AF352898

Current Trends and Perspectives in the Application of Polymeric Materials for Wastewater Treatment

Current Trends and Perspectives in the Application of Polymeric Materials for Wastewater Treatment

Editors

Marta Otero
Ricardo N. Coimbra

MDPI • Basel • Beijing • Wuhan • Barcelona • Belgrade • Manchester • Tokyo • Cluj • Tianjin

Editors
Marta Otero
Environment and Planning
CESAM
University of Aveiro
Aveiro
Portugal

Ricardo N. Coimbra
Environment and Planning
University of Aveiro
Aveiro
Portugal

Editorial Office
MDPI
St. Alban-Anlage 66
4052 Basel, Switzerland

This is a reprint of articles from the Special Issue published online in the open access journal *Polymers* (ISSN 2073-4360) (available at: www.mdpi.com/journal/polymers/special_issues/curr_tren_pers_appl_poly_mate_wastr_treat).

For citation purposes, cite each article independently as indicated on the article page online and as indicated below:

LastName, A.A.; LastName, B.B.; LastName, C.C. Article Title. *Journal Name* **Year**, *Volume Number*, Page Range.

ISBN 978-3-0365-2372-9 (Hbk)
ISBN 978-3-0365-2371-2 (PDF)

Contents

About the Editors . . . . **vii**

Preface to "Current Trends and Perspectives in the Application of Polymeric Materials for Wastewater Treatment" . . . . **ix**

Ricardo N. Coimbra and Marta Otero
Current Trends and Perspectives in the Application of Polymeric Materials to Wastewater Treatment
Reprinted from: *Polymers* **2021**, *13*, 1089, doi:10.3390/polym13071089 . . . . **1**

Mohammed Abdulsalam, Hasfalina Che Man, Pei Sean Goh, Khairul Faezah Yunos, Zurina Zainal Abidin, Aida Isma M.I. and Ahmad Fauzi Ismail
Permeability and Antifouling Augmentation of a Hybrid PVDF-PEG Membrane Using Nano-Magnesium Oxide as a Powerful Mediator for POME Decolorization
Reprinted from: *Polymers* **2020**, *12*, 549, doi:10.3390/polym12030549 . . . . **5**

Mohd Shaban Ansari, Kashif Raees, Moonis Ali Khan, M.Z.A. Rafiquee and Marta Otero
Kinetic Studies on the Catalytic Degradation of Rhodamine B by Hydrogen Peroxide: Effect of Surfactant Coated and Non-Coated Iron (III) Oxide Nanoparticles
Reprinted from: *Polymers* **2020**, *12*, 2246, doi:10.3390/polym12102246 . . . . **27**

Tjasa Gornik, Sudhirkumar Shinde, Lea Lamovsek, Maja Koblar, Ester Heath, Börje Sellergren and Tina Kosjek
Molecularly Imprinted Polymers for the Removal of Antide-Pressants from Contaminated Wastewater
Reprinted from: *Polymers* **2020**, *13*, 120, doi:10.3390/polym13010120 . . . . **43**

Hiroyuki Hoshina, Jinhua Chen, Haruyo Amada and Noriaki Seko
Chain Entanglement of 2-Ethylhexyl Hydrogen-2-Ethylhexylphosphonate into Methacrylate-Grafted Nonwoven Fabrics for Applications in Separation and Recovery of Dy (III) and Nd (III) from Aqueous Solution
Reprinted from: *Polymers* **2020**, *12*, 2656, doi:10.3390/polym12112656 . . . . **63**

Moonis Ali Khan, ¡void¿ Momina, Masoom Raza Siddiqui, Marta Otero, Shareefa Ahmed Alshareef and Mohd Rafatullah
Removal of Rhodamine B from Water Using a Solvent Impregnated Polymeric Dowex 5WX8 Resin: Statistical Optimization and Batch Adsorption Studies
Reprinted from: *Polymers* **2020**, *12*, 500, doi:10.3390/polym12020500 . . . . **75**

Aline M. F. Linhares, Cristiano P. Borges and Fabiana V. Fonseca
Investigation of Biocidal Effect of Microfiltration Membranes Impregnated with Silver Nanoparticles by Sputtering Technique
Reprinted from: *Polymers* **2020**, *12*, 1686, doi:10.3390/polym12081686 . . . . **87**

Geaneth Pertunia Mashile, Kgokgobi Mogolodi Dimpe and Philiswa Nosizo Nomngongo
A Biodegradable Magnetic Nanocomposite as a Superabsorbent for the Simultaneous Removal of Selected Fluoroquinolones from Environmental Water Matrices: Isotherm, Kinetics, Thermodynamic Studies and Cost Analysis
Reprinted from: *Polymers* **2020**, *12*, 1102, doi:10.3390/polym12051102 . . . . **105**

Md Shipan Mia, Biaobiao Yan, Xiaowei Zhu, Tieling Xing and Guoqiang Chen
Dopamine Grafted Iron-Loaded Waste Silk for Fenton-Like Removal of Toxic Water Pollutants
Reprinted from: *Polymers* **2019**, *11*, 2037, doi:10.3390/polym11122037 . . . . **131**

Leili Mohammadi, Abbas Rahdar, Razieh Khaksefidi, Aliyeh Ghamkhari, Georgios Fytianos and George Z. Kyzas
Polystyrene Magnetic Nanocomposites as Antibiotic Adsorbents
Reprinted from: *Polymers* **2020**, *12*, 1313, doi:10.3390/polym12061313 . . . . **145**

Liang Qiu, Guilaine Jaria, María Victoria Gil, Jundong Feng, Yaodong Dai, Valdemar I. Esteves, Marta Otero and Vânia Calisto
CoreShell Molecularly Imprinted Polymers on Magnetic Yeast for the Removal of Sulfamethoxazole from Water
Reprinted from: *Polymers* **2020**, *12*, 1385, doi:10.3390/polym12061385 . . . . **163**

Keyan Yang, Jingchen Xing, Pingping Xu, Jianmin Chang, Qingfa Zhang and Khan Muhammad Usman
Activated Carbon Microsphere from Sodium Lignosulfonate for Cr(VI) Adsorption Evaluation in Wastewater Treatment
Reprinted from: *Polymers* **2020**, *12*, 236, doi:10.3390/polym12010236 . . . . **185**

About the Editors

Marta Otero

Marta Otero received her Ph.D. in Chemical Engineering from University of León (Spain) in 2000, then occupying several lecturer and researcher positions, including a Marie Curie fellowship at the Faculty of Engineering of the University of Porto (FEUP, Portugal) and a Ramón y Cajal contract at University of León (Spain). In 2017, she moved to University of Aveiro (Portugal), where she is currently senior researcher at the Associate Laboratory CESAM (Centre for Environmental and Marine Studies) and Department of Environment and Planning. Her research is mainly focussed on two fields: (i) sustainable treatments and materials for water decontamination; and (ii) bio-wastes management and valorization.

Ricardo N. Coimbra

Ricardo N. de Coimbra is a chemical engineer graduated at the Faculty of Engineering of the University of Porto (FEUP, Portugal) in 1998. Since then, his professional activity has always been related to polymers, polymeric materials and their processing. In these fields, he has carried out engineering, consulting, research and lecturing activities. In addition, wastewater treatment was the focus of his doctorate studies, which allowed him to obtain his Ph.D. at the University of León (Spain) in 2017. After that, he has worked both in academia and industry.

Preface to "Current Trends and Perspectives in the Application of Polymeric Materials for Wastewater Treatment"

Water is indispensable to the functioning of most known life forms, and good water quality is essential to human health, social and economic development, and ecosystem functioning. Nonetheless, population growth has been leading to the degradation and depletion of fresh water resources around the world. Under these circumstances, ensuring sufficient and safe water supplies for everyone is one of the Sustainable Development Goals (SDGs) set by the United Nations General Assembly in 2015 for the year 2030. For this goal to be achieved, the development and implementation of appropriate and efficient wastewater treatments that allow us to reduce water pollution is a major challenge.

The application of polymers and polymeric materials in wastewater treatment is a research field that has largely developed. Conventional and novel approaches have been carried out by researchers from different areas, who have demonstrated that polymers and polymeric materials may have an important role in the removal of pollutants of different origin and nature from wastewater, in the disposal of sludge, in the recycling of materials, in the improved efficiency and economy of wastewater, etc.

In view of the relevant contribution that polymers and polymeric materials may have in the conservation of the aquatic environment, namely by their application in wastewater treatment, original research and review papers on "Current trends and perspectives in the application of polymeric materials for wastewater treatment" were here brought together. Authors of the here included works are deeply acknowledged for their outstanding contributions, which we hope may be helpful and inspiring for readers interested in this topic.

Marta Otero, Ricardo N. Coimbra
Editors

Editorial

Current Trends and Perspectives in the Application of Polymeric Materials to Wastewater Treatment

Ricardo N. Coimbra [1] and Marta Otero [1,2,*]

1 Department of Environment and Planning, University of Aveiro, Campus Universitário de Santiago, 3810-193 Aveiro, Portugal; ricardo.coimbra@ua.pt
2 Centre for Environmental and Marine Studies (CESAM), University of Aveiro, Campus Universitário de Santiago, 3810-193 Aveiro, Portugal
* Correspondence: marta.otero@ua.pt

Citation: Coimbra, R.N.; Otero, M. Current Trends and Perspectives in the Application of Polymeric Materials to Wastewater Treatment. *Polymers* **2021**, *13*, 1089. https://doi.org/10.3390/polym13071089

Received: 18 March 2021
Accepted: 24 March 2021
Published: 30 March 2021

Publisher's Note: MDPI stays neutral with regard to jurisdictional claims in published maps and institutional affiliations.

Water with the necessary quality is indispensable to the functioning of most of the known life forms, being essential to human health, social and economic development, and ecosystems functioning. However, only 2.5% of all water on Earth is freshwater, and less than 1% is accessible, its availability being actually affected by climate change and direct human impacts. Furthermore, population growth and industry expansion have been leading to the continuous degradation of freshwater quality around the world. Under these circumstances, ensuring sufficient and safe water supplies for everyone is one of the Sustainable Development Goals (SDGs) set by the United Nations General Assembly in 2015 for the year 2030. For this goal to be achieved, the development and implementation of appropriate and efficient wastewater treatments is a major challenge.

The application of polymers and polymeric materials in wastewater treatment is a research field that has greatly developed from the end of the last century. The very nature, structure, and versatility of polymers make them useful for many applications, including wastewater treatment processes. Conventional and novel approaches have been elaborated or refined by researchers from different areas, who have demonstrated that polymers and polymeric materials may have an important role not only in the removal of pollutants of different origin and nature from wastewater but also in the recycling of materials and the improvement of wastewater efficiency and economy.

In view of the relevant contribution of polymers and polymeric materials to the conservation of the aquatic environment, namely by their application in wastewater treatment, this Special Issue (SI) was launched for the publication of original research or review papers within this topic. The aim was to bring forth the challenges and discuss current trends and perspectives in the utilization of polymers and polymeric materials—either synthetic or natural—for the treatment or purification of wastewater.

Eleven research works [1–11] by distinguished international authors were published within this SI on "Current Trends and Perspectives in the Application of Polymeric Materials for Wastewater Treatment", which covered a wide range of issues related to this topic.

Abdulsalam et al. [1] developed hydrophilic hybrid polyvinylidene fluoride (PVDF)-polyethylene glycol (PEG) ultrafiltration membranes loaded with Nano-MgO (NMO) by using a phase inversion technique. These authors demonstrated that NMO incorporation not only improved the membranes' antifouling properties and permeation performance but also allowed for an enhanced color separation from palm oil mill effluent, which was related to the synergism between surface deprotonation and pore size screening. For their part, Linhares et al. [6] loaded silver nanoparticles (AgNPs) in microfiltration polymeric membranes (15 wt.% polyethersulfone and 7.5 wt.% polyvinylpyrrolidone in *N*,*N*-dimethylacetamide) by the sputtering technique. They demonstrated the efficiency of this technique to load AgNPs, which provided membranes with biocidal properties resistant to biofouling and made them proficient for water disinfection treatments.

Mia et al. [8] loaded iron on waste silk fibers previously grafted with polydopamine by oxidative polymerization to produce wSF-DA/Fe, which was tested for the catalytic removal of toxic dyes (Methylene Blue, Cationic Violet X-5BLN, and Reactive Orange GRN). The authors postulated that the dye removal was due to the synergistic effect of free radicals and reactive species, which resulted from a heterogeneous Fenton reaction and oxidized the dyes into colorless nontoxic substances. The catalytic performance of wSF-DA/Fe was affected by the H_2O_2 concentration, initial dye concentration, temperature, and presence of electrolytes (NaCl, Na_2SO_4). Aiming at the catalytic removal of Rhodamine B (RB) dye, Ansari et al. [2] synthesized Fe_3O_4 nanoparticles (NPs) and sodium dodecyl sulfate (SDS) coated Fe_3O_4 NPs (SDS@Fe_3O_4) by the co-precipitation method. RB degradation was tested in presence of H_2O_2, H_2O_2 and Fe_3O_4 NPs, and H_2O_2 and SDS@Fe_3O_4 NPs, the latter providing an increased catalytic removal, as the SDS coating avoided the aggregation of Fe_3O_4 and the associated efficiency reduction.

In the field of adsorptive treatments, Khan et al. [5] investigated RB adsorption by a novel solvent impregnated resin (SIR). SIR, which was produced by modifying the cationic polymeric resin Dowex 5WX8 with the solvent t-butyl phosphate, adsorbed RB mainly by electrostatic interactions and π–π bonding. The authors optimized the operational conditions, viz. pH, SIR dosage, and contact time, for the adsorptive removal of RB. The adsorptive removal of Cr(VI) from water was studied by Yang et al. [11], who produced for this purpose activated carbon microspheres (SLACM), using sodium lignosulfonate ((SL), a waste from the pulp and paper industry) as raw material. The synthesis of SLACM, which was to be attained with no binder addition, consisted of (i) amination of chloromethylstyrene-divinylbenzene-styrene copolymer (CMPS) with 1,3-diaminopropane; (ii) Mannich reaction of SL and amino CMPS to produce the adsorbent resin microsphere; (iii) impregnation with $ZnCl_2$ and pyrolysis at 600 °C. The adsorption of Cr (VI) by the produced SLACM was shown to be favored by decreasing pH (within pH 2 and 9) and increasing temperature (within 20 and 40 °C). Hoshima et al. [4] aimed at the adsorption of the rare earths dysprosium (Dy) and neodymium (Nd) from water as a previous necessary step for their subsequent recovery. With this objective, the authors produced a new adsorbent by the radiation-induced graft polymerization of methacrylate with a long alkyl chain on a PE/PP nonwoven fabric and the subsequent loading of 2-ethylhexyl hydrogen-2-ethylhexylphosphonate by hydrophobic interaction and chain entanglement between the alkyl chains. Four different methacrylate monomers were tested, namely butyl methacrylate (BMA), hexyl methacrylate (HMA), dodecyl methacrylate (DMA), and octadecyl methacrylate (OMA). Only the OMA-adsorbent was stable under subsequent uses, which was related to the suppression of EHEP losses due to the strong hydrophobic interaction and chain entanglement between the long alkyl chains. Moreover, OMA-adsorbent was efficient in the adsorption of Dy (III) and Nd (III) from water, being a promissory material for the recovery of rare-earth metals from NdFeB permanent magnet scraps.

Four papers in the SI dealt with the adsorptive removal of pharmaceuticals [3,7,9,10]. Mashile et al. [7] synthesized a magnetic mesoporous carbon/-cyclodextrin–chitosan (MMPC/Cyc-Chit) nanocomposite for the adsorption of fluoroquinolones (FQs), viz. danofloxacin, enrofloxacin, and levofloxacin, from different water samples, including a synthetic mixture of the FQs, the influent and effluent from a wastewater treatment plant, river water, and tap water. The authors found that incorporating biodegradable polymers such as chitosan and cyclodextrin into magnetic mesoporous carbon brought about a nanocomposite with large surface area and adsorption capacity. This was a very complete study, which included the optimization of operational parameters (pH, mass of MMPC/Cyc-Chit, and sonication power) by a response surface methodology, kinetic and equilibrium modeling, assessment of regeneration and reusability, thermodynamic parameters determination, and cost analysis for the synthesis of MMPC/Cyc-Chit. Mohammadi et al. [9] synthesized a poly(styrene-block-acrylic acid) diblock copolymer/Fe_3O_4 magnetic nanocomposite (P(St-b-AAc)/Fe_3O_4)) and tested it for the adsorption of antibiotic ciprofloxacin from synthetic wastewater. This work included the optimization of operational parameters, namely antibiotic concentration, pH, nanocomposite mass, and contact time, and showed that P(St-b-AAc)/Fe$_3$O$_4$ was efficient in the adsorptive removal of ciprofloxacin. Aiming at the adsorption of selective serotonin reuptake inhibitor (SSRI) antidepressants and their metabolites from water, Gornik et al. [3] optimized the synthesis of molecularly imprinted polymer (MIP) adsorbents in which the SSRI sertraline was used as template. Different MIPs were synthesized by varying the functional monomer, the porogen, and/or the template form, which were shown to largely affect the adsorbent performance of the resulting material, so the authors selected the most efficient. Even when the selected MIPs had a relatively lower surface area than conventional activated carbons, they displayed a larger adsorption capacity in real wastewater samples. Apart from the MIPs optimization and characterization, this work included the assessment of their reusability, occurrence of cross-reactivity, adsorption kinetics, matrix effects, upscale, and leaching, with authors pointing out the necessity of carrying out future work at a larger scale to confirm the advantages of the synthetized materials. In order to overcome the drawbacks of MIPs, mainly the small number of recognition sites per unit of volume and the low mass transfer, surface molecular imprinting on magnetic yeast (MY) was carried out by Qiu et al. [10], who synthesized highly selective magnetic yeast-molecularly imprinted polymers (MY@MIPs) for the adsorptive removal of antibiotic sulfamethoxazole (SMX). For the production of MY@MIPs, these authors started by preparing nano-Fe$_3$O$_4$ by an in situ one-step procedure, which was loaded onto yeast cells to obtain MY. Then, MY was used as core for the polymerization of MIPs using SMX as template to produce MY@MIPs. The authors compared MY@MIPs with MY and non-imprinted MY@NIPs (synthesized in the same way as MY@MIPs but in the absence of template) and evidenced the superior SMX adsorption capacity of MY@MIPs. Furthermore, besides the selectivity of MY@MIPs towards SMX in the presence of other pharmaceuticals and in real wastewater, the reutilization capability of this material was also proved, pointing to its possible application as an alternative adsorbent for SMX selective removal from wastewater.

Membrane filtration and catalytic applications together with the utilization of polymeric materials for the adsorptive removal of pollutants were discussed in this SI. Furthermore, it is worth highlighting the attention given to the adsorptive removal of pharmaceuticals from wastewater. Since the 1990s, a growing scientific concern about pharmaceuticals may be inferred from the number of related publications, and this SI was no exception. Such a concern is mainly related to (i) the analytic development that has made possible their detection at trace levels and the confirmation of their ubiquity in environment; and (ii) the ecological risks related to their potential to cause physiological responses in non-target individuals, including endocrine disruption and antibiotic resistance. An important benefit of pharmaceuticals removal by adsorption is that such a treatment does not result into the formation of by-products, which in some cases can be more hazardous than the parent compounds.

Polymers have been used for long in conventional wastewater treatment for the flocculation/coagulation of solids, so that they may be easily separated from water. However, the SI hereby presented makes evident that polymers and polymeric materials may have many new and varied applications in wastewater treatment. Most published works take advantage of polymers' versatility and capacity to be combined or modified to produce advanced materials with relevant features for water treatment. In fact, polymer modification is presently a hot topic, since it allows for the development of specific and even smart materials for target applications in different sectors, including wastewater treatment. Owing to the efforts of polymer engineers and scientists, many alternative and advanced polymeric materials are continuously developed. In the specific case of wastewater treatment, polymer applications have become very important due to the increased pollutant removal efficiencies that they offer, so this is a research field in great expansion. The progress in the last years and the successful applications of polymer and polymeric materials are reflected by the high quality works published within this SI.

Author Contributions: M.O. and R.N.C., as Guest Editors of the Special Issue entitled "Current Trends and Perspectives in the Application of Polymeric Materials for Wastewater Treatment", contributed to the preparation of this Editorial. Conceptualization, M.O. and R.N.C.; writing—original draft preparation, R.N.C. and M.O.; writing—review and editing, M.O.; Supervision, M.O. Both authors have read and agreed to the published version of the manuscript.

Funding: Thanks are due to the Portuguese Fundação para a Ciência e a Tecnologia/ Ministério da Ciência, Tecnologia e Ensino Superior (FCT/MCTES) for the financial support to the Associated Laboratory CESAM (UIDP/50017/2020+UIDB/50017/2020) through national funds. FCT is also acknowledged for funding through the Investigator Program (IF/00314/2015).

Institutional Review Board Statement: Not applicable.

Informed Consent Statement: Not applicable.

Data Availability Statement: Data sharing is not applicable to this article.

Acknowledgments: Contributions to this Polymers Special Issue (SI) are deeply acknowledged. As Guest Editors, we would like to sincerely thank the authors of these contributions for considering this SI for the publication of their outstanding works. Thanks are also due to all the anonymous reviewers who gently reviewed these works and sent their wise comments, corrections, and suggestions. Finally, our recognition to the editorial managers for their helpful assistance in the assembling of this SI.

Conflicts of Interest: The authors declare no conflict of interest.

References

1. Abdulsalam, M.; Man, H.C.; Goh, P.S.; Yunos, K.F.; Abidin, Z.Z.; Isma, M.I.A.; Ismail, A.F. Permeability and antifouling augmentation of a hybrid PVDF-PEG membrane using nano-magnesium oxide as a powerful mediator for POME decolorization. *Polymers* **2020**, *12*, 549. [CrossRef] [PubMed]
2. Ansari, M.S.; Raees, K.; Khan, M.A.; Rafiquee, M.Z.A.; Otero, M. Kinetic studies on the catalytic degradation of rhodamine b by hydrogen peroxide: Effect of surfactant coated and non-coated iron (III) oxide nanoparticles. *Polymers* **2020**, *12*, 2246. [CrossRef] [PubMed]
3. Gornik, T.; Shinde, S.; Lamovsek, L.; Koblar, M.; Heath, E.; Sellergren, B.; Kosjek, T. Molecularly imprinted polymers for the removal of antidepressants from contaminated wastewater. *Polymers* **2021**, *13*, 120. [CrossRef] [PubMed]
4. Hoshina, H.; Chen, J.; Amada, H.; Seko, N. Chain entanglement of 2-ethylhexyl hydrogen-2-ethylhexylphosphonate into methacrylate-grafted nonwoven fabrics for applications in separation and recovery of Dy (III) and Nd (III) from aqueous solution. *Polymers* **2020**, *12*, 2656. [CrossRef] [PubMed]
5. Khan, M.A.; Momina; Siddiqui, M.R.; Otero, M.; Alshareef, S.A.; Rafatullah, M. Removal of rhodamine b from water using a solvent impregnated polymeric Dowex 5WX8 resin: Statistical optimization and batch adsorption studies. *Polymers* **2020**, *12*, 500. [CrossRef] [PubMed]
6. Linhares, A.M.F.; Borges, C.P.; Fonseca, F.V. Investigation of biocidal effect of microfiltration membranes impregnated with silver nanoparticles by sputtering technique. *Polymers* **2020**, *12*, 1686. [CrossRef] [PubMed]
7. Mashile, G.P.; Dimpe, K.M.; Nomngongo, P.N. A biodegradable magnetic nanocomposite as a superabsorbent for the simultaneous removal of selected fluoroquinolones from environmentalwater matrices: Isotherm, kinetics, thermodynamic studies and cost analysis. *Polymers* **2020**, *12*, 1102. [CrossRef]
8. Mia, M.S.; Yan, B.; Zhu, X.; Xing, T.; Chen, G. Dopamine grafted iron-loaded waste silk for Fenton-like removal of toxic water pollutants. *Polymers* **2019**, *11*, 2037. [CrossRef] [PubMed]
9. Mohammadi, L.; Rahdar, A.; Khaksefidi, R.; Ghamkhari, A. Polystyrene Magnetic Nanocomposites as Antibiotic Adsorbents. *Polymers* **2020**, *12*, 1313. [CrossRef] [PubMed]
10. Qiu, L.; Jaria, G.; Gil, M.V.; Feng, J.; Dai, Y.; Esteves, V.I.; Otero, M.; Calisto, V. Core-shell molecularly imprinted polymers on magnetic yeast for the removal of sulfamethoxazole from water. *Polymers* **2020**, *12*, 1385. [CrossRef] [PubMed]
11. Yang, K.; Xing, J.; Xu, P.; Chang, J.; Zhang, Q.; Usman, K.M. Activated carbon microsphere from sodium lignosulfonate for Cr(VI) adsorption evaluation in wastewater treatment. *Polymers* **2020**, *12*, 236. [CrossRef]

 *polymers*

Article

Permeability and Antifouling Augmentation of a Hybrid PVDF-PEG Membrane Using Nano-Magnesium Oxide as a Powerful Mediator for POME Decolorization

Mohammed Abdulsalam [1,2], Hasfalina Che Man [1,*], Pei Sean Goh [3], Khairul Faezah Yunos [4], Zurina Zainal Abidin [5], Aida Isma M.I. [6] and Ahmad Fauzi Ismail [3]

1 Department of Biological and Agricultural Engineering, Faculty of Engineering, Universiti Putra Malaysia, UPM Serdang 43400, Selangor, Malaysia; m.abdul_22@yahoo.com
2 Department of Agricultural and Bioresources, Ahmadu Bello University, Zaria 810107, Nigeria
3 Advanced Membrane Technology Research Centre (AMTEC), School and Chemical and Energy Engineering, Faculty of Engineering, Universiti Teknologi Malaysia, UTM Skudai 81310, Johor, Malaysia; peisean@petroleum.utm.my (P.S.G.); afauzi@utm.my (A.F.I.)
4 Departments of Food and Process Engineering, Faculty of Engineering, Universiti Putra Malaysia, UPM Serdang 43400, Selangor, Malaysia; kfaezah@upm.edu.my
5 Departments of Chemical and Environmental Engineering, Faculty of Engineering, Universiti Putra Malaysia, UPM Serdang 43400, Selangor, Malaysia; zurina@upm.edu.my
6 Departments of Chemical Engineering, Segi University, Kota Damansara Selangor 47810, Malaysia; aidaisma@segi.edu.my
* Correspondence: hasfalina@upm.edu.my

Received: 19 December 2019; Accepted: 7 January 2020; Published: 3 March 2020

Abstract: This study focused on developing a hydrophilic hybrid polyvinylidene fluoride (PVDF)-polyethylene glycol (PEG) hollow membrane by incorporating Nano-magnesium oxide (NMO) as a potent antifouling mediator. The Nano-hybrid hollow fibers with varied loading of NMO (0 g; 0.25 g; 0.50 g; 0.75 g and 1.25 g) were spun through phase inversion technique. The resultants Nano-hybrid fibers were characterized and compared based on SEM, EDX, contact angle, surface zeta-potential, permeability flux, fouling resistance and color rejection from palm oil mill effluent (POME). Noticeably, the permeability flux, fouling resistance and color rejection improved with the increase in NMO loading. PVDF-PEG with 0.50 g-NMO loading displayed an outstanding performance with 198.35 L/m^2·h, 61.33 L/m^2·h and 74.65% of water flux, POME flux and color rejection from POME, respectively. More so, a remarkable fouling resistance were obtained such that the flux recovery, reversible fouling percentage and irreversible fouling percentage remains relatively steady at 90.98%, 61.39% and 7.68%, respectively, even after 3 cycles of continuous filtrations for a total period of 9 h. However, at excess loading of 0.75 and 1.25 g-NMO, deterioration in the flux and fouling resistance was observed. This was due to the agglomeration of nanoparticles within the matrix structure at the excessive loading.

Keywords: nano-MgO; structural modification; permeability; antifouling; color rejection; POME

---

1. Introduction

Owing to the exceptional physical, mechanical and chemical stability of polyvinylidene difluoride (PVDF) polymer, its applications for bio-system/tissue engineering [1], pervaporation [2,3] and separation technology [4,5] have gained a considerable attentions. In term of solubility, PVDF polymer can easily dissolve in most of the organic solvents such as *N,N*-dimethylformamide (DMF),

N,N-dimethylacetatamide (DMAc), Triethylphosphate (TEP) and *N*-methyl-*N*-pyrolidinone (NMP) [6,7]. As a result of this flexibility, PVDF polymer is suitable for fabricating polymeric membrane [6]. It has likewise been noted that PVDF membrane requires a relatively low pressure and minimal energy demand during filtration. This feature is most desirable for the microfiltration and ultrafiltration separation processes [8]. The aforementioned separation process using PVDF polymeric membrane have been applied on several wastewater, such as palm oil mill effluent [9], dye wastewater generated from textile industry [10], saline-water [11] and endocrine compounds [12]. Some drawbacks were reported which include continuous diminishing in permeation, low rejection as well as fouling [13]. The reduction in the membrane flux and susceptibility to fouling was collectively attributed to the hydrophobic nature of the polymer [14,15]. Thus, this justified the significant research conducted to subdue the hydrophobicity by improving the water-liking property along with permeability flux of the polymeric membrane using inorganic nanoparticles as an additive [16].

The commonly applied nanoparticles to modified polymeric membranes includes ZnO [17,18], TiO_2 [9], Ag_2O_3 [19,20], Al_2O_3 [21], graphene oxide [22,23], SiO_2 and CuO [24]. Tan et al. [17] enhanced the permeation and antifouling properties of a bared PVDF membrane by incorporating synthesized Zn-Fe oxide (ZIO) into the matrix structure. The nanocomposite membrane with 0.5 wt % ZIO loading demonstrated a significant increase in permeate flux to the magnitude of 25% increment compare to the bared PVDF membrane. Also, Shen et al. [25] modified the matrix structure of polymeric membrane using ZnO nano-additive. The result showed that the additive has sizeable influences on the pores structure as well as the hydrophilicity of the membrane. Over 254% improvement in permeability flux was reported at the loading of 0.3 g Nano-ZnO. Furthermore, the authors validated that the porosity of the modified composite membrane increased with the presence of the nano-additive, which ultimately justify the reported high flux. Nano-ZnO has large surface area that expedite the formation of hydroxyl (–OH) functional group on the surface [26,27]. The presence of –OH enhanced hydrophilicity of the PVDF polymeric membrane, thus mitigating fouling rate [26]. Analogously, TiO_2 Ag_2O_3 and Al_2O_3 also exhibit similar features when incorporated into the matrix structure of polymeric membrane. Subramanium et al. [9] synthesized titanate nanotubes using nano-TiO_2 as a precursor to modify the PVDF membrane. The results substantiated that under photo-catalytic condition, the involvement of TiO_2 in the dope formulation improved the permeation consistency (35.8 L/m^2·h) of the resultant nanocomposite membrane. Negligible fouling was observed at 0.5 wt % TNT loading throughout the 4 h continuous filtration. Another report has shown strong agreement with this observation, and the results affirmed that the antifouling performance of TiO_2 was due to generated –OH under UV spectrum and its aptitude to exhibit self-cleaning [28]. Also, attentions are long drifted to the use of nano-Ag_2O_3 to modify polymeric membrane [19,20,29]. The nano-additive released Ag ions to inhibit the metabolism of the microbial-foulants, thereby preventing the generation of the extracellular polymeric substance (EPS). Thus, this ultimately curtailed the most serious type of fouling, which are organic and bio-fouling [22]. Maximous et al. [30] investigated the antifouling impact of Al_2O_3 and the results showed that at optimum 0.05 wt % loading, the modified composite polymeric membrane was less prone to fouling. Besides, the presence of Al_2O_3 in the matrix structure of a polymeric membrane not only reduces the fouling but also improves the flux consistency.

However, most of the widely used nanoparticles, particularly as mentioned above, are photo-catalytic driven to effectively address the fouling issue [31,32]. This implies that the presence of ultraviolent radiation is a prerequisite to precede the antifouling performance. More so, the issue continues releasing antifouling radicals and superoxide could seriously jeopardize the stability of the composite matrix structure. In line with this, Tan et al. [17] reported that the structural instability of the modified nanocomposite PVDF-ZIO membrane was discernible after four filtration cycles due to the collapse of the incorporated nanoparticles. This could significantly undermine the overall antifouling performance and also re-exposing the modified membrane to inconsistencies in permeation flux along with frequent fouling challenges. Moreover, the nano- ZnO, TiO_2, Ag_2O_3, Al_2O_3 and CuO are relatively expensive and the running cost could be unsustainable for industrial application [33,34].

Meanwhile, nano-MgO has remained one of the antimicrobial and super hydrophilic nanomaterials yet to be fully explored in improving filtration and antifouling performance of polymeric membrane [35]. The nano-MgO has the ability to release reactive oxygen species (ROS) which directly extract lipid from the cells of the microbial-foulants. This ultimately disrupted the metabolic activities and hindered biofilm formation [31,36]. More interestingly, the nano-MgO not only averts bio-fouling formation but also capable of evincing self-cleaning mechanism [31,36]. The nano-MgO precursor is readily available and comparably cheaper than other nanomaterials (such as TiO_2, Ag_2O_3, Al_2O_3, and ZnO) [31]. Therefore, MgO-nanoparticles intimate a promising additive capable of enhancing antifouling properties and permeation performance of polymeric membranes for industrial filtration purposes. Currently, application of a modified Nano-MgO (NMO) composite PVDF-PEG membrane for separation of color pigment from POME has not been reported.

In view of these, the present study focuses on modifying the structure of an in-house fabricated PVDF-PEG ultrafiltration membrane at various loadings of NMO. The impacts of the incorporated NMO were examined based on the morphological changes, hydrophilicity, the permeability flux and fouling resistance of the resultant membrane. Furthermore, the rejection and color separation efficiency was studied, and also the used membranes were characterized using Fourier transform infrared spectroscopy. The outcome of the study demonstrated that the involvement of the NMO in the dope formulation significantly improved the antifouling properties and the permeation performance, alongside with color separation from palm oil mill effluent (POME) via the synergy of surface deprotonation and pore size screening mechanism.

2. Experimental Methods

2.1. Chemicals and Materials

Nano-MgO (NMO: particles size (BET) < 50 nm; MW = 40.30 g/Mol; purification ≥ 99.9%), Pellets PVDF (Kynar 740) and polyethylene glycol (PEG: 12,000 g/mol) were procured from Sigma Aldrich (M) Sdn Bhd, Selangor, Malaysia. The PVDF was applied as the major membrane matric polymer, while the PEG as co-polymer to enhance pore formation. The *N,N*-dimethylformamide (DMF: ≥99.9%; 87.12 g/mol), ethanol (≥99.98%; 46.07 g/Mol) and glycerol (≥99%; 92.09 g/Mol) were also obtained from Sigma Aldrich (M) Sdn Bhd, Selangor, Malaysia, and respectively used as doping and post-treatment solvents without any further purification. $LiCl_2.H_2O$ (MW 42.39 g/Mol; ≥99%) was purchased from Acros Organic Industry, Chemicals and Reagents, Semenyih, Selangor, Malaysia and it was applied to improve the hydrophilicity of the polymers (PVDF/PEG). Also, high strength POME with initial color concentration of 8570 ADMI was collected from an Oil Palm Milling industry, Malaysia. Prior to the usage, the POME sample was filtered to remove all the visible debris and then diluted using a factor of 2 to give 4285 ADMI. This procedure was to represent the industrial final discharged color concentration ranges [9].

2.2. Synthesis of Nano-Hybrid PVDF/PEG-NMO

2.2.1. Dope formulation

The dope formulation was preceded by adding NMO into DMF and then subjected to sonication using digital ultrasonic water bath (VWR 142-0300) at 75 °C for a period of 20 min. Subsequently, $LiCl_2 \cdot H_2O$ was added into the mixture under steady agitation of 350 rpm and 75 °C for a period of 24 h. Essentially, this procedure assists in dispersion of the NMO as well as ensuring good blending of the mixtures. The mixture was followed by adding the dehydrated PVDF pellets and the co-polymer, PEG. The combination was stirred continuously at 350 rpm and 100 °C for another 24 h using a hot-plate stirrer (Monotaro; C-MAG HS7, Malaysia) to achieve a homogenous solution. The amount of the NMO contained in the dope solutions were varied as contained in Table 1.

Table 1. Chemical formulation of Neat PVDF-PEG and modified Nano-hybrid membranes.

Samples	PVDF/PEG (g)		DMF (g)	LiCl.H₂O (g)	NMO (g)	Mass of Dope (g)
Neat PVDF/PEG	30/10	(3:1)	158	2.0	–	200.00
NMO-0.25 g	30/10	(3:1)	158	2.0	0.25	200.25
NMO-0.50 g	30/10	(3:1)	158	2.0	0.50	200.50
NMO-0.75 g	30/10	(3:1)	158	2.0	0.75	200.75
NMO-1.25 g	30/10	(3:1)	158	2.0	1.25	201.25

2.2.2. Spinning of Nano-Hybrid PVDF/PEG-NMO Hollow Membrane

It is important to note that same spinning parameters were applied for all of the samples. The dopes were spun through dry-jet wet swirling technique employing an annular spinneret. The inner and outside diameter of annular spinneret was 0.55 and 1.15 mm, respectively. During the fabrication of the membrane, tap water and distilled water was used as the external and internal coagulant. In addition, the pick speed control, collecting drum speed, extrusion rate, air gap, external coagulant temperature, room temperature and room humidity remains constant at 7 rpm, 11 rpm, 5 mL/min, 10 cm, 25 °C, 29.5 ± 1 °C and 72.7%, respectively. The spun fibers were drench in a continues flow water-bath at least for a period of 24 h to dislodged solvent remnants. In order to minimize shrinkage, post-treatments were applied on the fibers by immersing in ethanol for 12 h, then followed by drenching in glycerol for 5 h, respectively. The post-treated fibers were air-dried for at least 24 h to ensure complete dehydration.

2.3. Characterization of Nano-Hybrid PVDF/PEG-NMO Fibre

2.3.1. Morphology Analysis

The cross section and surface morphology of the synthesized Nano-hybrid fiber were examined using scanning-electron-microscope, (SEM: S-3400N). Erstwhile, the composite fibers were immersed into liquid nitrogen for a period of 5 min. This procedure ensures sharp breaking of the fiber to reveal the cross sectional structure. Then, the fractured samples were sputtered coated with gold thin layer, and the voltage acceleration was maintained constant at 20 kV during the image capturing.

Also, SEM/EDX (scanning electron microscope/energy dispersive X-ray) was engaged for examining the NMO particle distribution and the dimension, as well as elemental composition in the Nano-hybrid fiber. The fractured membrane samples were place on the adhesive carbon tape of a metal plate and examined at 20 kV accelerating voltage using SEM-Thermo Scientific (Hitachi & S-3400N) for the analysis and imaging.

2.3.2. Hydrophilic and Porosity Analysis

The hydrophilicity of the synthesized membrane was analyzed based on static contact angle using goniometer (GmbH OCA 15pro, Data-Physics). The goniometer uses RO water as the probing liquid, such that the liquid droplets were captured with the equip camera after 30 s stand. The measured contact angles were analyzed using SCA20 software. In order to minimize error, each of the measurement was repeated ten times on different spots of the membrane and the average was determined as the contact angle. Throughout the analysis, the measurements were conducted at ambient temperature.

The procedure employed for the determination of porosity is based on gravimetric method [9]. Ten pieces of the synthesized membrane of 20 cm of equal length were hermetically sealed at both ends using glue-resin. The prepared samples were submersed in distilled water for 5 h under ambient temperature and humidity. Afterwards, the superficial water drops on the surface of the samples were eliminated using dry tissue paper, and then weighed as wet membrane (M_w). The wet membrane samples were air dried overnight at 70 °C and reweighed as dried membrane (M_d). Hence, the porosity (ε) for each sample was determined using Equation (1):

$$\varepsilon,\ (\%) = \frac{1}{\rho_w} \times \left(\frac{M_w - M_d}{V}\right)100 \tag{1}$$

where ε Symbolises the membrane porosity in %, ρ_w is density of water, $(M_w - M_d)$ is the quantity of pores water in gram and V is the volume of the membrane sample.

2.3.3. Surface Charge Analysis

The surface charge of the membranes was examined based on the potential of streaming analysis conducted using Anton SurPass Analyser (Paar Inc. Ireland). The samples were placed on the sample-holder by mean of a carbon-adhesive tape, and then the analyzer was pre-set to a maximum pressure of 400 mbar. This procedure was to ensure laminar flow throughout the analysis [17]. Also, a solution of 1 mM KCl was applied as contextual electrolyte and the pH varied between 2 and 10 was accomplished using aqueous 0.1 M HCl or NaOH. The Helmholts Smoluchowski procedure was adopted for the calculation of the Zeta-Potential surface charge.

2.4. Membrane Performance

2.4.1. Permeability Analysis

The filtration performance was examined using a fabricated dead-end permeation system, equipped with membrane module cell and a peristaltic pump to provide the required suction pressure. Each of the modules comprises of 10 units of the membrane with equal length of 20 cm. initially, the membrane was compacted at a pressure of 0.4 MPa for a period of 30 min to ensure steady flux, while the subsequent filtration operations were performed at lower pressure of 0.3 MPa. The pure water (J_w) and permeates flux (J_p) were determined using Equation (2):

$$J = \frac{V}{A_s \Delta t} \tag{2}$$

where J denotes the flux in L/m^2·h, V is the volume of permeate (L), A_s is membrane surface area (m^2) and Δt is filtration time in h.

2.4.2. POME Decolorization

The membrane fibers were further subjected to filtration of diluted POME with 4285 ADMI color concentration using same set-up as applied when pure water was used as feed. Initially, the membranes fibers were engrossed in the POME solution for 90 min to initiate adhesion of thin layer color pigments on the fibers. This procedure assists in achieving accurate color rejection performances of the fibers [9]. The apparent color content of the feed POME and permeate were analyze in ADMI using UV-spectrophotometric technique (DR4000U, HACH) at an absorbance wavelength of 400 nm. The POME decolorization efficiency was determined using Equation (3):

$$\%\ C_{removal} = \left(1 - \frac{C_{Permeate}}{C_{Feed\ POME}}\right) \times 100 \tag{3}$$

where, $C_{removal}$ is percentage of color removal in %, $C_{permeate}$ is permeate color concentration in ADMI, and $C_{Feed\ POME}$ is feed POME color concentration in ADMI.

2.4.3. Fouling Analysis

In order to mimic practical situation and to evaluate the reusability, the membranes were subjected 3 cycle of continues filtration for a total period of 9 h using a known color concentration (4285 ADMI) of a diluted POME as feed. In each of the completed cycle, the antifouling performances of the membrane were evaluated after 180 min continues filtration without interruption. The indices used for the antifouling analysis were as stated in Equations (4)–(6). At every completion of each filtration cycle, the volume of POME permeates, flux and color rejected were determined. Then, the used membranes were physically cleaned under running tap-water for a period of 15 min. The washed and cleaned membranes were reapplied for the 2nd cycle of POME filtration for another 180 min, following same procedure as highlighted above. In this study, the POME filtration cycle was repeated 3 times to determine the antifouling performance of the membranes using percentage of flux recovery (*FR*), reversible fouling (*RF*) and irreversible fouling (*IF*) as indices Equations (4)–(6):

$$Percentage\ of\ flux\ recovery,\ (\%FR) = \frac{J_{w2}}{J_w} \times 100 \tag{4}$$

$$Percentage\ of\ reversible\ fouling,\ (\%RF) = \%FR - \frac{J_P}{J_w} \times 100 \tag{5}$$

$$Percentage\ of\ irreversible\ fouling,\ (\%IF) = (1 - \%FR) \times 100 \tag{6}$$

where, J_{w2} is the water flux after the POME filtration, L/m^2·h.

2.4.4. Characterization of used Membranes by FTIR Analysis

The surface chemical functional groups and transformation of the neat and modified membrane (0.5 g-NMO) after used were characterized using Fourier transform infrared spectroscopy (FTIR-Perkin Elmer spectrum 100 Series). The spectra analysis was taken over a wide range from 400 to 4000 cm^{-1}. Essentially, the analysis involves shining a beam of light rays with variable frequencies, and then measures how much of that rays got absorbed by the specimen (i.e., the membrane samples). This ensure high signal-to-noise ratio of the spectra; thus accurate analysis of fouling level based on functional group is achievable [37].

3. Results

3.1. Effect of Nano-MgO on Membrane Characteristic

3.1.1. Morphological Studies

Figure 1 presents the scanning electron microscope (SEM) images of cross-sectional view of the neat and modified hybrid Nano-MgO (NMO) PVDF-PEG membranes. It is obvious that the neat fiber had 3 distinctive layers with thin layers both at inner and outer section of the membrane. The middle layer constitutes majorly of sandwich-like morphology containing short finger-like pores at both sides toward the ultra-thin layers. However, different scenarios were observed with the modified Nanocomposite membranes. With the increasing NMO loading, the finger-like pores became longer and the number of the micro-pores structure increased considerably Figure 1b–d [38]. In addition, some spongy macrovoids were also noticed towards the inner ultra-thin layer at the higher NMO loading. This observation is in agreement with previous studies reported [38–40]. Though at 1.25 g-NMO loading, the pores structure was significantly suppressed and this may be due to the inhomogeneity dispersion of the Nanoparticles [25]. The uneven dispersion of the NMO at higher dosage resulted in the formation of agglomerated particles within the matrix structure [41]. Thus, the resultants membrane comprised of dense structure with suppressed pore sizes as shown in Figure 1e. From Figure 1b,c, homogenous dispersion of NMO can be observed, and this indicates a good compatibility at the loadings within the matrix structure.

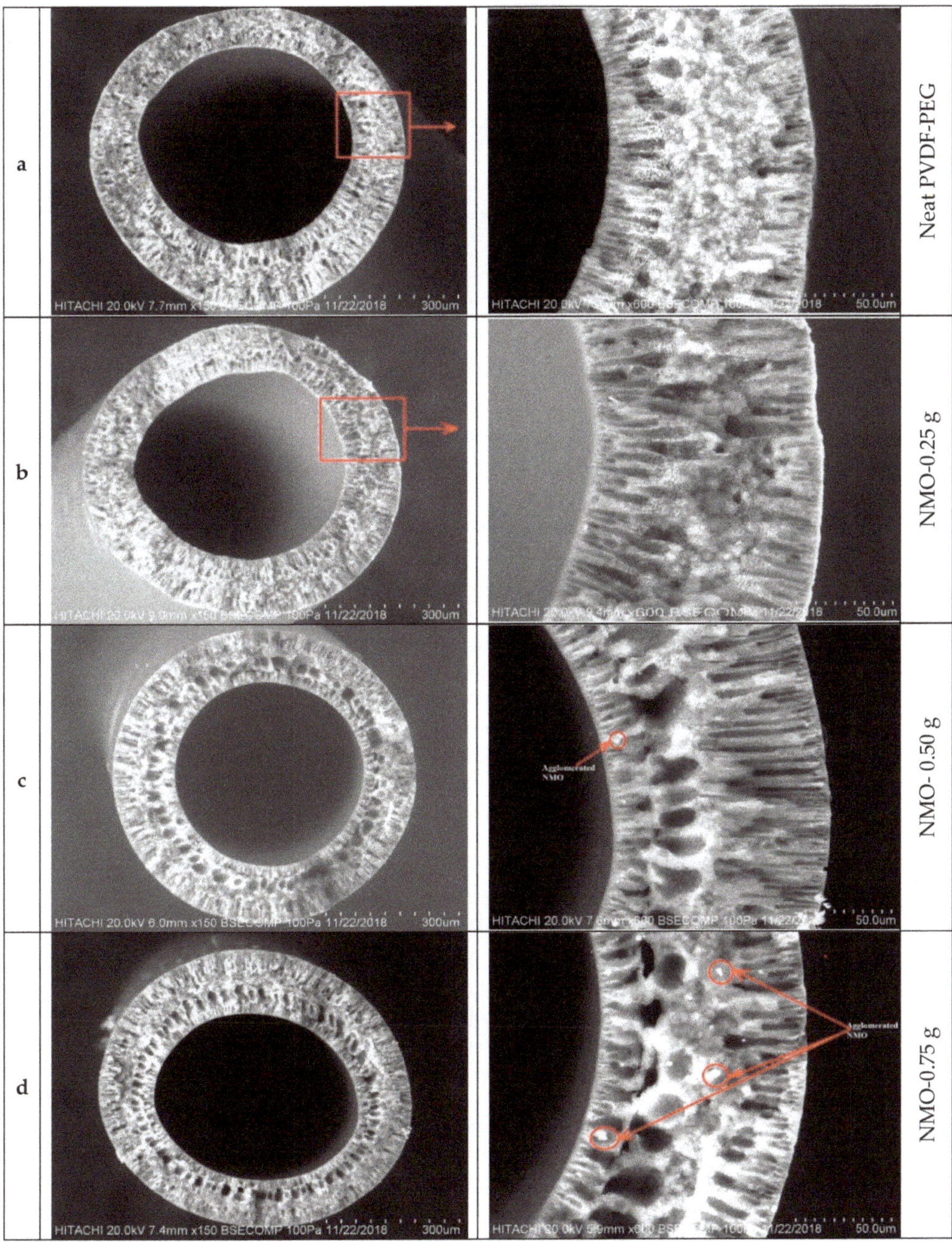

Figure 1. *Cont.*

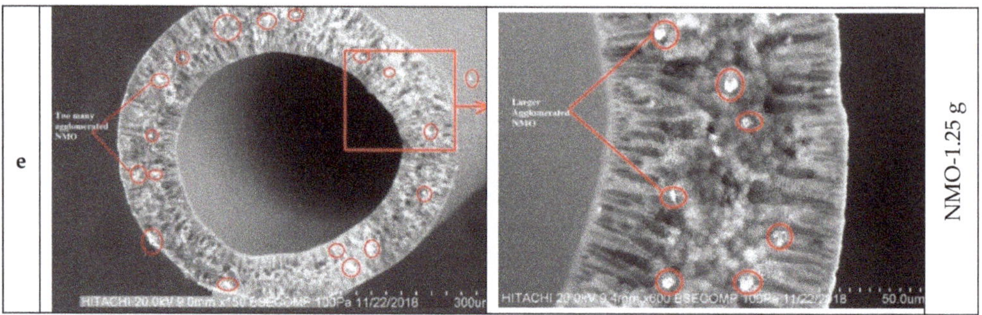

Figure 1. SEM Micro-structure of cross-sectional view of the Spun Fiber: (**a–e**) with different NMO loading.

In recap, NMO loading has demonstrated a notable effect on the shape and magnitude of the pores formation. An increase in the dosage of NMO results in continuously suppression of the sandwich structure and accompanied with the increase of spongy-finger like pores at the inner and outer walls of the fiber, respectively, as indicated in Figure 1b–d. However, a contrary scenario was observed in the Figure 1e, which presented dense structure even with higher NMO loading of 1.25 g. This is indicating that a relatively uniformly distributed NMO within the polymeric matric structure were achieved at a loading range of 0.25–0.75 g to give larger surface interaction with O–H in the coagulating water bath [42]. This phenomenon often leads to formation of finger-like pore substructure and/or combination with spongy voids, fine gravimetric, porosity and thin skin layers [43]. Conversely, the tendency of non-uniformly distribution of the augmented nanoparticles (NMO) advances with increase in the dosage (1.25 g), since the spinning parameters were maintained constant for all the samples [44]. Thus, the excessive NMO got agglomerated, thereby skewing the surface interaction with O–H during crystallization. This effect in conjunction with the increase in the viscosity of the dope solution due to the higher NMO dosage, jointly delayed the demixing process, and consequently, the formation of denser-sandwich structure alongside with suppressed finger-like pore structure [25,41,43]. In addition, high NMO content dopes are in meta-stable states that are extremely supersaturated with respect to polymer crystallization [43], and as such there may exhibit a considerable amount of pre-nucleation embryos along with several aggregated Nano-particles in the dope [25,43].

The results of the porosity analysis concurred with the morphological structure observed from the SEM characterization. Figure 2 displayed the porosity analysis of the spun neat and modified hollow fibers (a–e). The neat membrane (a) had the least porosity of 64.12%, while the modified membranes recorded higher values of 67.38%, 78.96%, 81.51% and 70.17% for b, c, d and e fiber, respectively. Similar remarks have been previously reported on the use of NMO to modify polymeric membranes [39,40,45]. The blended NMO accelerated the exchange speed between the non-solvent (water) and solvent (DMF) through the gelation process, thus, the porosity was increased considerably. The positive effect of the NMO on the porosity property could improve the membrane permeate flux significantly [45,46]. However, the diminished porosity percentage noticed with 1.25 g NMO loading might be due to the retardation in the crystallization which resulted from the high viscosity of the dope solution and presence of agglomerated particle at the excessive loading [47]. This effect resulted in the formation of a denser structure with suppressed pores and several visible aggregated particles, Figure 1e. Therefore, it can be deduced that the excessive NMO undermine the pores formation which could significantly diminish the porosity along with permeability of the resultant membrane.

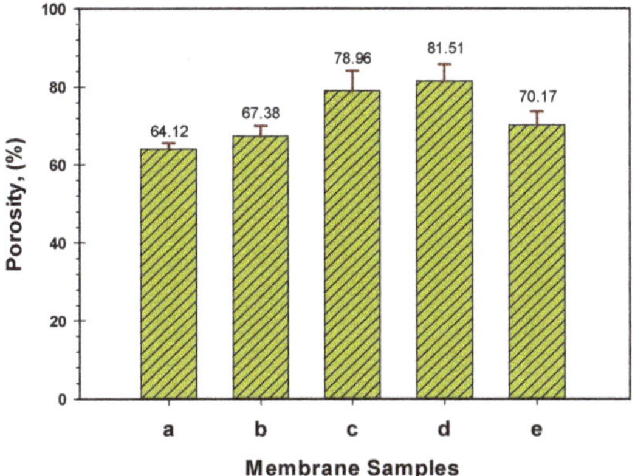

Figure 2. Porosity properties of neat (**a**), and modified membrane with varied NMO loading (**b–e**).

3.1.2. Membrane Hydrophilicity

The hydrophilic properties of the neat and modified membranes were examined based on surface contact angle analysis. Principally, surface with a lower contact angle is said to be more hydrophilic and water-liking [48]. Results of the contact angle analysis along with the respective goniometric images for the fabricated membranes are presented in Figure 3. Essentially, the goniometric image of the dropped probing water on the horizontally-positioned membrane samples were captured after 30 s. The membrane surface hydrophilicity increased with the increasing NMO loading. As can be seen from the Figure, the neat membrane presents the highest contact angle to the magnitude of 87.34°. However, considerable reductions in the contact angle were noticed with 0.50 g-NMO and 0.75 g-NMO loading to 60.01° and 57.19°, respectively. Essentially, the decrease in the contact angles is an indication of improvements in the hydrophilicity [24]. However, at higher NMO loading of 1.25 g, the contact angle upsurges to 75° and probably this might be due to the inhomogeneity dispersion and aggregation of the NMO particles which essentially reduced its overall effect. In a whole, the hydrophilicity of the modified membrane has shown a correlation with the dosage of the NMO, which also influences the pores morphology [40], formation of hydroxyl functional group [17,39] and surface charges [42].

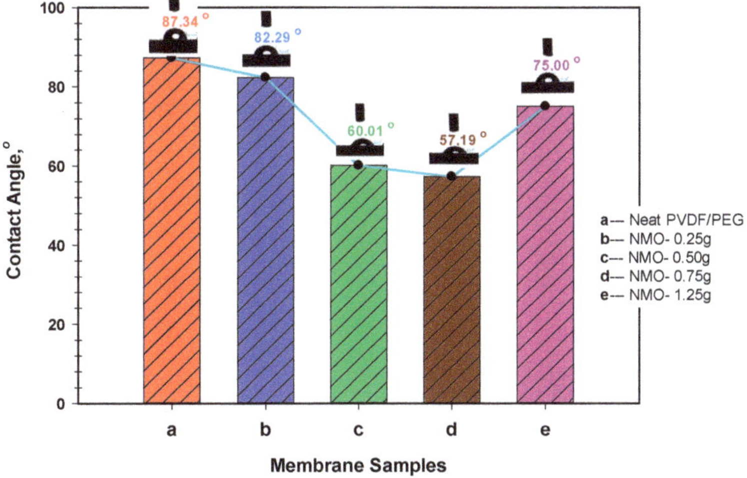

Figure 3. Hydrophilicity of neat PVDF-PEG (**a**), and modified Nano-hybrid membranes (**b–e**) with the respective goniometric images after 30 second of contact.

3.1.3. Elemental Analysis

EDX mapping was used to examine the presence of NMO within the matrix structure of the spun hollow fibers. Figure 4a presents the EDX spectrum of the neat membrane, and it clearly shows no trace of NMO in the composition. The major constituted elements are the C and F, which were the primary elements of the polymeric materials (PVDF/PEG). However, Figure 4b–e has evidenced the presence of the modified Nanocomposite membranes with various loadings of NMO particles. As expected, Mg composition was observed in the spectrum of all the modified fibers at distinct intensities with respect to the NMO dosage. The mapping of Mg in EDX spectra indicates successful blending of the NMO into the matrix structure as well as on the surface of the membranes. Despite the compatibility, only the 0.25 g-NMO and 0.5 g-NMO modified membranes present a free agglomeration with the matrix as depicted in Figure 4b,c. However, at higher NMO loading (0.75 and 1.25 g), an inhomogeneity dispersion and cluster of particles were observed (Figure 4d,e). Reports have shown that presence agglomerated particles in dope solution increases the viscosity, and the this results to irregular nucleation along with uneven crystallization process during the phase separation process [36].

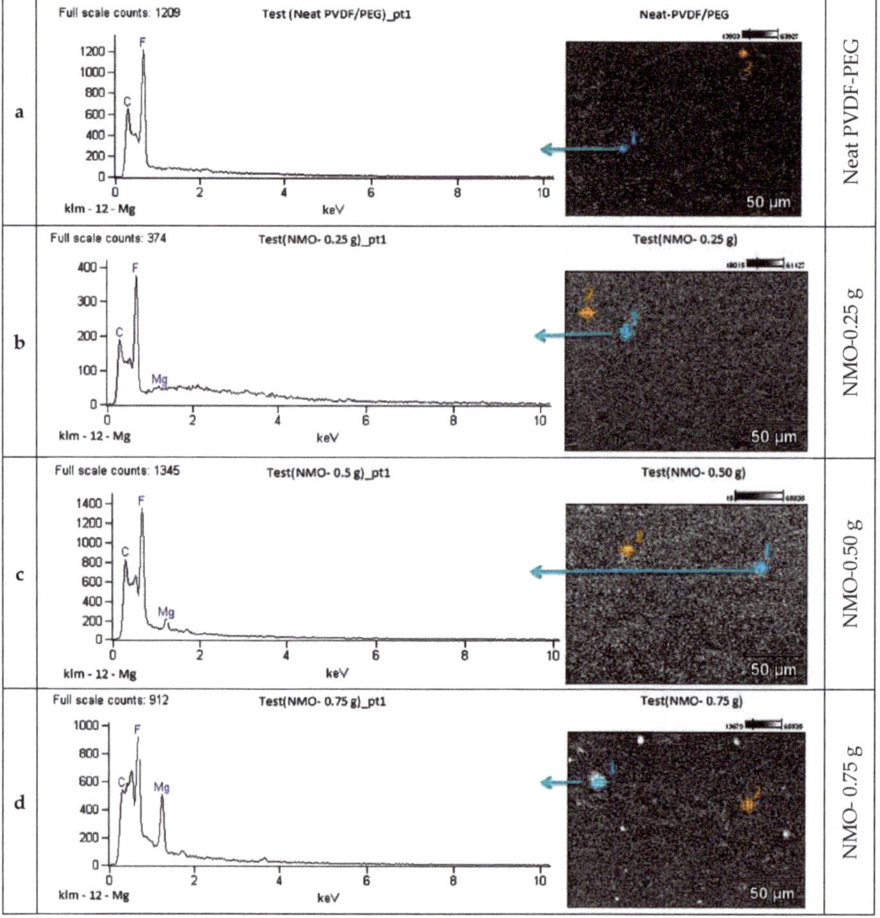

Figure 4. *Cont.*

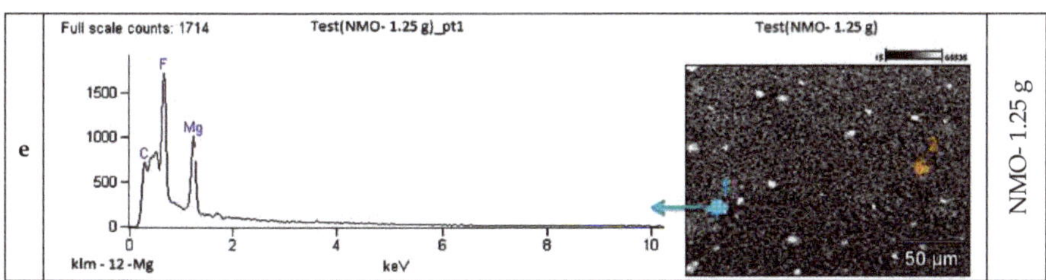

Figure 4. SEM/EDX spectra Mapping the Presence of NMO in (**a–e**) membrane.

3.1.4. Surface Charge of Nano-MgO Membrane

The surface charges of the neat and modified membrane were examined using surface zeta-potential analyzer. The analysis of the surface zeta-potentials for the fabricated hollow fibers with respect to the varied wide range of pH (2 to 10) are presented in Figure 5. The equilibrium isoelectric point of the neat membrane was observed at the pH 3.6, which is in agreement with the previous study [49]. However, the addition of NMO into the dope formulation significantly influenced the surface negativity of the membranes due to the oxidation effect and the formation of the acidic oxides [17]. As shown in Figure 5, the surface of the membranes became more negatively charged with the increasing amount of NMO. This indicates that the NMO became hydrated when added into DMF solvent [17,47]. During the spinning process, the NMO exposed to the membrane surface were hydrolyzed to form some hydroxyl functional groups in the presence of water. The protonation of the NMO has resulted in the deprotonation of membrane surface [42]. Thus, the membrane surface became more negatively charged [17,25]. The fiber with 0.75 g NMO loading presented the highest negative zeta potential within the pH range of 5.5 and 9 with value of −41.99 and −57.44, respectively (Figure 5). The lower surface charges in the 1.25 g NMO loading might be due to excessive agglomeration which resulted in the uneven dispersion and reduction in overall active-surface-area [42]. The membrane with 0.5 g NMO loading recorded higher zeta potential than the fiber with 0.75 g loading, (Figure 5). From same figure, it is obvious that the neat membrane had the least zeta potential at both extreme of the pH range (5.5 and 9) with value of −18.43 and −38.17, respectively. Based on this result, it can be deduced that to derive intensive surface negativity, NMO loading should be maintained between 0.5 and 0.75 g. One of the important applications of negatively surface charged membrane is in the separation of like-charged color pigments; such as the lignin and tannin substances in a typical oily wastewater that contains POME [9].

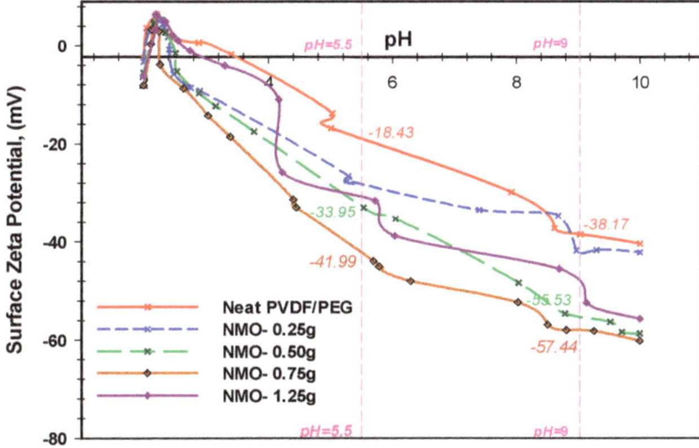

Figure 5. Surface zeta potential of neat PVDF-PEG, and modified Nano-hybrid membranes.

More so, researchers have reported that the main effect of the deprotonation process of a Nano-composite polymeric membrane is the resulted hydroxyl functional groups (OH–) formed on the membrane interacting surface in aqueous media, and this accounted for the surface negatively charge zeta potential [17,25]. More interestingly, the presences of this OH– on the membrane interacting interface improves its water-liking properties (hydrophilicity), permeability as well as repulsion of hydrophobic foulants, such as extracellular polymeric substance (EPS) [9,17]. This remark is strongly in agreement with other studies [17,39,46].

3.2. Effect of Nano-MgO on Membrane Performance

3.2.1. Permeability

Permeability results of the spun fibers both in pure water and diluted POME (4285 ADMI concentration) are shown in Figure 6. The modified fibers demonstrated higher permeate flux performance in both pure water and the diluted-POME. The modified membrane with 0.5 g-NMO and 0.75 g-NMO loading recorded almost equal water flux with 198.35 L/m^2·h and 201.22 L/m^2·h, respectively. The 1.25 g-NMO and 0.25 g NMO modified membrane had 97.40 L/m^2·h and 83.02 L/m^2·h water flux, respectively, whereas, the neat membrane presented the least flux of 80.72 L/m^2·h. Based on the flux pattern, it shows that the addition of NMO has a positive impact on the permeability, and this may be due to its ability to enhance the hydrophilicity through the protonation process. However, this effect seems otherwise at higher NMO loading of 1.25 g, Figure 6. This may be attributed to the suppressed pores and dense structure which were the consequential effect of the excessive aggregated particles present within the matrix structure.

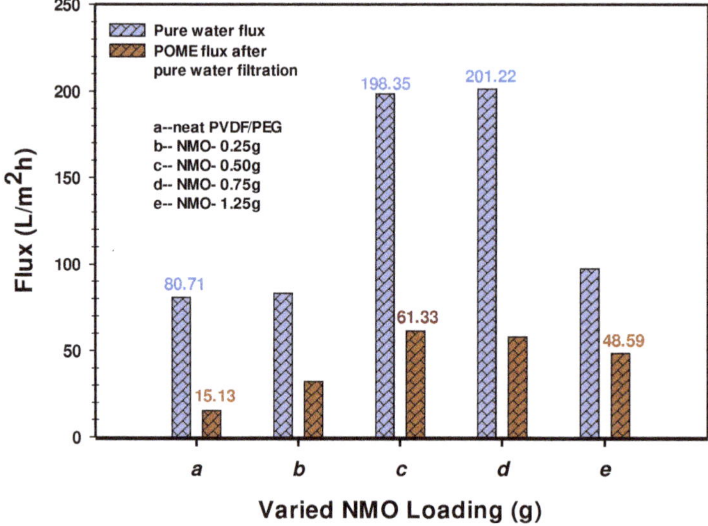

Figure 6. Flux of pure water and POME for neat PVDF-PEG (**a**), and modified Nano-hybrid fibers (**b–e**).

Similar trend was observed in the filtration of the diluted-POME. Both 0.5 g-NMO and 0.75 g-NMO modified membranes presents higher flux of 61.33 L/m^2·h and 57.97 L/m^2·h, respectively. On the contrary, the neat membrane had the least flux of 15.13 L/m^2·h, while the 1.25 g-NMO modified membrane recorded 48.59 L/m^2·h. Collectively, the magnitudes of the fluxes in POME filtration were noticeably lower compared to the pure water filtration. This could be due to the presence of contaminants (such as suspended solids, color pigments, organic and inorganic substances) in the POME, which apparently restrict the free flow of permeate through the membranes. Basically, the selectivity and rejection of the contaminants is based on size differences in the pores as well as the surface zeta potential of the membrane [17]. In addition, the results of the pure water flux after POME filtration indicated that membrane with 0.50 g NMO loading had an outstanding recovery with flux of 183.11 L/m^2·h.

3.2.2. Rejection of Color Pigment from POME

Figure 7A presents the color removal efficiency for both neat and modified membranes. Noticeably, the membrane with 0.50 g and 0.75 g NMO loading demonstrated outstanding color removal with 74.65% and 72.94%, respectively. It can be noticed that despite the high loading of 1.25 g-NMO, the modified fiber "e" recorded a lower rejection of 47.18% compared to the membrane with 0.50 g-NMO loading. The reason may not be devoid of the inhomogeneity dispersion of the NMO particles which instigated diminution in the effective interacting surface area required to repel the color pigments [36]. Also, Figure 7B display the pictorial visual color differences in feed POME, permeate of the neat and modified membrane as well as pure water. Principally, pore size is a key factor in the separation of color pigment from the POME [17]. This shows that the membrane with pores sizes smaller than the size of the contaminants is capable of giving higher rejection efficiency, though the permeability performance may diminish [50,51]. Besides, the distinct upturn rejection efficiency obtained was not only influenced by the pore sizes but majorly due to the strong negatively charge surface zeta potential. Basically, the surface negativity of the membrane was developed as a result of NMO protonation effect in the presence of water, as explained earlier. In this study, the fed POME has a pH of 8.45 and according to the previous studies, the pH ranging from 8 to 9 was due to the contained color pigments (lignin and tannin) [17,47]. The pigments became negatively charged when the pH of feed was adjusted to acidic conditions (ranging from pH 5.5), this assist in the aggregation of the color pigments [17]. Thus, under this condition of likes negatively charges of color pigments and membrane surface, repulsion of the color pigment prevails during the filtration. This phenomenon considerably improved the color removal efficiency, as observed in the modified fiber (c) and (d).

Figure 8 depicts a simple representation of the color rejection mechanism based on surface negativity of the membrane. As mentioned earlier, the principle of the color rejection was based on the protonation process of both the color pigments present in the diluted POME and NMO to form like charges on the surfaces [47]. During filtration, the hydrated and negatively charged color pigments in the POME migrated towards the surface of the membrane. The presence of water in the medium preceded the protonation of the NMO contained on the surface within the matrix of the membrane, thus resulted in the formation of negative charges. The surging negatively charged pigments were repelled when they approached the membrane surface with like charges [45]. The repulsion of the pigments prevented the deposition of hydrophobic substances (tannin and lignin) on the membrane surface and pore walls. On this basis, the rejection of color pigments not only improves the separation efficiency but also enhance fouling control. Though, some of the un-repelled color pigments with smaller sizes were still able to navigates through the membrane pores along with permeate.

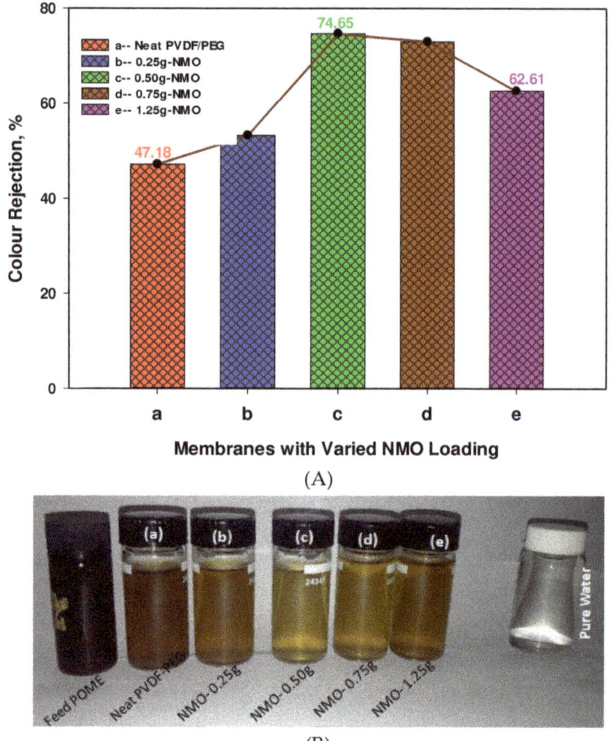

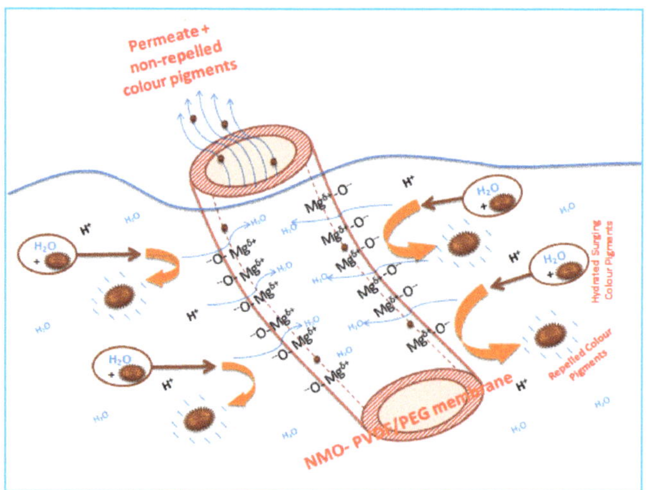

Figure 7. (**A**) Color rejection performance of neat PVDF-PEG (**a**) and modified Nano-hybrid membrane (**b–e**). (**B**). Pictorial view of color differences in the feed POME, permeate of (**a**) neat, (**b–e**) modified membranes, and pure water.

Figure 8. Schematic of color rejection and antifouling mechanism of modified Nanocomposite NMO-PVDF/PEG membranes during filtration.

3.2.3. Antifouling Performance and Membrane Reusability

Fouling involves deposition of foulants on the surface and pore walls of a polymeric membrane. Several reports have shown that hydrophobic membrane are more prone to severe fouling due to the strong adhesive attraction that exist between the interacting interface [41,52]. This implies that suppressing the hydrophobic properties through the incorporation of hydrophilic Nanoparticles (NMO) into the membrane matrix structure has the potential to curtail foulants deposition, hence the improvement in antifouling properties. On this note, 3 cycles of POME filtration analysis were conducted in accordance to Subramaniam et al. [9] procedure. Throughout, the filtration time and color concentration of the feed were maintained at 180 min and 4285 ADMI for each of the cycles. At the end of each filtration cycle, the membranes were only physically cleaned under running tap water for 15 min. Essentially, this experiment gives the basis to examine the membrane reusability and antifouling performance using FR, RF and IF as the indicators.

Figure 9 presents all the 3 filtration cycles' antifouling performance of the spun membranes. From the cycle 1, 0.5 g-NMO modified membrane recorded the highest FR of 92.32%, while the neat membrane had 27.18%. The FR of 1.25 g, 0.75 g and 0.25 g NMO modified membranes were 69.32%, 86.45% and 61.05%, respectively. Furthermore, the fouling resistance performances of the modified membranes were also superior to the neat membrane. The RF% and IF% of the 0.25 g-NMO, 0.5 g-NMO, 0.75 g-NMO and 1.25 g-NMO were (22.36, 38.95), (61.39, 7.68), (57.64, 13.55) and (19.44, 30.68), respectively, as against (8.44, 72.82) for that of the neat membrane. Even at the end of the third filtration cycle 3, the membrane (c) with 0.50 g NMO loading recorded an FR over 90% with good fouling resistance RF% and IF% at 60.64 and 8.77%, respectively. On the contrary, the flux recovery and antifouling resistance of the neat membrane deteriorated significantly with FR%, RF% and IF% of 10.33%, 4.14% and 91.73%, respectively.

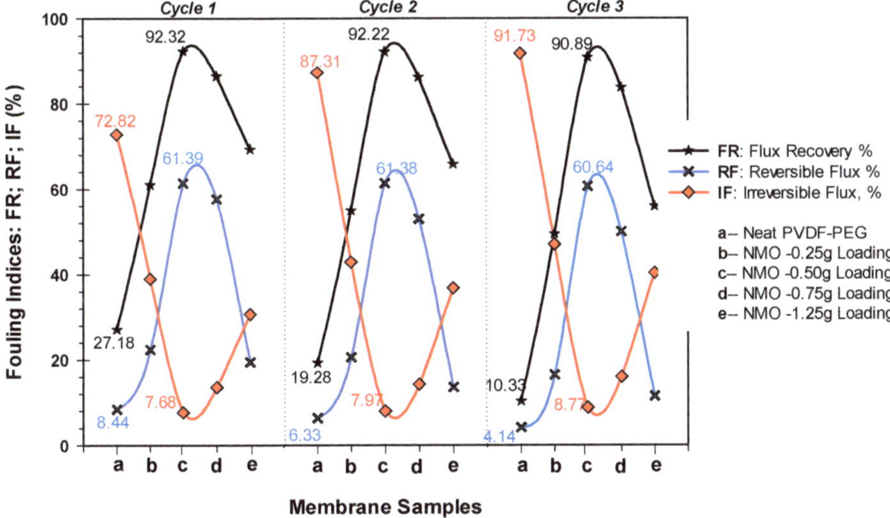

Figure 9. Fouling characteristics (FR, RF and IF) of membrane fibers during three (3) consecutive filtration cycles using POME as feed.

Overall, the better antifouling performance of the resultant modified membranes was due to the NMO which essentially improved the hydrophilicity alongside with the negatively surface zeta-potential [25]. Particularly, membrane with 0.50 g loading showed a remarkable improved antifouling performance, and only loss of 1.43% FR with relatively steady RF and IF were observed at the end of the third filtration cycles-3 using diluted-POME as feed. This excellent antifouling performance of the membrane at this loading may be due to the homogenous dispersion of the Nanoparticles, thereby creating superior interacting surface for effective repellence of foulants [36]. More so, apart from the protonation purpose of the NMO in the presence of water to develop a negative charges on the membrane surface, it is also capable of generating free reactive oxygen species (ROS) to inhibit bio-fouling [36]. Thus, preventing the formation of bio-film and cake layer on the membrane surface and pore walls. In addition, Hikku et al. [42] reported that the NMO antifouling activity has a correlation with magnitude of surface area of the Nanoparticle contacting with the microbes (foulants). This implies that evenly dispersed NMO within the membrane matrix presents larger interacting surface area, thus advancing antifouling as well as flux recoverability [25].

3.2.4. FTIR Analysis of used Membranes

Figure 10 compares the FTIR spectra of the fouled neat membrane (red spectrum) and the modified membrane (blue spectrum) after-used. The modified-membrane had a broad absorbance at 3440, 2926–2304 and 725 cm^{-1}, and these can be attributed to the stretch disturbance of O–H, C–H and Mg–O, respectively [40]. The first functional group (O–H) often indicate the degree of surface hydrophilicity and antifouling properties [24,40,53], as accounted by the broad bands of O–H (blue spectrum). In addition, the FTIR spectra suggest that Nano-MgO (NMO) was successfully incorporated into the PVDF-PEG matrix structure using phase inversion technique. On the contrary, no similar band observed in the neat membrane spectra (red spectrum). The noticeable bands in the neat membrane spectra appeared at 1663, 1397, 1175, 1067 and 883 cm^{-1} which can be attributed to stretch vibrations of amide-I, methyl bonds, CF$_2$, polysaccharides and complex aromatic functional group, respectively [54,55]. It was noticed that the stretch band at 1175 cm^{-1} was peculiar in both spectra which resulted due to intrinsic vibration of asymmetric CF$_2$ of the PVDF-PEG composition [56]. As suggested by the spectra (red spectrum) of the neat membrane, among the most prevailing functional groups the protein (amide-I 1663 cm^{-1}) and polysaccharides (1067 cm^{-1}) were both included. These are major source of EPS and SMP substance that are usually responsible for the initiation and formation of biofilm which eventually degenerates into cake layer [57]. Therefore, this shows that the neat membrane is highly prone to fouling. On the contrary, the deprotonating process of the Nano-MgO (NMO) improves the intensity of the reactive O–H (blue spectrum) in the modified membrane. Thus, generated O–H denaturize the deposited EPS/SMP as well as improving the hydrophilicity of the polymeric membrane [40]. This phenomenon effectively averts formation of the cake layer on the modified membrane, as observed in Figure 10.

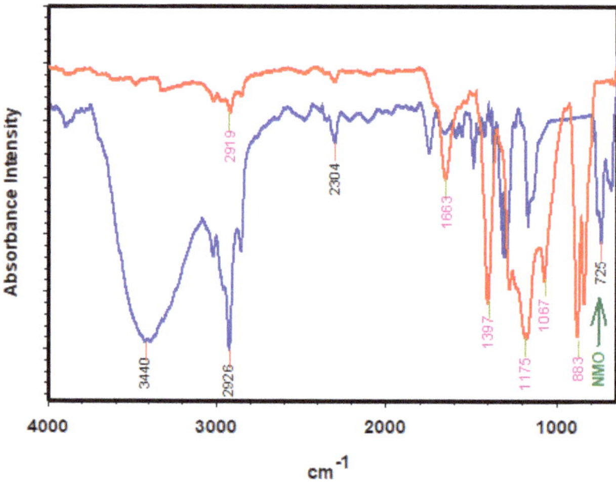

Figure 10. FTIR spectra of used neat membrane (red spectrum), and the modified membrane (blue spectrum).

3.3. Performance Appraisal with Literatures

This section presents a concise comparison of this study with previous works that uses MgO Nanoparticle (NMO) to modified polymeric membranes. In this study, NMO was successfully incorporated into the PVDF-PEG matrix using phase inversion technique. This resulted in considerable augmentation of the hydrophilicity and surface negativity of the spun hollow fibers. At 0.50 g NMO loading, a remarkable water flux, and POME permeability flux along with the color rejection of 198.35 L/m^2·h, 61.33 L/m^2·h and 74.65% were obtained, respectively. Even after the 3 filtration cycle of a total period of 9 h, the flux recovery percentage remains relatively steady at 92.32%. Previously, Parvizian et al. [39] reported that at a varied loading of NMO blended into PVC matrix, the flux improves and the highest flux was obtained at 0.5 wt % dosage. However, a comprehensive fouling analysis and color rejection of the PVC membrane were not considered. Arumugham et al. [58] used NMO to modify sulfonated polyphenyl sulfone (SPPSU) polymeric membrane and applied to treat oily wastewater. The NMO loading was fixed at 25 wt % while the SPPSU and organic solvent were varied. From the results, a good hydrophilic performance (99.00% of flux recovery) was achieved. However, the improvement in the hydrophilicity may not be attributed to incorporated NMO since the dosage was not varied [58]. Based on this appraisal, it can be deduced that this study not only bridge the information gap, but also reported the significant role of the NMO at varied dosage in improving performance of the hybrid PVDF-PEG membrane.

4. Conclusions

Hybrid PVDF-PEG hollow fibers blended with NMO at various loading of 0–1.25 g were fabricated using phase inversion technique. The increasing NMO loading has demonstrated a significant effect on the morphology, hydrophilicity, permeability and antifouling properties of the resultants composite hollow fibers. The loading at 0.50 g-NMO presented the most auspicious performance with 198.35 and 61.33 L/m^2·h of pure water and POME filtration flux, respectively. After a continues 3 filtration cycles of diluted-POME for a total period of 9 h, the 0.50 g-NMO composite membrane recorded the best flux recovery (FR), reversible fouling (RF) and least irreversible fouling (IF) percentages of 90.98%, 61.39% and 7.68%, respectively. The outstanding performance was due to the homogenous distribution and compatibility of the membrane matrix with the NMO particles at 0.50 g loading. Conversely, after the third filtration cycle-3, a significant deterioration in the FR and fouling resistance (RF and IF) were noticed at higher loading of 1.25 g-NMO with a values of 55.98%, 11.44% and 40.38%, respectively. This was due to the excessive NMO loading which resulted in the formation of numerous aggregated particles within the matrix structure. The agglomerated particles skewed the nucleation and protonation process during the phase separation and crystallization, as well as creating weak spots in the resultant membranes. Therefore, it can be deduced that Nano-hybrid PVDF-PEG membrane with 0.50 g-NMO loading presents the best performance. Overall, the modified fibers presented better performances compared to the neat membrane.

5. Patents

Malaysian Patent No. PI 2019006570, 2019: A Hybrid System and Method for Treating Palm Oil Mill Effluent.

Author Contributions: Conceptualization, M.A., and H.C.M.; methodology, M.A., H.C.M. and P.S.G.; software, M.A.; validation, M.A., H.C.M., P.S.G., K.F.Y., and Z.Z.A.; formal analysis, M.A., H.C.M. and P.S.G.; investigation, M.A.; resources, H.C.M., P.S.G., A.I.M.I. and A.F.I.; data curation, M.A.; writing—original draft preparation, M.A.; writing—review and editing, M.A., H.C.M., P.S.G., K.F.Y., Z.Z.A. and A.I.M.I.; visualization, M.A., H.C.M., and P.S.G.; supervision, H.C.M., K.F.Y., Z.Z.A. and A.I.M.I.; project administration, H.C.M., P.S.G. and A.F.I.; funding acquisition, H.C.M. and M.A. All authors have read and agreed to the published version of the manuscript.

Funding: This work was supported by the Universiti Putra Malaysia (PUTRA Impak: 9530900), and Tertiary Education Trust Fund (TETF/UNIV/ZARIA/ASTD/2017).

Conflicts of Interest: The authors declare no conflict of interest, and also the funders had no role in the design of the study; in the collection, analyses, or interpretation of data; in the writing of the manuscript, or in the decision to publish the results.

References

1. Nunes-Pereira, J.; Ribeiro, S.; Ribeiro, C.; Gombek, C.J.; Gama, F.M.; Gomes, A.C.; Patterson, D.A.; Lanceros-Méndez, S. Poly(vinylidene fluoride) and copolymers as porous membranes for tissue engineering applications. *Polym. Test.* **2015**, *44*, 234–241. [CrossRef]

2. Marquez, J.A.D.; Ang, M.B.M.Y.; Doma, B.T.; Huang, S.H.; Tsai, H.A.; Lee, K.R.; Lai, J.Y. Application of cosolvent-assisted interfacial polymerization technique to fabricate thin-film composite polyamide pervaporation membranes with PVDF hollow fiber as support. *J. Memb. Sci.* **2018**, *564*, 722–731. [CrossRef]

3. Gao, L.; Alberto, M.; Gorgojo, P.; Szekely, G.; Budd, P.M. High-flux PIM-1/PVDF thin film composite membranes for 1-butanol/water pervaporation. *J. Memb. Sci.* **2017**, *529*, 207–214. [CrossRef]

4. Subramaniam, M.N.; Goh, P.S.; Lau, W.J.; Ng, B.C.; Ismail, A.F. AT-POME colour removal through photocatalytic submerged filtration using antifouling PVDF-TNT nanocomposite membrane. *Sep. Purif. Technol.* **2018**, *191*, 266–275. [CrossRef]

5. Dharupaneedi, S.P.; Nataraj, S.K.; Nadagouda, M.; Reddy, K.R.; Shukla, S.S.; Aminabhavi, T.M. Membrane-based separation of potential emerging pollutants. *Sep. Purif. Technol.* **2019**, *210*, 850–866. [CrossRef]

6. Liu, T.; Ren, C.; Zhang, Y.; Wang, Y.; Lei, L.; Chen, F. Solvent effects on the morphology and performance of the anode substrates for solid oxide fuel cells. *J. Power Sources* **2017**, *363*, 304–310. [CrossRef]

7. Arefi-Oskoui, S.; Khataee, A.; Vatanpour, V. Effect of solvent type on the physicochemical properties and performance of NLDH/PVDF nanocomposite ultrafiltration membranes. *Sep. Purif. Technol.* **2017**, *184*, 97–118. [CrossRef]

8. Padaki, M.; Surya Murali, R.; Abdullah, M.S.; Misdan, N.; Moslehyani, A.; Kassim, M.A.; Hilal, N.; Ismail, A.F. Membrane technology enhancement in oil-water separation. A review. *Desalination* **2015**, *357*, 197–207. [CrossRef]

9. Subramaniam, M.N.; Goh, P.S.; Lau, W.J.; Tan, Y.H.; Ng, B.C.; Ismail, A.F. Hydrophilic hollow fiber PVDF ultrafiltration membrane incorporated with titanate nanotubes for decolourization of aerobically-treated palm oil mill effluent. *Chem. Eng. J.* **2017**, *316*, 101–110. [CrossRef]

10. Zeng, G.; Ye, Z.; He, Y.; Yang, X.; Ma, J.; Shi, H.; Feng, Z. Application of dopamine-modified halloysite nanotubes/PVDF blend membranes for direct dyes removal from wastewater. *Chem. Eng. J.* **2017**, *323*, 572–583. [CrossRef]

11. Ngang, H.P.; Ahmad, A.L.; Low, S.C.; Ooi, B.S. Preparation of thermoresponsive PVDF/SiO 2 -PNIPAM mixed matrix membrane for saline oil emulsion separation and its cleaning efficiency. *Desalination* **2017**, *408*, 1–12. [CrossRef]

12. Zhu, Y.; Wei, J.; Zhang, H.; Liu, K.; Kong, Z.; Dong, Y.; Jin, G.; Tian, J.; Qin, Z. Fabrication of composite membrane with adsorption property and its application to the removal of endocrine disrupting compounds during filtration process. *Chem. Eng. J.* **2018**, *352*, 53–63. [CrossRef]

13. Liao, Y.; Bokhary, A.; Maleki, E.; Liao, B. A review of membrane fouling and its control in algal-related membrane processes. *Bioresour. Technol.* **2018**, *264*, 343–358. [CrossRef] [PubMed]

14. Guo, J.; Farid, M.U.; Lee, E.J.; Yan, D.Y.S.; Jeong, S.; Kyoungjin An, A. Fouling behavior of negatively charged PVDF membrane in membrane distillation for removal of antibiotics from wastewater. *J. Memb. Sci.* **2018**, *551*, 12–19. [CrossRef]

15. Rabuni, M.F.; Nik Sulaiman, N.M.; Aroua, M.K.; Yern Chee, C.; Awanis Hashim, N. Impact of in situ physical and chemical cleaning on PVDF membrane properties and performances. *Chem. Eng. Sci.* **2015**, *122*, 426–435. [CrossRef]

16. Zin, G.; Wu, J.; Rezzadori, K.; Petrus, J.C.C.; Di Luccio, M.; Li, Q. Modification of hydrophobic commercial PVDF microfiltration membranes into superhydrophilic membranes by the mussel-inspired method with dopamine and polyethyleneimine. *Sep. Purif. Technol.* **2019**, *212*, 641–649. [CrossRef]

17. Tan, Y.H.; Goh, P.S.; Ismail, A.F.; Ng, B.C.; Lai, G.S. Decolourization of aerobically treated palm oil mill effluent (AT-POME) using polyvinylidene fluoride (PVDF) ultrafiltration membrane incorporated with coupled zinc-iron oxide nanoparticles. *Chem. Eng. J.* **2017**, *308*, 359–369. [CrossRef]

18. Zinadini, S.; Rostami, S.; Vatanpour, V.; Jalilian, E. Preparation of antibiofouling polyethersulfone mixed matrix NF membrane using photocatalytic activity of ZnO/MWCNTs nanocomposite. *J. Memb. Sci.* **2017**, *529*, 133–141. [CrossRef]

19. Li, J.H.; Shao, X.S.; Zhou, Q.; Li, M.Z.; Zhang, Q.Q. The double effects of silver nanoparticles on the PVDF membrane: Surface hydrophilicity and antifouling performance. *Appl. Surf. Sci.* **2013**, *265*, 663–670. [CrossRef]

20. Mauter, M.S.; Okemgbo, K.C.; Osuji, C.O.; Elimelech, M.; Wang, Y.; Giannelis, E.P. Antifouling ultrafiltration membranes via post-fabrication grafting of biocidal nanomaterials. *ACS Appl. Mater. Interfaces* **2011**, *3*, 2861–2868. [CrossRef]

21. Jhaveri, J.H.; Murthy, Z.V.P. A comprehensive review on anti-fouling nanocomposite membranes for pressure driven membrane separation processes. *Desalination* **2016**, *379*, 137–154. [CrossRef]

22. Meng, F.; Zhang, S.; Oh, Y.; Zhou, Z.; Shin, H.S.; Chae, S.R. Fouling in membrane bioreactors: An updated review. *Water Res.* **2017**, *114*, 151–180. [CrossRef]

23. Chae, H.R.; Lee, J.; Lee, C.H.; Kim, I.C.; Park, P.K. Graphene oxide-embedded thin-film composite reverse osmosis membrane with high flux, anti-biofouling, and chlorine resistance. *J. Memb. Sci.* **2015**, *483*, 128–135. [CrossRef]

24. Nasrollahi, N.; Vatanpour, V.; Aber, S.; Mahmoodi, N.M. Preparation and characterization of a novel polyethersulfone (PES) ultrafiltration membrane modified with a CuO/ZnO nanocomposite to improve permeability and antifouling properties. *Sep. Purif. Technol.* **2018**, *192*, 369–382. [CrossRef]

25. Shen, L.; Bian, X.; Lu, X.; Shi, L.; Liu, Z.; Chen, L.; Hou, Z.; Fan, K. Preparation and characterization of ZnO/polyethersulfone (PES) hybrid membranes. *Desalination* **2012**, *293*, 21–29. [CrossRef]

26. Mao, C.; Xiang, Y.; Liu, X.; Cui, Z.; Yang, X.; Yeung, K.W.K.; Pan, H.; Wang, X.; Chu, P.K.; Wu, S. Photo-Inspired Antibacterial Activity and Wound Healing Acceleration by Hydrogel Embedded with Ag/Ag@AgCl/ZnO Nanostructures. *ACS Nano* **2017**, *11*, 9010–9021. [CrossRef]

27. Elango, M.; Deepa, M.; Subramanian, R.; Mohamed Musthafa, A. Synthesis, characterization of polyindole/Ag[sbnd]ZnO nanocomposites and its antibacterial activity. *J. Alloys Compd.* **2017**, *696*, 391–401. [CrossRef]

28. Wu, H.; Liu, Y.; Mao, L.; Jiang, C.; Ang, J.; Lu, X. Doping polysulfone ultrafiltration membrane with TiO2-PDA nanohybrid for simultaneous self-cleaning and self-protection. *J. Memb. Sci.* **2017**, *532*, 20–29. [CrossRef]

29. Zodrow, K.; Brunet, L.; Mahendra, S.; Li, D.; Zhang, A.; Li, Q.; Alvarez, P.J.J. Polysulfone ultrafiltration membranes impregnated with silver nanoparticles show improved biofouling resistance and virus removal. *Water Res.* **2009**, *43*, 715–723. [CrossRef]

30. Maximous, N.; Nakhla, G.; Wong, K.; Wan, W. Optimization of Al2O3/PES membranes for wastewater filtration. *Sep. Purif. Technol.* **2010**, *73*, 294–301. [CrossRef]

31. Yemmireddy, V.K.; Hung, Y.C. Using Photocatalyst Metal Oxides as Antimicrobial Surface Coatings to Ensure Food Safety—Opportunities and Challenges. *Compr. Rev. Food Sci. Food Saf.* **2017**, *16*, 617–631. [CrossRef]

32. Ong, C.B.; Ng, L.Y.; Mohammad, A.W. A review of ZnO nanoparticles as solar photocatalysts: Synthesis, mechanisms and applications. *Renew. Sustain. Energy Rev.* **2018**, *81*, 536–551. [CrossRef]

33. Raghunath, A.; Perumal, E. Metal oxide nanoparticles as antimicrobial agents: a promise for the future. *Int. J. Antimicrob. Agents* **2017**, *49*, 137–152. [CrossRef]

34. Van Thuan, T.; Quynh, B.T.P.; Nguyen, T.D.; Ho, V.T.T.; Bach, L.G. Response surface methodology approach for optimization of Cu2+, Ni2+ and Pb2+ adsorption using KOH-activated carbon from banana peel. *Surf. Interfaces* **2017**, *6*, 209–217. [CrossRef]

35. Baruah, A.; Chaudhary, V.; Malik, R.; Tomer, V.K. *Nanotechnology Based Solutions for Wastewater Treatment*; Elsevier Inc.: Amsterdam, The Netherlands, 2018; ISBN 9780128139028.

36. Hikku, G.S.; Jeyasubramanian, K.; Vignesh Kumar, S. Nanoporous MgO as self-cleaning and anti-bacterial pigment for alkyd based coating. *J. Ind. Eng. Chem.* **2017**, *52*, 168–178. [CrossRef]

37. Fakayode, S.O.; Baker, G.A.; Bwambok, D.K.; Bhawawet, N.; Elzey, B.; Siraj, N.; Macchi, S.; Pollard, D.A.; Perez, R.L.; Duncan, A.V.; et al. Molecular (Raman, NIR, and FTIR) spectroscopy and multivariate analysis in consumable products analysis 1. *Appl. Spectrosc. Rev.* **2019**, 1–77. [CrossRef]

38. Han, S.; Mao, L.; Wu, T.; Wang, H. Author's Accepted Manuscript inversion method. *J. Memb. Sci.* **2016**, *516*, 47–55. [CrossRef]

39. Parvizian, F.; Sadeghi, Z.; Hosseini, S.M. PVC Based Ion-Exchange Membrane Blended with Magnesium Oxide Nanoparticles for Desalination: Fabrication, Characterization and Performance. *J. Appl. Membr. Sci. Technol.* **2017**, *21*, 11–24. [CrossRef]

40. Dong, C.; He, G.; Li, H.; Zhao, R.; Han, Y.; Deng, Y. Antifouling enhancement of poly (vinylidene fluoride) microfiltration membrane by adding $Mg(OH)_2$ nanoparticles. *J. Memb. Sci.* **2012**, *387–388*, 40–47. [CrossRef]

41. Zhu, Z.; Jiang, J.; Wang, X.; Huo, X.; Xu, Y.; Li, Q.; Wang, L. Improving the hydrophilic and antifouling properties of polyvinylidene fluoride membrane by incorporation of novel nanohybrid GO@SiO2 particles. *Chem. Eng. J.* **2017**, *314*, 266–276. [CrossRef]

42. Mukherjee, A.; Mohammed Sadiq, I.; Prathna, T.C.; Chandrasekaran, N. Antimicrobial activity of aluminium oxide nanoparticles for potential clinical applications. *Sci. against Microb. Pathog. Commun. Curr. Res. Technol. Adv.* **2011**, *1*, 245–251.

43. Lin, D.J.; Chang, C.L.; Huang, F.M.; Cheng, L.P. Effect of salt additive on the formation of microporous poly(vinylidene fluoride) membranes by phase inversion from LiClo 4 /water/DMF/PVDF system. *Polymer (Guildf).* **2002**, *44*, 413–422. [CrossRef]

44. Vani, C.V.; Karuppasamy, K.; Sridevi, N.A.; Balakumar, S.; Shajan, X.S. Effect of electron beam irradiation on the mechanical and electrochemical properties of plasticized polymer electrolytes dispersed with nanoparticles. *Adv. Mater. Res.* **2013**, *678*, 229–233. [CrossRef]

45. Younas, H.; Bai, H.; Shao, J.; Han, Q.; Ling, Y.; He, Y. Super-hydrophilic and fouling resistant PVDF ultrafiltration membranes based on a facile prefabricated surface. *J. Memb. Sci.* **2017**, *541*, 529–540. [CrossRef]

46. Emadzadeh, D.; Ghanbari, M.; Lau, W.J.; Rahbari-Sisakht, M.; Rana, D.; Matsuura, T.; Kruczek, B.; Ismail, A.F. Surface modification of thin film composite membrane by nanoporous titanate nanoparticles for improving combined organic and inorganic antifouling properties. *Mater. Sci. Eng. C* **2017**, *75*, 463–470. [CrossRef]

47. Liu, C.; Lee, J.; Small, C.; Ma, J.; Elimelech, M. Comparison of organic fouling resistance of thin-film composite membranes modified by hydrophilic silica nanoparticles and zwitterionic polymer brushes. *J. Memb. Sci.* **2017**, *544*, 135–142. [CrossRef]

48. Ho, K.C.; Teow, Y.H.; Ang, W.L.; Mohammad, A.W. Novel GO/OMWCNTs mixed-matrix membrane with enhanced antifouling property for palm oil mill effluent treatment. *Sep. Purif. Technol.* **2017**, *177*, 337–349. [CrossRef]

49. Shi, B.-L.; Su, X.; He, J.; Wang, L.-L. Surface hydrophilicity modification of PVDF membranes with an external electric field in the phase inversion process. *Membr. Water Treat.* **2015**, *6*, 351–363. [CrossRef]

50. Wang, X.M.; Li, X.Y.; Shih, K. In situ embedment and growth of anhydrous and hydrated aluminum oxide particles on polyvinylidene fluoride (PVDF) membranes. *J. Memb. Sci.* **2011**, *368*, 134–143. [CrossRef]

51. Lim, S.K.; Goh, K.; Bae, T.H.; Wang, R. Polymer-based membranes for solvent-resistant nanofiltration: A review. *Chinese J. Chem. Eng.* **2017**, *25*, 1653–1675. [CrossRef]

52. Abed, M.M.R.; Kumbharkar, S.C.; Groth, A.M.; Li, K. Ultrafiltration PVDF hollow fibre membranes with interconnected bicontinuous structures produced via a single-step phase inversion technique. *J. Memb. Sci.* **2012**, *407–408*, 145–154. [CrossRef]

53. Zuo, G.; Wang, R. Novel membrane surface modification to enhance anti-oil fouling property for membrane distillation application. *J. Memb. Sci.* **2013**, *447*, 26–35. [CrossRef]

54. Ashfaq, M.Y.; Al-Ghouti, M.A.; Qiblawey, H.; Zouari, N. Evaluating the effect of antiscalants on membrane biofouling using FTIR and multivariate analysis. *Biofouling* **2019**, *35*, 1–14. [CrossRef]

55. Rahman, M.M.; Al-Sulaimi, S.; Farooque, A.M. Characterization of new and fouled SWRO membranes by ATR/FTIR spectroscopy. *Appl. Water Sci.* **2018**, *8*, 1–11. [CrossRef]

56. Wang, Z.; Wu, Z.; Yin, X.; Tian, L. Membrane fouling in a submerged membrane bioreactor (MBR) under sub-critical flux operation: Membrane foulant and gel layer characterization. *J. Memb. Sci.* **2008**, *325*, 238–244. [CrossRef]

57. Gao, D.; Fu, Y.; Ren, N. Tracing biofouling to the structure of the microbial community and its metabolic products: A study of the three-stage MBR process. *Water Res.* **2013**, *47*, 6680–6690. [CrossRef]
58. Arumugham, T.; Kaleekkal, N.J.; Rana, D.; Doraiswamy, M. Separation of oil/water emulsions using nano MgO anchored hybrid ultrafiltration membranes for environmental abatement. *J. Appl. Polym. Sci.* **2016**, *133*, 1–12. [CrossRef]

 *polymers*

Article

Kinetic Studies on the Catalytic Degradation of Rhodamine B by Hydrogen Peroxide: Effect of Surfactant Coated and Non-Coated Iron (III) Oxide Nanoparticles

Mohd Shaban Ansari [1], Kashif Raees [1], Moonis Ali Khan [2], M.Z.A. Rafiquee [1,*] and Marta Otero [3,*]

1   Department of Applied Chemistry, Zakir Hussain College of Engineering and Technology, Aligarh Muslim University, Aligarh 202002, UP, India; shabanansari126@gmail.com (M.S.A.); raeeskashif@gmail.com (K.R.)
2   Chemistry Department, College of Science, King Saud University, Riyadh 11451, Saudi Arabia; mokhan@ksu.edu.sa
3   CESAM—Centre for Environmental and Marine Studies, Department of Environment and Planning, University of Aveiro, Campus de Santiago, 3810-193 Aveiro, Portugal
*   Correspondence: drrafiquee@yahoo.com (M.Z.A.R.); marta.otero@ua.pt (M.O.)

Received: 3 September 2020; Accepted: 26 September 2020; Published: 29 September 2020

Abstract: Iron (III) oxide (Fe_3O_4) and sodium dodecyl sulfate (SDS) coated iron (III) oxide (SDS@Fe_3O_4) nanoparticles (NPs) were synthesized by the co-precipitation method for application in the catalytic degradation of Rhodamine B (RB) dye. The synthesized NPs were characterized using X-ray diffractometer (XRD), vibrating sample magnetometer (VSM), scanning electron microscopy (SEM), transmission electron microscopy (TEM), and Fourier transform infra-red (FT-IR) spectroscopy techniques and tested in the removal of RB. A kinetic study on RB degradation by hydrogen peroxide (H_2O_2) was carried out and the influence of Fe_3O_4 and SDS@Fe_3O_4 magnetic NPs on the degradation rate was assessed. The activity of magnetic NPs, viz. Fe_3O_4 and SDS@Fe_3O_4, in the degradation of RB was spectrophotometrically studied and found effective in the removal of RB dye from water. The rate of RB degradation was found linearly dependent upon H_2O_2 concentration and within 5.0×10^{-2} to 4.0×10^{-1} M H_2O_2, the observed pseudo-first-order kinetic rates (k_{obs}, s^{-1}) for the degradation of RB (10 mg L^{-1}) at pH 3 and temperature 25 ± 2 °C were between 0.4 and 1.7×10^4 s^{-1}, while in presence of 0.1% *w/v* Fe_3O_4 or SDS@Fe_3O_4 NPs, k_{obs} were between 1.3 and 2.8×10^4 s^{-1} and between 2.6 and 4.8×10^4 s^{-1}, respectively. Furthermore, in presence of Fe_3O_4 or SDS@Fe_3O_4, k_{obs} increased with NPs dosage and showed a peaked pH behavior with a maximum at pH 3. The magnitude of thermodynamic parameters E_a and ΔH for RB degradation in presence of SDS@Fe_3O_4 were 15.63 kJ mol^{-1} and 13.01 kJ mol^{-1}, respectively, lowest among the used catalysts, confirming its effectiveness during degradation. Furthermore, SDS in the presence of Fe_3O_4 NPs and H_2O_2 remarkably enhanced the rate of RB degradation.

Keywords: magnetite; co-precipitation method; Rhodamine B; sodium dodecyl sulfate; wastewater treatment

---

1. Introduction

Mushrooming industrialization and urbanization are primarily responsible for deteriorating the surface and sub-surface water quality, causing hazardous effects on both aquatic organisms and human health. Among water contaminants, dyes and pigments, which are widely discharged from textile, pharmaceutical, paint, rubber, cosmetic, and confectionary industries effluents [1,2], produce unwanted color to water bodies, resulting in intoxication of ecosystems. Rhodamine B (RB) is a synthetic cationic dye, containing a multi-ring aromatic xanthene core planar structure [3]. It is widely used for dyeing and printing applications [4]. The carcinogenic, mutagenic, and toxic effects of RB have been well reported [5–7], evidencing the need of RB contaminated effluents treatment prior to their discharge. Various treatment methodologies, such as reverse osmosis, ion-exchange, precipitation, adsorption, ozone treatment, catalytic reduction, biodegradation, ultrasonic decomposition, coagulation, electrocoagulation, chemical oxidation, and nano-filtration, have been used for the removal of RB and other dyes from water [8–10]. However, high-cost, long process duration, large energy consumption, regeneration difficulties, and pollutants transfer from one phase to another are the major demerits of the aforementioned processes. Thus, advanced oxidation processes (AOPs) are considered comparatively advantageous since they possess favorable decolorizing ability for reactive dyes [11]. Fenton reaction, which is one of the most effective AOPs, has attracted widespread attention. It is operated at acidic pH in the presence of hydrogen peroxide (H_2O_2) and ferrous ions while yielding hydroxyl radical with powerful oxidation capacity leading to complete decomposition of organic dyes, thus, converting them into non-toxic lower molecular weight products [12]. In this sense, the Fe^{2+}-H_2O_2 Fenton system has been widely used for the oxidative removal of RB from water [11–13].

Recently, iron (III) oxide (Fe_3O_4) nanoparticles (NPs) have been used for removing various dyes and heavy metals from water [14–16]. These NPs are inert, economical, possess unique magnetic properties, and can be easily separated from reaction medium through an external magnetic field [17–19]. Additionally, Fe_3O_4 magnetic NPs exhibit a high surface area, depending on the particle size, and show the ability for surface modification. Furthermore, the interaction of Fe_3O_4 NPs with H_2O_2 generates hydroxyl and peroxyl radicals, which are able to undergo the oxidative degradation of organic pollutants [20–22]. However, bare Fe_3O_4 NPs suffer some shortcomings such as agglomeration, limited adsorption ability, and limited working pH range. Coating of Fe_3O_4 NPs with surfactants, polymers, silica, starch, polyelectrolytes, etc., render an enhancement in their surface properties and chemical stability, making them suitable for industrial wastewater treatment and catalytic applications [23–29]. Among surfactants used for coating, the anionic sodium dodecyl sulfate (SDS) is known to enhance the ability of NPs to remove pollutants from wastewater, which has been related with the binding and chelating efficiency of its functional groups [30]. Although no studies were found on the specific case of RB, Fe_3O_4 NPs modified with SDS have been successfully used for the adsorptive removal of several dyes from water, including tolonium chloride [31], Basic Blue 41 [32], or Brilliant Green [33].

The present work was undertaken with the aim of developing an efficient, eco-friendly, and economical treatment for the removal of cationic dyes from water. For this purpose, Fe_3O_4 NPs were synthesized, coated with SDS, and tested as catalyst for the degradation of RB under the presence of H_2O_2. The synthesized Fe_3O_4 and SDS-coated Fe_3O_4 (SDS@Fe_3O_4) NPs were thoroughly characterized through XRD, VSM, SEM, TEM, and FT-IR techniques. The main novelty of this work was the comparative study of the dye degradation by H_2O_2 under three different situations, namely, in absence of ferrous NPs, in the presence of Fe_3O_4 NPs, and in the presence of SDS@Fe_3O_4 NPs. Kinetic experiments were carried out to explore the influence of these catalysts dosages, H_2O_2, SDS, and solution pH on the RB degradation rate.

2. Materials and Methods

2.1. Chemicals and Reagents

Rhodamine B (RB: AR grade 80%; CDH, New Delhi, India), hydrochloric acid (HCl: AR grade 36%; Fisher Scientific, Mumbai, India), hydrogen peroxide (H_2O_2: 35% v/v, Merck, Mumbai, India), sodium dodecyl sulphate (SDS: 99%; CDH, New Delhi, India), Ammonia solution (NH$_4$OH: 25% with purity index 99%, Thermo Fisher Scientific, Mumbai, India), ferrous chloride dihydrate (FeCl$_2$. 2 H$_2$O: 99%; CDH, New Delhi), ferric Chloride (FeCl$_3$: 97.0%; CDH, New Delhi, India), and sodium hydroxide pellets (NaOH: 97%, Merck, Mumbai, India) were used as supplied. All the other reagents used during the experimental work were of reagent grade. All the solutions were prepared in deionized (DI) water. The stock solutions of NaOH (1.0 M) and SDS (1.0×10^{-2} M) were prepared in DI water. The stock solution (500 mgL^{-1}) of dye was prepared by dissolving 50 mg RB in 100 mL DI water. Likewise, 250 mL stock solution of H_2O_2 was prepared by dissolving 25 mL of H_2O_2 in DI water. The stock solution of HCl (0.1 M) was prepared in 100 mL DI water.

2.2. Synthesis and Surfactant Coating of Fe₃O₄ Magnetic NPs

Magnetic nanoparticles were synthesized by adopting the co-precipitation method as described in the literature [34]. Briefly, Fe_3O_4 NPs were synthesized by mixing 20.0 g of $FeCl_3$ (0.4 M) and 10.0 g of $FeCl_2.2H_2O$ (0.2 M) into a 1.0 L conical flask. These iron salts were dissolved in 300 mL DI water. The mixture was purged with N_2 gas and stirred for about an hour. Then, liquor ammonia (25%) was added drop-wise in the flask. The pH of the solution in flask was further increased to ~10 by adding 2.0 M of NaOH solution. The temperature of the solution was then raised to 70 °C with stirring and purging of N_2 gas for 5 h. Black precipitate was formed in the flask. It was filtered, washed with acetone, and thereafter with DI water to a neutral pH value. The precipitate was then dried at 70 °C in a vacuum oven. The synthesis of Fe_3O_4 NPs can be given by the following reaction:

$$Fe^{2+} + 2Fe^{3+} + 8OH^- \rightarrow Fe_3O_4 + 4H_2O$$

To prepare the SDS-coated Fe_3O_4 NPs, $FeCl_3.6H_2O$ (20 g, 0.40 M), $FeCl_2.4H_2O$ (10 g, 0.20 M), and SDS (8.64 g, 0.10 M) were taken into the conical flask of 1.0 L capacity containing 300 mL DI water. The overhead stirrer was used to mix the reactants properly. The solution was stirred vigorously for 45 min under the N_2 gas atmosphere. Then, 200 mL of 25% ammonium hydroxide solution was added drop-wise into the above solution until the pH of the resulting solution reached 9–11. The pH of the reaction medium was further raised to 14 by adding 2.0 M NaOH solution drop-wise. The mixture was then stirred vigorously under N_2 gas purging for 5 h. The black precipitate that formed was filtered and washed with acetone and DI water until the pH came to a neutral value.

2.3. Characterization

The crystallinity and phase composition of Fe_3O_4 and SDS@Fe_3O_4 NPs were studied by X-ray diffraction (XRD: MiniFlex II, Rigaku, Tokyo, Japan) analysis equipped with a Cu K$_\alpha$ radiation source (with λ = 1.5406 nm). The surface functionalities present over Fe_3O_4 and SDS@Fe_3O_4 NPs surface were determined by Fourier infra-red spectrometer (FT-IR: Nicolet iS50, Thermo Fisher Scientific, Madison, WI, USA). The surface morphology and particle size were analyzed by scanning electron microscopy (SEM: JSM-5600LV, JEOL, Tokyo, Japan) and transmission electron microscopy (TEM: CM120, Philips, Amsterdam, The Netherlands). The magnetic properties of Fe_3O_4 and SDS@Fe_3O_4 NPs were determined using a vibrating sample magnetometer (VSM: 7307, Lakeshore, Westerville, OH, USA).

2.4. Degradation Kinetic Experiments

A Genesys 10S UV–visible spectrophotometer (Thermo Fisher Scientific, Madison, WI, USA) was used to monitor the change in the absorbance intensity of RB during its degradation under the varying reaction conditions. The spectrophotometer was provided with multiple cell holders in which a 3.0 mL quartz cuvette with a path length of 10 mm was used to measure absorbance. All the kinetic experiments were performed at a constant temperature of 25.0 ± 0.2 °C by using a thermostatic water-bath. A 0.1% *w/v* of magnetic Fe_3O_4 NPs was put together with RB solution with an initial concentration of 10 mg L^{-1} into a three necked round bottom flask of 100 mL capacity. Solution pH was adjusted by adding hydrochloric acid or sodium hydroxide solution and monitored by using a pH meter. The reaction vessel containing RB solution and magnetic Fe_3O_4 NPs was kept in the water-bath to equilibrate with the required temperature. The reaction was started with the addition of 5.0×10^{-2} to 4.0×10^{-1} M H_2O_2 and zero time was taken when the half of the amount of H_2O_2 was added. The concentration of RB was spectrophotometrically analyzed at its maximum absorbance wavelength (λ_{max}: 554 nm) at constant time intervals. All the kinetic experiments were carried out under pseudo-first-order conditions in which H_2O_2 was kept in excess over RB. The progress of the reaction gradually resulted in the decrease of RB concentration and the values of the pseudo-first-order rate constants were obtained from the slopes of the plots of ln (absorbance) versus time. Each kinetic run was carried out in triplicate to check their repeatability and the rate constant was observed to be within the error limits of ~5%.

3. Results and Discussion

3.1. Characterization of Fe₃O₄ and SDS@Fe₃O₄ NPs

3.1.1. X-ray Diffraction (XRD)

Figure 1A shows the XRD patterns obtained for the synthesized Fe_3O_4 NPs and it confirms the nanocrystal structure and phase purity of Fe_3O_4 NPs. The diffraction peaks appeared at $2\theta = 30.26°$, $35.5°$, $43.12°$, $53.74°$, $57.10°$, and $62.92°$ corresponding to planes (220), (311), (400), (422), (511), and (440), respectively [35], consistent with standard magnetite database (JCPDS-19-0629), indicating a highly crystalline nature of Fe_3O_4 NPs. Figure 1B shows the XRD patterns for SDS@Fe_3O_4 NPs with reduced peak intensity due to the SDS coating over Fe_3O_4 surface. This confirms crystalline-to-amorphous transition of Fe_3O_4 NPs due to SDS coating during SDS@Fe_3O_4 NPs synthesis [36].

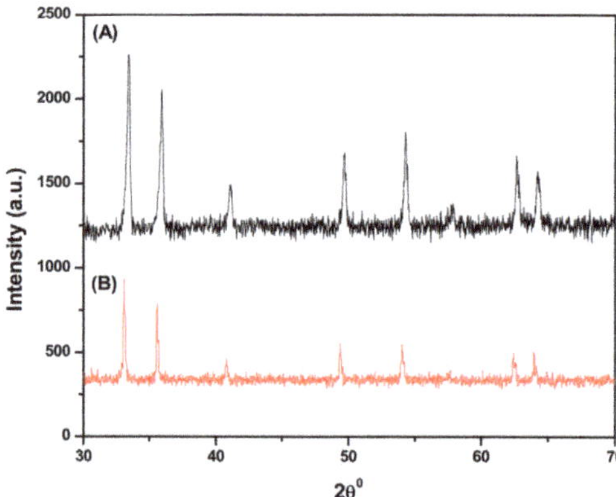

Figure 1. X-ray diffraction patterns for synthesized iron (III) oxide (Fe_3O_4) (**A**) and sodium dodecyl sulfate (SDS) coated iron (III) oxide (SDS@Fe_3O_4) (**B**).

3.1.2. Fourier Transform Infrared Spectroscopy (FTIR)

The FT-IR spectra of Fe_3O_4 and SDS@Fe_3O_4 NPs are shown in Figure 2.

The two peaks at 585 and 435 cm^{-1}, as shown in Figure 2A, correspond to the Fe-O bond vibrations of Fe_3O_4 NPs [37]. From these observations, it is confirmed the spinel structure of Fe_3O_4 NPs and also inferred the existence of the difference in the bond length in Fe-O. The peak at 3424 cm^{-1} in Figure 2A was associated to the O-H stretching vibrations arising from the hydroxyl group due to the presence of water molecules associated with Fe_3O_4 [38]. The H-O-H bending of water molecules in Figure 2A is observed at 1631 cm^{-1} in Fe_3O_4 NPs [39]. The FTIR spectrum of SDS@Fe_3O_4 NPs is shown in Figure 2B, which displayed a new absorption peak at 1252 cm^{-1} due to the stretching vibration of S=O groups of SDS and the presence of peaks at 2929 cm^{-1} and 2842 cm^{-1}, which were assigned to the stretching mode for aliphatic C-H groups of SDS [40]. The peak at 1635 cm^{-1} in SDS@Fe_3O_4 (Figure 2B) was attributed to the H-O-H bending of water molecules and that at 3431 cm^{-1} was due to stretching vibration of hydroxyl group on the surface of the NPs. The presence of two peaks at 547 cm^{-1} and at 474 cm^{-1} in Figure 2B is attributed to Fe-O bonds in SDS-modified Fe_3O_4 [41]. Thus, the FTIR results confirmed successful synthesis of Fe_3O_4 NPs and their surface modifications through the adsorption of SDS molecules.

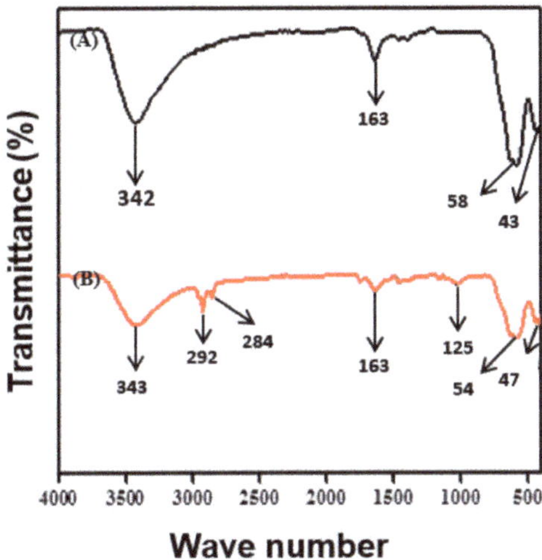

Figure 2. FTIR spectra of Fe_3O_4 (**A**) and SDS@Fe_3O_4 (**B**).

3.1.3. Scanning Electron Microscopy (SEM)

The SEM micrograph of the synthesized magnetite (Fe_3O_4) NPs is shown in Figure 3A. It can be observed that the NPs exhibit spherical surface morphology, having a particle size lower than 100 nm scale with low polydispersity. The SEM image of the SDS@Fe_3O_4 NPs is shown in Figure 3D on the scale of up to 5 μm. The image depicts successful functionalization of Fe_3O_4 by SDS and the larger dispersion of SDS@Fe_3O_4 as compared with Fe_3O_4 NPs.

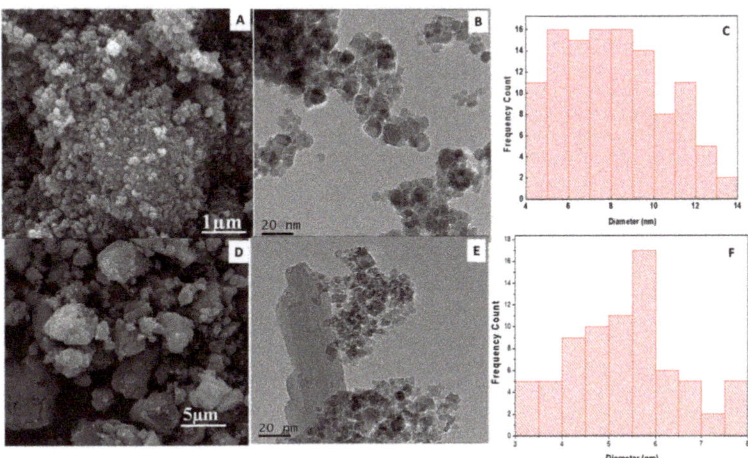

Figure 3. SEM image of Fe_3O_4 (**A**); TEM image of Fe_3O_4 (**B**); histogram images of Fe_3O_4 (**C**); SEM image of SDS@Fe_3O_4 (**D**); TEM image of SDS@Fe_3O_4 (**E**); histogram images of SDS@Fe_3O_4 (**F**). Note that scales of figures A and B are different.

3.1.4. Transmission Electron Microscopy (TEM)

The TEM micrograph of pristine Fe_3O_4 NPs (Figure 3B) on the scale of up to 20 µm shows their spherical shape with a narrow range particle size distribution centered at 9 ± 2 nm, as demonstrated by the histogram in Figure 3C. The TEM image of the SDS@Fe_3O_4 NPs is illustrated in Figure 3E. After coating with SDS, the size of SDS@Fe_3O_4 NPs appears to be smaller, as shown by the histogram (Figure 3F). This might be due to the coating of Fe_3O_4 NPs with SDS, which hinders NPs agglomeration.

3.1.5. Vibrating Sample Magnetometer (VSM)

The magnetic behavior of Fe_3O_4 and SDS@Fe_3O_4 NPs was studied by using VSM. As it may be seen in Figure 4, both Fe_3O_4 and SDS@Fe_3O_4 showed superparamagnetic behavior with different magnetic saturations level. The specific magnetic saturation magnitudes for Fe_3O_4 and SDS@Fe_3O_4 NPs were 60.0 and 50.0 $emug^{-1}$, respectively, as displayed in Figure 4. Comparatively lower magnetic saturation of SDS@Fe_3O_4 NPs might be due to their coating with SDS [42]. In order to avoid aggregation of Fe_3O_4 NPs, which may severely reduce their catalytic efficiency, coating with SDS was executed in this work. In any case, as for the large magnetic saturation and superparamagnetic property of SDS@Fe_3O_4 NPs (Figure 4), such a coating did not affect the high efficiency in magnetic separation and recovery.

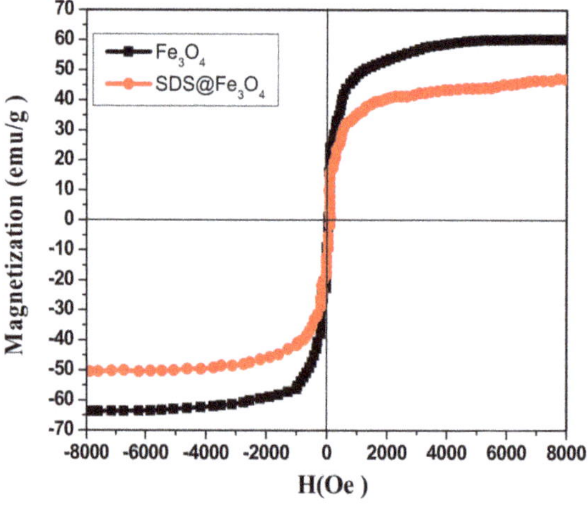

Figure 4. Magnetization curve for Fe_3O_4 and for SDS@Fe_3O_4 at room temperature.

3.2. Degradation of RB by H_2O_2

The repetitive scans of the reactant mixture containing RB (10 mg L^{-1}) and H_2O_2 (2×10^{-1} M) were recorded at constant time intervals of ten minutes in the visible region (460–600 nm). The temperature and pH were kept constant at 25 ± 0.2 °C and 3, respectively. These spectra, which are shown in Figure 5A, indicated that the absorbance intensities at λ_{max} (554 nm) progressively decreased with time. A decrease in the absorbance intensities was due to the degradation of RB by H_2O_2.

The degradation of RB can be represented by the following representative reaction and rate Equation (1):

$$\text{RB} + H_2O_2 \rightarrow \text{Oxidized Products} + H_2O, \text{ Rate } = \frac{-d[\text{RB}]}{dt} = k_{obs}[\text{RB}], \tag{1}$$

where k_{obs} is the observed value of the rate constant and was calculated from the slope of the plot of $\ln \frac{[\text{RB}]_0}{[\text{RB}]_t}$ versus t. The terms $[\text{RB}]_0$ and $[\text{RB}]_t$ are the concentrations of RB at time zero and at any time t, respectively. The observed rate constant depends upon the concentration of H_2O_2 as given by Equation (2). The order of the reaction is assumed to be x with respect to the concentration of H_2O_2.

$$k_{obs} = k\,[H_2O_2]^x \tag{2}$$

where k is the specific rate constant with respect to H_2O_2 concentration. The values of k and x were respectively obtained from the intercept and slope of the plot of log k_{obs} versus log $[H_2O_2]$.

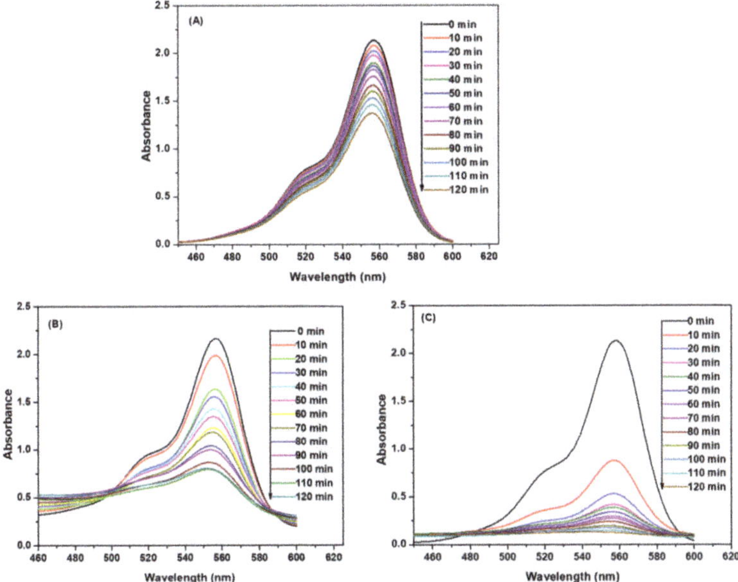

Figure 5. UV–visible spectra of Rhodamine B (RB) at various degradation times. In absence of Fe_3O_4 (**A**); in presence of Fe_3O_4 (**B**); and in presence of SDS@Fe_3O_4 (**C**). (Reaction conditions: 10 mg L^{-1} RB, 2.0×10^{-1} M H_2O_2, 0.1% *w/v* Fe_3O_4 NPs, 0.1% *w/v* SDS@Fe_3O_4, pH 3, and temperature 25 ± 2 °C).

The rate of RB degradation was studied at varied concentrations of H_2O_2 in the range from 5.0×10^{-2} to 4.0×10^{-1} M while keeping a RB concentration of 10 mg L^{-1} at pH 3 and temperature 25 ± 0.2 °C. The values of rate constants were calculated and the plot of rate constant versus H_2O_2 concentration (Figure 6) shows a linear dependence of the rate constant values on H_2O_2 concentration.

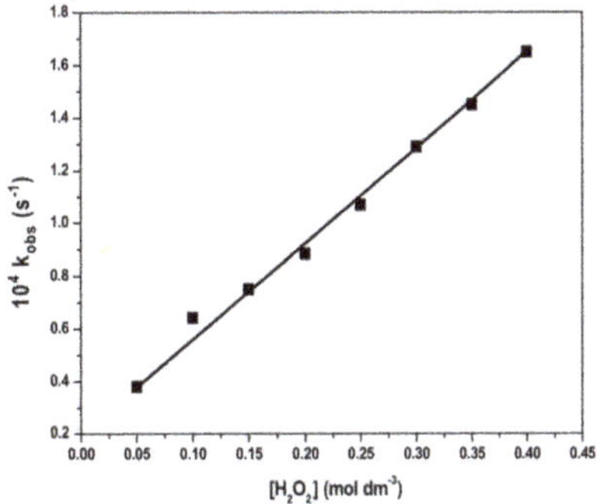

Figure 6. Plots of k_{obs} versus hydrogen peroxide (H_2O_2) concentration for the degradation of RB. (Reaction conditions: 10 mg L^{-1} RB, 5.0×10^{-2} to 4.0×10^{-1} M H_2O_2, pH 3, and temperature 25 ± 2 °C).

3.3. Degradation of RB in the Presence of Fe₃O₄ and SDS@Fe₃O₄ NPs

The addition of 0.1% *w/v* of Fe_3O_4 NPs to the solution containing RB and H_2O_2 increased the rate of degradation of RB, as is evident from the decrease in the rate of absorbance intensities with time, which is presented in Figure 5). The increase in the degradation rate of RB can be attributed to the catalytic role of Fe_3O_4 NPs. The degradation rate was further increased in the presence of 0.1% *w/v* SDS@Fe_3O_4 NPs as displayed in Figure 5C.

In order to assess the effect of pH, the degradation rate of RB was studied in the pH range 1–10 by adjusting it with HCl/NaOH solutions. The observed results are presented in Figure 7. The plot of the rate constant versus pH (Figure 7) demonstrates that the values of the rate constant increase with pH until pH 3. Thereafter, on further increasing the pH beyond 3, the values of the rate constant decreased. Thus, a peaked behavior plot was obtained with the maximum degradation rate at pH 3.

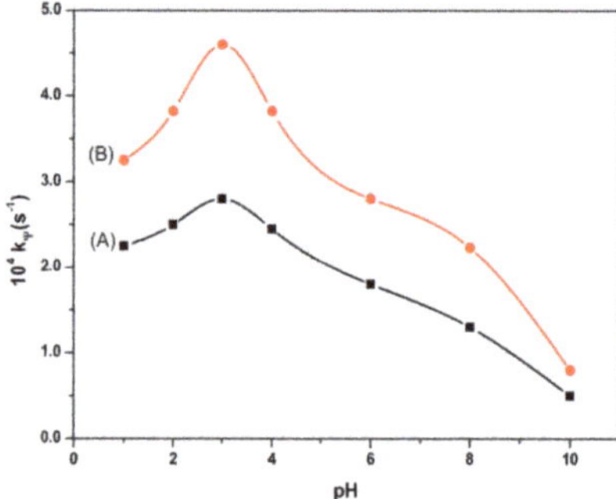

Figure 7. Effect of pH on the Rhodamine B (RB) degradation process. In presence of Fe_3O_4 (**A**); and in presence of SDS@Fe_3O_4 (**B**). (Reaction conditions: 10 mg L^{-1} RB, 2.0 × 10^{-1} M H_2O_2, 0.1% *w/v* Fe_3O_4, 0.1% *w/v* SDS@Fe_3O_4, and temperature 25 ± 2 °C).

The influence of the magnetic Fe_3O_4 NPs dosage on the RB degradation rate was studied in the range between 0.02% and 0.2% *w/v* Fe_3O_4. The respective concentrations of H_2O_2 and RB were set at 2.0 × 10^{-1} M and 10 mg L^{-1}, while the pH and temperature of the solution were 3 and 25 ± 0.2 °C, respectively. The increase in the amount of Fe_3O_4 increased the RB degradation rate, as shown by data graphically presented in Figure 8A. Furthermore, as it may be seen in Figure 8B, the influence of SDS@Fe_3O_4 concentration on the RB degradation rate showed the same pattern observed for Fe_3O_4, but with higher values of the rate constant.

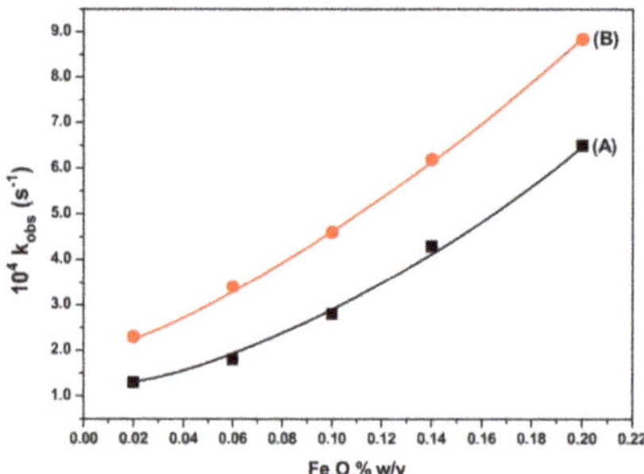

Figure 8. Plots of k_{obs} versus varying concentration of magnetic nanoparticles (NPs) for the degradation of the RB in presence of Fe_3O_4 (**A**), and in presence of SDS@Fe_3O_4 (**B**). (Reaction conditions: 10 mg L^{-1} RB, 2.0 × 10^{-1} M H_2O_2, 0.1% *w/v* Fe_3O_4, 0.1% *w/v* SDS@Fe_3O_4, pH 3, and temperature 25 ± 2 °C).

The observed enhancement in the rate of the RB degradation in presence of Fe_3O_4 can be described through the production of highly reactive hydroxyl radicals due to the interaction between the NPs and H_2O_2 [43,44], followed by the formation of peroxyl radicals and the subsequent oxidation of RB by these radicals, as described by the following reactions:

(i) $Fe^{2+} + H_2O_2 \rightarrow Fe^{3+} + HO^{\bullet} + OH^-$ The possible reactions of free radicals are:

(ii) $Fe^{2+} + HO^{\bullet} \rightarrow Fe^{3+} + OH^-$,

(iii) $H_2O_2 + HO^{\bullet} \rightarrow H_2O + HO_2^{\bullet}$,

(iv) $HO_2^{\bullet} + HO^{\bullet} \rightarrow H_2O + O_2$,

(v) $HO^{\bullet} + HO^{\bullet} \rightarrow H_2O_2$,

(vi) $HO^{\bullet} + RB \rightarrow Products$,

(vii) $HO_2^* + RB \rightarrow Products$.

Thus, the oxidation of RB by HO^* and HO_2^* radicals leads to a decrease in its concentration. As for the RB degradation in the presence and absence of Fe_3O_4 and $SDS@Fe_3O_4$, results shown in Figure 5 to 8 allow to state that RB degradation was enlarged under the presence of Fe_3O_4 and $SDS@Fe_3O_4$ NPs. As for the degradation rate of RB, it was linearly dependent on the initial concentration of H_2O_2 in the absence of NPs. The RB degradation was comparatively higher in presence of $SDS@Fe_3O_4$ than Fe_3O_4 and increased with the increase in the dosage of either Fe_3O_4 or $SDS@Fe_3O_4$ NPs. The variations in pH displayed a similar influence on the RB degradation rate in the presence of Fe_3O_4 and $SDS@Fe_3O_4$ NPs, the rate showing a peaked behavior. From these observations, it was confirmed that the reaction proceeded through the formation of highly reactive free radicals in the presence of Fe_3O_4 and $SDS@Fe_3O_4$ NPs due to the interaction between Fe^{2+} and H_2O_2, as described by reactions (i) to (vii). The increase in the amount of the NPs increases the production of HO^* radicals (step (i)) and, therefore, an enhancement in the RB degradation rate was observed with the increase in the NPs dosage, which is coincident with previous observations [45]. As shown in Figure 6, the RB degradation rate increased from 0.4 to 1.7×10^4 s^{-1} with increasing H_2O_2 concentration in the absence of NPs. In Figure 9, under the presence of NPs, larger degradation rates are represented, varying between 1.3 and 2.8×10^4 s^{-1} in the case of Fe_3O_4 and between 2.6 and 4.8×10^4 s^{-1} in the case of $SDS@Fe_3O_4$, which compare rather well with published rate constants for the catalytic degradation of RB (Table S1 in the Supplementary Materials). However, as it may be seen in Figure 9, in the presence of Fe_3O_4 and $SDS@Fe_3O_4$ NPs, the rate constant value increased with the concentration of H_2O_2 until it was 2.5×10^{-1} M H_2O_2, and thereafter decreased with H_2O_2 concentration. After the maximum, this decreasing effect in the rate constant with the increase in the H_2O_2 concentration was due to the other free radical reactions taking place in steps (ii) to (v). Thus, at higher concentrations of H_2O_2, the side reactions scavenged the $HO^{\bullet}$ radicals and decreased the concentration of free radicals available to oxidize the dye and, therefore, the rate of the reaction decreased [46].

The rate of degradation of RB was highly pH-dependent and, as it is shown in Figure 7, the maximum rate of degradation was observed at pH 3 in the presence and the absence of NPs. At high concentrations of H^+ ions (pH < 3), peroxide gets solvated to form stable oxonium ions, which enhanced the activity of H_2O_2 and restricted the generation of hydroxyl radicals [47–49]. Moreover, the excess of H^+ ions acts as hydroxyl radical scavenger and, with the increase in H^+ ions, the concentration of $HO^{\bullet}$ radicals decreases, thus, decreasing the rate of reaction [48]. Furthermore, the strong electrostatic interaction between the anionic surfactant head groups and cationic dye molecules at lower pH also decreases the rate of RB degradation. The observed lower rate of reaction at higher pH may be related to the formation of the Fe^{3+}-complexes, which decreases the dissolved Fe^{2+} ions that were available to generate free radicals [49].

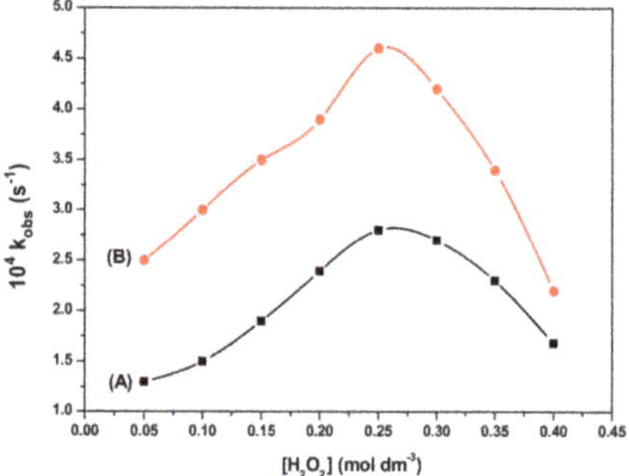

Figure 9. Effect of H_2O_2 concentration on RB degradation. In presence of Fe_3O_4 (**A**); and in presence of SDS@Fe_3O_4 (**B**). (Reaction conditions: 10 mg L^{-1} RB, 5.0×10^{-2} to 4.0×10^{-1} M H_2O_2, 0.1% *w/v* Fe_3O_4, 0.1% *w/v* SDS@Fe_3O_4, pH 3, and temperature 25 ± 2 °C).

The higher degradation rate of RB in the presence of SDS@Fe_3O_4 NPs in comparison with bare Fe_3O_4 NPs that is observed in Figure 8, might be due to the larger capture of RB by SDS@Fe_3O_4 than by Fe_3O_4. Thus, the generated free radicals at the NPs surface can readily attack the attached RB and thus leading to the increase in RB degradation rate. Binding of RB to the SDS@Fe_3O_4 surface can be explained by the electrostatic interaction between the anionic surfactant and protonated cationic dye at pH 3 [50].

3.4. Effect of SDS Concentration and Fe_3O_4 NPs Dosage on RB Degradation

The addition of SDS at varied concentrations (5.0×10^{-4} to 5.0×10^{-2} M) to a solution containing RB (10 mgL^{-1}), H_2O_2 (2.0×10^{-1} M), and Fe_3O_4 (0.1% *w/v*) NPs at pH 3 resulted in an increase in the rate of the degradation reaction, as shown in Figure 10. On the other hand, an increase in the amount of Fe_3O_4 NPs from 0.02% to 0.2% *w/v* at a fixed concentration of SDS (2.0×10^{-2} M) also increased the rate of RB degradation, as shown in Figure 11.

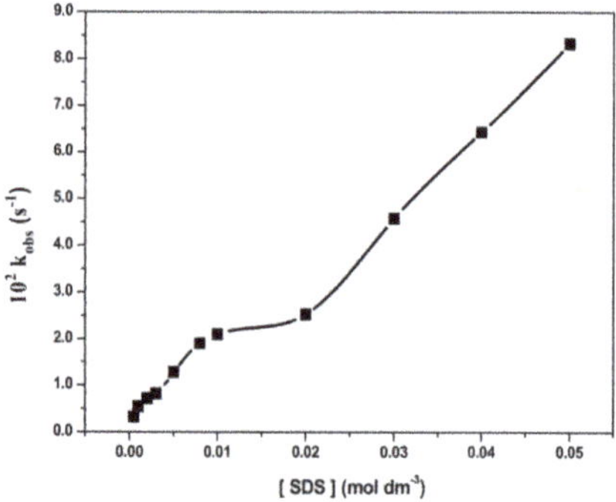

Figure 10. Effect of sodium dodecyl sulfate (SDS) concentration on RB degradation. (Reaction conditions: 10 mg L^{-1} RB, 2.0×10^{-1} M H_2O_2, 0.1% *w/v* Fe_3O_4, pH 3, and temperature 25 ± 2 °C).

The respective degradation rates of RB in presence of SDS and Fe_3O_4 can be represented by Equations (3) and (4).

$$RB + D_n \overset{K_s}{\rightleftharpoons} RB_{mic} \overset{H_2O_2}{\rightarrow} Products, \tag{3}$$

$$Fe_3O_4 + RB_{mic} \overset{K_c}{\rightleftharpoons} RB_{mic} - Fe_3O_4 \overset{H_2O_2}{\rightarrow} Products. \tag{4}$$

The presence of SDS micelles (D_n) partitions RB into micellar (RB_{mic}) and aqueous pseudo-phases resulting into the retardation of RB oxidation with H_2O_2 in the presence of SDS, which may be related to the electrostatic repulsion and, therefore, separation between the species involved in the reaction. However, in the presence of Fe_3O_4, micellised RB (RB_{mic}) is incorporated to the NPs surface to form $RB_{mic}-Fe_3O_4$ where H_2O_2 interacts to form reactive $HO^\bullet$ radicals readily available to oxidize RB at the same site. Therefore, RB degradation is catalyzed and the rate of the reaction increases with increasing SDS concentration in the presence of Fe_3O_4 NPs (Figure 10) and also with increasing the amount of Fe_3O_4 NPs in the presence of SDS (Figure 11). In Figure 10, a two steps increase of the degradation may be observed, which may be related to the formation of premicellar aggregates below the critical micelle concentration (cmc) of SDS and micelles above cmc [8,10,51,52], then increasing micelles formation with SDS concentration. Regarding Figure 11, at a SDS concentration above cmc, an increasing degradation rate occurred under increasing Fe_3O_4 concentration, as previously observed in Figure 8 and explained by reactions (i) to (vii). These results are in agreement with previous studies on RB photocatalytic degradation [53,54].

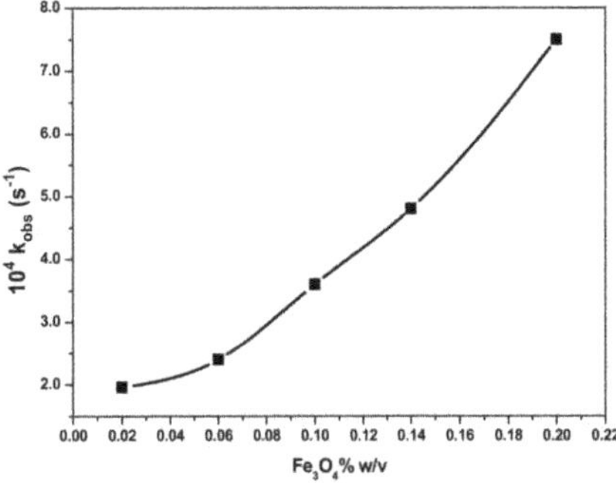

Figure 11. Effect of Fe_3O_4 concentration on RB degradation. (Reaction conditions: 10 mg L^{-1} RB, 0.02% to 0.2% *w/v* Fe_3O_4, 2.0×10^{-2} M SDS, pH 3, and temperature $25 \pm 2\ ^\circ$C).

3.5. Effect of Temperature on RB Degradation

The effect of temperature on RB degradation (10 mg L^{-1}) in aqueous solutions in the presence of H_2O_2 (2.5×10^{-1} M) and at pH 3 was studied at varied temperatures ranging from 25 to 60 $^\circ$C (because above 60 $^\circ$C, due to thermal disintegration of H_2O_2 and free radicals, the rate of RB degradation slowed down) in the absence or in the presence of Fe_3O_4 NPs (0.1% *w/v*), Fe_3O_4 NPs (0.1% *w/v*) together with SDS (2.0×10^{-2} M) or SDS@Fe_3O_4 (0.1% *w/v*).

The energy of activation was calculated using the Arrhenius equation Equation (5), which gave a straight line plot for log k versus 1/T.

$$\log k_{obs} = -\frac{E_a}{2.303\ RT} + \log A_o \tag{5}$$

where E_a is the activation energy (kJ mol^{-1}), R (8.314 J mol^{-1}K^{-1}) is the universal gas constant, T is the temperature in Kelvin (K), A_o is the frequency factor, and k_{obs} is the measured first-order rate constant. The E_a was determined from the slope and values are given in Table 1.

The value of ΔH (enthalpy of activation) and ΔS (entropy of activation) were calculated using the Erying equation Equation (6).

$$\ln\left(\frac{k_{obs}}{T}\right) = -\frac{\Delta H}{R} \times \frac{1}{T} + \ln\frac{k_B}{h} + \frac{\Delta S}{R}, \tag{6}$$

where k_B is the Bolzmann's constant and h is the Plank's constant. A plot of ln (k_{obs}/T) versus 1/T produces a straight line and the values of ΔH and ΔS may be obtained from the slope and the intercept, respectively. The so determined ΔH and ΔS values are given in Table 1.

As it may be seen in Table 1, good fittings ($R^2 > 0.94$) to the Erying equation were obtained within the temperature range here considered. The largest E_a and ΔH determined for RB degradation were those in the absence of NPs. These values progressively decreased in the presence of Fe_3O_4 NPs, Fe_3O_4 NPs together with SDS and SDS@Fe_3O_4, which provided the lowest E_a and ΔH. Regarding the ΔS, although the effect was not so remarkable as for E_a and ΔH, slightly lower values were also determined under the presence of NPs. These results point to the energetically favorable effect of Fe_3O_4 and SDS@Fe_3O_4 NPs, which confirms that these are efficient catalysts.

Table 1. Activation parameters determined for RB degradation by H_2O_2 in the absence and presence of Fe_3O_4 and SDS and SDS@Fe_3O_4.

Reaction Media	E_a (kJ mol^{-1})	ΔH (kJ mol^{-1})	ΔS (J mol^{-1}K^{-1})	R^2
H_2O_2	69.47	66.85	−132.66	0.949
$H_2O_2 + Fe_3O_4$	28.47	25.86	−188.18	0.953
$H_2O_2 + Fe_3O_4 + SDS$	32.47	29.86	−194.42	0.945
$H_2O_2 + SDS@Fe_3O_4$	15.63	13.01	−149.00	0.953

Reaction conditions: 10 mg L^{-1} RB, 2.0×10^{-1} M H_2O_2, 0.1% *w/v* Fe_3O_4 (when present), 0.1% *w/v* SDS@Fe_3O_4 (when present), 2.0×10^{-2} M SDS, pH 3, and varied temperatures (between 25 and 60 ± 2 °C).

4. Conclusions

In this work, Fe_3O_4 NPs were synthesized, coated with SDS to synthesize SDS@Fe_3O_4 NPs, and both tested as catalysts for the oxidation of RB under H_2O_2. The main novelty was to compare the dye degradation under three different situations, namely, in presence of just H_2O_2, of H_2O_2 and Fe_3O_4 NPs, and H_2O_2 and SDS@Fe_3O_4 NPs. Observed pseudo-first-order kinetic rates (k_{obs}, s^{-1}) for the degradation of RB (10 mg L^{-1}) at pH 3 and temperature 25 ± 2 °C were between 0.4 and 1.7×10^4 s^{-1}, linearly dependent upon H_2O_2 concentrations within 5.0×10^{-2} to 4.0×10^{-1} M. Under identical experimental conditions, except for the presence of 0.1% *w/v* NPs, the observed rates increased to values between 1.3 and 2.8×10^4 s^{-1} in the case of Fe_3O_4 and between 2.6 and 4.8×10^4 s^{-1} in the case of SDS@Fe_3O_4. Fe_3O_4 NPs with H_2O_2 gave readily the highly reactive hydroxyl radicals, which enhanced the rate of RB degradation. Furthermore, an increased catalytic effect was observed for SDS@Fe_3O_4 because the SDS coating avoided Fe_3O_4 aggregation and the consequent efficiency depletion. However, under the presence of Fe_3O_4 and SDS@Fe_3O_4 NPs, k_{obs} did not increase linearly with H_2O_2 concentration but just until 2.5×10^{-1} M H_2O_2, then decreased with increasing H_2O_2 concentration, which was associated to free radical competitive reactions. On the other hand, it was verified that the addition of SDS molecules to the dye solution containing Fe_3O_4 also increased the rate of reaction, which was related to the incorporation of micellized RB ions onto the Fe_3O_4 NPs surface. Overall, this work demonstrated that the application of Fe_3O_4 and SDS@Fe_3O_4 along with H_2O_2 can be an efficient method for the rapid removal of cationic dyes from wastewater in line with the green chemistry principles.

Supplementary Materials: The following are available online at http://www.mdpi.com/2073-4360/12/10/2246/s1, Table S1: Published results on the rate constant (k_{obs}, s^{-1}) for the catalytic degradation of RB using different catalysts.

Author Contributions: Conceptualization, M.Z.A.R. and M.A.K.; methodology, M.Z.A.R. and M.S.A.; software, M.S.A. and K.R.; validation, M.S.A. and K.R.; formal analysis, M.S.A.; investigation, M.S.A. and K.R.; resources, M.Z.A.R.; data curation, M.Z.A.R. and M.S.A.; writing—original draft preparation, M.S.A.; writing—review and editing, M.A.K. and M.O.; visualization, M.O.; supervision, M.Z.A.R.; project administration, M.Z.A.R.; funding acquisition, M.A.K. and M.O.; revision, M.A.K. and M.O. All authors have read and agreed to the published version of the manuscript.

Funding: Marta Otero is thankful to the Portuguese "Fundação para a Ciência e a Tecnologia" (FCT) for the Investigator Program (IF/00314/2015). We would also like to thank FCT/Ministério da Ciência, Tecnologia e

Ensino Superior (MCTES) for the financial support to CESAM (UIDP/50017/2020+UIDB/50017/2020) through national funds.

Conflicts of Interest: The authors declare no conflict of interest. Furthermore, the funders had no role in the design of the study; in the collection, analyses, or interpretation of data; in the writing of the manuscript, or in the decision to publish the results.

References

1. El-Refaie, K.; Ghfar, A.A.; Wabaidur, S.M.; Khan, M.A.; Siddiqui, M.R.; Alothman, Z.A.; Alqadami, A.A.; Hamid, M. Cetyltrimethylammonium bromide intercalated and branched polyhydroxystyrene functionalized montmorillonite clay to sequester cationic dyes. *J. Environ. Manag.* **2018**, *219*, 285–293.

2. Khan, M.A.; Wabaidur, S.M.; Siddiqui, M.R.; Alqadami, A.A.; Khan, A.H. Silico-Manganese Fumes Waste Encapsulated Cryogenic Alginate Beads for Aqueous Environment De-colorization. *J. Clean. Prod.* **2020**, *244*, 118867. [CrossRef]

3. Goyal, P.; Chakraborty, S.; Misra, S.K. Multifunctional Fe3O4-ZnO nanocomposites for environmental remediation applications. *Environ. Nanotechnol. Monit. Manag.* **2018**, *10*, 28–35. [CrossRef]

4. Khan, M.A.; Siddiqui, M.R.; Otero, M.; Alshareef, S.A.; Rafatullah, M. Removal of rhodamine B from water using a solvent impregnated polymeric Dowex 5WX8 resin: Statistical optimization and batch adsorption studies. *Polymers* **2020**, *12*, 500. [CrossRef] [PubMed]

5. Zhou, Y.; Lu, J.; Zhou, Y.; Liu, Y. Recent advances for dyes removal using novel adsorbents: A review. *Environ. Pollut.* **2019**, *252*, 352–365. [CrossRef]

6. Youssef, N.A.; Shaban, S.A.; Ibrahim, F.A.; Mahmoud, A.S. Degradation of methyl orange using Fenton catalytic reaction. *Egypt. J. Pet.* **2016**, *25*, 317–321. [CrossRef]

7. Ranjbari, E.; Hadjmohammadi, M.R.; Kiekens, F.; De Wael, K. Mixed Hemi/Ad-Micelle Sodium Dodecyl Sulfate-Coated Magnetic Iron Oxide Nanoparticles for the Efficient Removal and Trace Determination of Rhodamine-B and Rhodamine-6G. *Anal. Chem.* **2015**, *87*, 7894–7901. [CrossRef]

8. Pham, T.D.; Pham, T.T.; Phan, M.N.; Ngo, T.M.V.; Dang, V.D.; Vu, C.M. Adsorption characteristics of anionic surfactant onto laterite soil with differently charged surfaces and application for cationic dye removal. *J. Mol. Liq.* **2020**, *301*, 112456. [CrossRef]

9. Ngo, T.M.V.; Truong, T.H.; Nguyen, T.H.L.; Duong, T.T.A.; Vu, T.H.; Pham, T.D. Surface modified laterite soil with an anionic surfactant for the removal of a cationic dye (crystal violet) from an aqueous solution. *Water Air Soil Pollut.* **2020**, *231*, 1–15. [CrossRef]

10. Chu, T.P.M.; Nguyen, N.T.; Vu, T.L.; Dao, T.H.; Dinh, L.C.; Nguyen, H.L.; Hoang, T.H.; Le, T.S.; Pham, T.D. Synthesis, characterization, and modification of alumina nanoparticles for cationic dye removal. *Materials* **2019**, *12*, 450. [CrossRef]

11. Wu, J.M.; Zhang, T.W.; Wilhelm, P.; Stephan, D.; Carp, O.; Huisman, C.L.; Reller, A.; Topare, N.S.; Bisen, N.; Shrivastava, P.; et al. Photocatalytic Degradation of Rhodamine B by using UV/TiO2 and Nb2O5 Process: A Kinetic Study. *J. Photochem. Photobiol. A Chem.* **2004**, *162*, 33–177. [CrossRef]

12. Yan, P.; Gao, L.B.; Li, W.T. Microwave-enhanced fenton-like system, Fe3O4/H2O2, for Rhodamine B wastewater degradation. *Appl. Mech. Mater.* **2013**, *448*, 834–837. [CrossRef]

13. Mehrdad, A.; Hashemzadeh, R. Ultrasonic degradation of Rhodamine B in the presence of hydrogen peroxide and some metal oxide. *Ultrason. Sonochem.* **2010**, *17*, 168–172. [CrossRef] [PubMed]

14. Giraldo, L.; Erto, A.; Moreno-Piraján, J.C. Magnetite nanoparticles for removal of heavy metals from aqueous solutions: Synthesis and characterization. *Adsorption* **2013**, *19*, 465–474. [CrossRef]

15. Dalali, N.; Khoramnezhad, M.; Habibizadeh, M.; Faraji, M. Magnetic Removal of Acidic Dyes from Waste Waters Using Surfactant- Coated Magnetite Nanoparticles: Optimization of Process by Taguchi Method. In Proceedings of the 2011 International Conference on Environmental and Agriculture Engineering IPCBEE, Chengdu, China, 29–31 July 2011; pp. 89–93.

16. Jiaqi, Z.; Yimin, D.; Danyang, L.; Shengyun, W.; Liling, Z.; Yi, Z. Synthesis of carboxyl-functionalized magnetic nanoparticle for the removal of methylene blue. *Colloids Surf. A Physicochem. Eng. Asp.* **2019**, *572*, 58–66. [CrossRef]

17. Gupta, A.K.; Gupta, M. Synthesis and surface engineering of iron oxide nanoparticles for biomedical applications. *Biomaterials* **2005**, *26*, 3995–4021. [CrossRef]

18. Alishiri, T.; Oskooei, H.A.; Heravi, M.M. Fe 3 O 4 Nanoparticles as an Efficient and Magnetically Recoverable Catalyst for the Synthesis of α,β-Unsaturated Heterocyclic and Cyclic Ketones under Solvent-Free Conditions. *Synth. Commun.* **2013**, *43*, 3357–3362. [CrossRef]
19. Godoi, M.; Liz, D.G.; Ricardo, E.W.; Rocha, M.S.T.; Azeredo, J.B.; Braga, A.L. Magnetite (Fe3O4) nanoparticles: An efficient and recoverable catalyst for the synthesis of alkynyl chalcogenides (selenides and tellurides) from terminal acetylenes and diorganyl dichalcogenides. *Tetrahedron* **2014**, *70*, 3349–3354. [CrossRef]
20. Wu, X.; Nan, Z. Degradation of rhodamine B by a novel Fe_3O_4 /SiO_2 double-mesoporous-shelled hollow spheres through photo-Fenton process. *Mater. Chem. Phys.* **2019**, *227*, 302–312. [CrossRef]
21. Jiao, Y.; Wan, C.; Bao, W.; Gao, H.; Liang, D.; Li, J. Facile hydrothermal synthesis of Fe3O4@cellulose aerogel nanocomposite and its application in Fenton-like degradation of Rhodamine B. *Carbohydr. Polym.* **2018**, *189*, 371–378. [CrossRef]
22. AlHamedi, F.H.; Rauf, M.A.; Ashraf, S.S. Degradation studies of Rhodamine B in the presence of UV/H_2O_2. *Desalination* **2009**, *239*, 159–166. [CrossRef]
23. Zhu, N.; Ji, H.; Yu, P.; Niu, J.; Farooq, M.; Akram, M.; Udego, I.; Li, H.; Niu, X. Surface Modification of Magnetic Iron Oxide Nanoparticles. *Nanomaterials* **2018**, *8*, 810. [CrossRef] [PubMed]
24. Rajabi, A.A.; Yamini, Y.; Faraji, M.; Nourmohammadian, F. Modified magnetite nanoparticles with cetyltrimethylammonium bromide as superior adsorbent for rapid removal of the disperse dyes from wastewater of textile companies. *Nanochem. Res.* **2016**, *1*, 49–56. [CrossRef]
25. Nguyen, T.H.; Nguyen, T.T.L.; Pham, T.D.; Le, T.S. Removal of lindane from aqueous solution using aluminum hydroxide nanoparticles with surface modification by anionic surfactant. *Polymers* **2020**, *12*, 960. [CrossRef]
26. Sun, C.; Yang, S.T.; Gao, Z.; Yang, S.; Yilihamu, A.; Ma, Q.; Zhao, R.S.; Xue, F. Fe3O4 /TiO2 /reduced graphene oxide composites as highly efficient Fenton-like catalyst for the decoloration of methylene blue. *Mater. Chem. Phys.* **2019**, *223*, 751–757. [CrossRef]
27. Kim, D.K.; Mikhaylova, M.; Wang, F.H.; Kehr, J.; Bjelke, B.; Zhang, Y.; Tsakalakos, T.; Muhammed, M. Starch-Coated Superparamagnetic Nanoparticles as MR Contrast Agents. *Chem. Mater.* **2003**, *15*, 4343–4351. [CrossRef]
28. Zhang, J.L.; Srivastava, R.S.; Misra, R.D.K. Core−Shell Magnetite Nanoparticles Surface Encapsulated with Smart Stimuli-Responsive Polymer: Synthesis, Characterization, and LCST of Viable Drug-Targeting Delivery System. *Langmuir* **2007**, *23*, 6342–6351. [CrossRef]
29. Kievit, F.M.; Veiseh, O.; Bhattarai, N.; Fang, C.; Gunn, J.W.; Lee, D.; Ellenbogen, R.G.; Olson, J.M.; Zhang, M. PEI-PEG-Chitosan-Copolymer-Coated Iron Oxide Nanoparticles for Safe Gene Delivery: Synthesis, Complexation, and Transfection. *Adv. Funct. Mater.* **2009**, *19*, 2244–2251. [CrossRef]
30. El-kharrag, R.; Amin, A.; Greish, Y.E. Synthesis and characterization of mesoporous sodium dodecyl sulfate-coated magnetite nanoparticles. *J. Ceram. Sci. Technol.* **2011**, *2*, 203–210. [CrossRef]
31. Abedi, M.H.; Ahmadmoazzam, M.; Jaafarzadeh, N. Removal of cationic tolonium chloride dye using Fe3O4 nanoparticles modified with sodium dodecyl sulfate. *Chem. Biochem. Eng. Q.* **2018**, *32*, 205–213. [CrossRef]
32. Yamini, Y.; Faraji, M.; Rajabi, A.A.; Nourmohammadian, F. Ultra efficient removal of Basic Blue 41 from textile industry's wastewaters by sodium dodecyl sulphate coated magnetite nanoparticles: Removal, kinetic and isotherm study. *Anal. Bioanal. Chem. Res.* **2018**, *5*, 205–215. [CrossRef]
33. Zolgharnein, J.; Feshki, S. Solid-phase extraction and separation of brilliant green by Fe3O4 magnetic nano-particles functionalized by sodium dodecyl sulphate from aqueous solution: Multivariate optimization and adsorption characterization. *Desalin. Water Treat.* **2017**, *75*, 58–69. [CrossRef]
34. Raees, K.; Ansari, M.S.; Rafiquee, M.Z.A. Influence of surfactants and surfactant-coated IONs on the rate of alkaline hydrolysis of procaine in the presence of PEG. *J. King Saud Univ. Sci.* **2020**, *32*, 1182–1189. [CrossRef]
35. Raees, K.; Ansari, M.S.; Rafiquee, M.Z.A. Inhibitive effect of super paramagnetic iron oxide nanoparticles on the alkaline hydrolysis of procaine. *J. Nanostruct. Chem.* **2019**, *9*, 175–187. [CrossRef]
36. Khan, M.A.; Alqadami, A.A.; Wabaidur, S.M.; Siddiqui, M.R.; Jeon, B.-H.; Alshareef, S.A.; Alothman, Z.A.; Hamedelniel, A.E. Oil industry waste based non-magnetic and magnetic hydrochar to sequester potentially toxic post-transition metal ions from water. *J. Hazard. Mater.* **2020**, *400*, 123247. [CrossRef]
37. Saranya, T.; Parasuraman, K.; Anbarasu, M.; Balamurugan, K. XRD, FT-IR and SEM Study of Magnetite (Fe3O4) Nanoparticles Prepared by Hydrothermal Method. *Nano Vis.* **2015**, *5*, 149–154.
38. Aliramaji, S.; Zamanian, A.; Sohrabijam, Z. Characterization and Synthesis of Magnetite Nanoparticles by Innovative Sonochemical Method. *Procedia Mater. Sci.* **2015**, *11*, 265–269. [CrossRef]

39. Saif, B.; Wang, C.; Chuan, D.; Shuang, S. Synthesis and Characterization of Fe3O4 Coated on APTES as Carriers for Morin-Anticancer Drug. *J. Biomater. Nanobiotechnol.* **2015**, *6*, 267–275. [CrossRef]

40. Azari, Z.; Pourbasheer, E.; Beheshti, A. Mixed hemimicelles solid-phase extraction based on sodium dodecyl sulfate (SDS)-coated nano-magnets for the spectrophotometric determination of Fingolomid in biological fluids, Spectrochim. *Acta Part A Mol. Biomol. Spectrosc.* **2016**, *153*, 599–604. [CrossRef]

41. Tabrizi, A.B.; Teymurlouie, N.D. Application of Sodium Dodecyl Sulfate Coated Iron Oxide Magnetic Nanoparticles for the Extraction and Spectrofluorimetric Determination of Propranolol in Different Biological Samples. *J. Mex. Chem. Soc.* **2018**, *60*, 108–116. [CrossRef]

42. Alqadami, A.A.; Khan, M.A.; Otero, M.; Siddiqui, M.R.; Jeon, B.-H.; Batoo, K.M. A magnetic nanocomposite produced from camel bones for an efficient adsorption of toxic metals from water. *J. Clean. Prod.* **2018**, *178*, 293–304. [CrossRef]

43. Nacéra, Z. Comparative study of discoloration of mono-azo dye by catalytic oxidation based on wells-dawson heteropolyanion catalyst. *Environ. Nanotechnol. Monit. Manag.* **2018**, *10*, 10–16. [CrossRef]

44. Wang, B.; Yin, J.-J.; Zhou, X.; Kurash, I.; Chai, Z.; Zhao, Y.; Feng, W. Physicochemical Origin for Free Radical Generation of Iron Oxide Nanoparticles in Biomicroenvironment: Catalytic Activities Mediated by Surface Chemical States. *J. Phys. Chem. C* **2013**, *117*, 383–392. [CrossRef]

45. Jiang, J.; Zou, J.; Zhu, L.; Huang, L.; Jiang, H.; Zhang, Y. Degradation of methylene blue with H2O2 activated by peroxidase-like Fe3O4 magnetic nanoparticles. *J. Nanosci. Nanotechnol.* **2011**, *11*, 4793–4799. [CrossRef] [PubMed]

46. Lucas, M.; Peres, J. Decolorization of the azo dye Reactive Black 5 by Fenton and photo-Fenton oxidation. *Dye Pigment* **2006**, *71*, 236–244. [CrossRef]

47. Ma, Y.S.; Chang, C.N.; Chao, C.R. Decolorization of Rhodamine B by a Photo-Fenton Process: Effect of System Parameters and Kinetic Study. *Int. J. Environ. Resour.* **2012**, *1*, 73–80.

48. Gulkaya, I.; Surucu, G.; Dilek, F. Importance of H_2O_2/Fe^{2+} ratio in Fenton's treatment of a carpet dyeing wastewater. *J. Hazard. Mater.* **2006**, *136*, 763–769. [CrossRef]

49. Reza, K.M.; Kurny, A.; Gulshan, F. Photocatalytic Degradation of Methylene Blue by Magnetite+H2O2+UV Process. *Int. J. Environ. Sci. Dev.* **2016**, *7*, 325–329. [CrossRef]

50. Keyhanian, F.; Shariati, S.; Faraji, M.; Hesabi, M. Magnetite nanoparticles with surface modification for removal of methyl violet from aqueous solutions. *Arab. J. Chem.* **2016**, *9*, S348–S354. [CrossRef]

51. Pham, T.D.; Tran, T.T.; Le, V.A.; Pham, T.T.; Dao, T.H.; Le, T.S. Adsorption characteristics of molecular oxytetracycline onto alumina particles: The role of surface modification with an anionic surfactant. *J. Mol. Liq.* **2019**, *287*, 110900. [CrossRef]

52. Rather, M.A.; Bhat, S.A.; Pandit, S.A.; Bhat, F.A.; Rather, G.M.; Bhat, M.A. As Catalytic as Silver Nanoparticles Anchored to Reduced Graphene Oxide: Fascinating Activity of Imidazolium Based Surface Active Ionic Liquid for Chemical Degradation of Rhodamine B. *Catal. Lett.* **2019**, *149*, 2195. [CrossRef]

53. Khan, A.M.; Mehmood, A.; Sayed, M.; Nazar, M.F.; Ismail, B.; Khan, R.A.; Ullah, H.; Rehman, H.M.A.; Khan, A.Y.; Khan, A.R. Influence of acids, bases and surfactants on the photocatalytic degradation of a model dye rhodamine B. *J. Mol. Liq.* **2017**, *236*, 395–403. [CrossRef]

54. Oliveira, E.G.L.; Rodrigues, J.J.; de Oliveira, H.P. Influence of surfactant on the fast photodegradation of rhodamine B induced by TiO2 dispersions in aqueous solution. *Chem. Eng. J.* **2011**, *172*, 96–101. [CrossRef]

Article

Molecularly Imprinted Polymers for the Removal of Antide-Pressants from Contaminated Wastewater

Tjasa Gornik [1,2], Sudhirkumar Shinde [3,4], Lea Lamovsek [5], Maja Koblar [2,6], Ester Heath [1,2], Börje Sellergren [3] and Tina Kosjek [1,2,*]

[1] Department of Environmental Sciences, Jozef Stefan Institute, Jamova 39, 1000 Ljubljana, Slovenia; tjasa.gornik@ijs.si (T.G.); ester.heath@ijs.si (E.H.)
[2] Jozef Stefan International Postgraduate School, Jamova 39, 1000 Ljubljana, Slovenia; maja.koblar@ijs.si
[3] Department of Biomedical Sciences and Biofilms-Research Center for Biointerfaces (BRCB), Faculty of Health and Society, Malmö University, 20506 Malmö, Sweden; sudhirshinde1@gmail.com (S.S.); borje.sellergren@mau.se (B.S.)
[4] School of Chemistry and Chemical Engineering, Queens University Belfast, Belfast BT9 5AG, UK
[5] Department of Biopharmacy and Pharmacokinetics, Faculty of Pharmacy, University of Ljubljana, Askerceva 7, 1000 Ljubljana, Slovenia; lealamovsek@gmail.com
[6] Center for Electron Microscopy and Microanalysis (CEMM), Jamova 39, 1000 Ljubljana, Slovenia
* Correspondence: tina.kosjek@ijs.si; Tel.: +386/1-477-3288

Abstract: Selective serotonin reuptake inhibitors (SSRIs) are a class of antidepressants regularly detected in the environment. This indicates that the existing wastewater treatment techniques are not successfully removing them beforehand. This study investigated the potential of molecularly imprinted polymers (MIPs) to serve as sorbents for removal of SSRIs in water treatment. Sertraline was chosen as the template for imprinting. We optimized the composition of MIPs in order to obtain materials with highest capacity, affinity, and selectivity for sertraline. We report the maximum capacity of MIP for sertraline in water at 72.6 mg g^{-1}, and the maximum imprinting factor at 3.7. The MIPs were cross-reactive towards other SSRIs and the metabolite norsertraline. They showed a stable performance in wastewater-relevant pH range between 6 and 8, and were reusable after a short washing cycle. Despite having a smaller surface area between 27.4 and 193.8 m$^2 \cdot$g^{-1}, as compared to that of the activated carbon at 1400 m$^2 \cdot$g^{-1}, their sorption capabilities in wastewaters were generally superior. The MIPs with higher surface area and pore volume that formed more non-specific interactions with the targets considerably contributed to the overall removal efficiency, which made them better suited for use in wastewater treatment.

Keywords: molecular imprinting; polymer; wastewater treatment; sertraline; cross-reactivity; SSRI; template; sorbent

Citation: Gornik, T.; Shinde, S.; Lamovsek, L.; Koblar, M.; Heath, E.; Sellergren, B.; Kosjek, T. Molecularly Imprinted Polymers for the Removal of Antide-Pressants from Contaminated Wastewater. *Polymers* **2020**, *13*, 120. https://doi.org/10.3390/polym13010120

Academic Editor: Marta Otero
Received: 24 November 2020
Accepted: 23 December 2020
Published: 30 December 2020

1. Introduction

The fast population growth, advances in industry, and increased agricultural activity have greatly influenced the environment. In order to continue with the current pace, we need solutions in environmental management, especially wastewater (WW) reuse. The development in the area of sample preparation and instrumentation has put the removal of trace-level emerging contaminants in the forefront of environmental research [1,2]. Among them, pharmaceuticals are a very problematic group, since they are particularly designed to have a pharmacological effect on humans or animals, thus potentially yielding adverse effects in living organisms [3] after they have entered the aquatic environment.

The selective serotonin reuptake inhibitors (SSRIs) are members of the most prescribed class of antidepressants in the USA and Europe [4–6]. They have been repeatedly detected in WW, surface waters, sediments, and aquatic organisms [7–12], and are thus part of different monitoring programs [13]. In aquatic organisms, SSRIs cause changes in biochemical processes, feeding behavior, survivorship behavior, growth, and potential changes in their genetic material [7,14–17]. Hence, it is crucial to improve their removal from WW before they are introduced into the environment. Among the existing WW treatment techniques, advanced oxidation processes and biological treatment are most successful in removing SSRIs from WW [18,19]. While during the former, the leading process of removal is degradation, sorption to activated sludge seems to be responsible for the removal of the majority of SSRIs during biological treatment [19]. Hence, other adsorption-based treatment techniques have been considered. Among them, activated carbon (AC) is by far the most researched material for SSRI removal [20,21]. AC as a treatment technique is technologically simple, has relatively fast kinetics, and removes a high variety of contaminants. Its main disadvantages are a high initial investment, the non-selectivity of the process, and the need for frequent regeneration due to fouling, which is expensive, time-consuming, and results in the loss of material in each regeneration cycle [22,23]. Greener alternatives, such as using products of pyrolysis of primary and secondary paper mill sludge, spent coffee grounds, and pine bark have been reported [24,25]. However, there is a lack of literature investigating modified synthetic composite materials, such as carbon-based nanomaterials, different types of membranes, and other forms of modified polymers, which, however, present promising alternatives to achieve superior SSRI removal from WWs [19,26–29]. On the basis of this knowledge gap, we investigated molecularly imprinted polymers (MIPs) as an alternative sorption material to AC [23,30,31].

MIPs are polymers that have been imprinted by a chosen template during the polymerization step in order to create selective recognition sites and are therefore often referred to as artificial antibodies or synthetic receptors [32]. After the template is removed from the MIP, the same or similar molecule can be rebound. They have already been commercially used for solid-phase extraction (SPE) [33,34] and researched for several other applications, such as catalysis, chromatography, and drug delivery [35,36]. In the last few years, the number of studies considering MIPs for water treatment has increased. Thus far, they have been utilized to remove non-steroidal anti-inflammatory drugs, antibiotics, antimicrobials, endocrine-disrupting compounds, herbicides, phenols, and beta-blockers from contaminated WW [30,35,37–41]. The advantages of using MIPs for water treatment are their high selectivity and affinity for their targets. Hence, we expect to be able to regenerate the material after longer intervals compared to AC, since slower fouling rates are expected. Literature reports MIPs as mechanically and chemically stable, and thus they should withstand several regeneration cycles unchanged, making the treatment more cost-effective [35,38]. The main disadvantage of MIPs is, however, the initial investment into the production of the polymers. Among multiple polymerization procedures available today, we chose bulk polymerization as one of the simplest and cheapest one for MIP production [2].

The aim of this work was to develop a MIP that could be used for removal of not only our targeted template, but for the whole class of SSRIs. We evaluated the affinity, capacity, and selectivity of the synthetized MIPs for sertraline (SER) and chose the best performing materials. Further characterization included cross-reactivity towards other antidepressants fluoxetine (FLU), paroxetine (PXT), escitalopram (ESC), bupropion (BUP), two SER metabolites—norsertraline (NS) and sertraline ketone (SEK) [9,10,42], and structurally related compound bupivacaine (BUC) (Figure 1). Potential parameters influencing the removal were considered and the performance of the MIPs in WW was tested in order to evaluate their applicability for WW treatment. The composition of the polymers was confirmed using Fourier transform infrared spectroscopy (FTIR) and elemental analysis. Surface properties and pore volume were calculated on the basis of the obtained Brunauer–Emmett–Teller (BET) isotherms, and scanning electron microscopy images of materials were taken for morphological characterization.

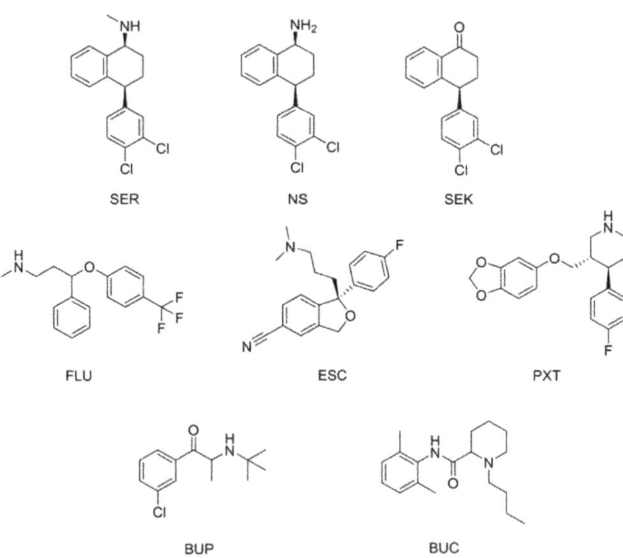

Figure 1. Chemical structures of the tested compounds.

2. Materials and Methods

The list of chemicals, materials, and the description of standard solution preparation and pre-preparation of the polymerization ingredients are reported in the Supplementary Material (SM) Section 1.

2.1. The Synthesis of MIP

The polymers were prepared via bulk radical polymerization with the ingredients in ratios specified in Table 1.

Table 1. Polymer compositions (molar ratio) and ingredients used for the synthesis of molecularly imprinted polymers (MIPs).

Material	Template	MAA	mMA	HEMA	EGDMA	Initiator (V-65)	Porogen
MIP1	SER×HCl (1)	4	/	/	20		CHCl$_3$
MIP2	SER×HCl (1)	4	8	/	12		CHCl$_3$
MIP3	SER×HCl (1)	4	/	8	12		CHCl$_3$
MIP4	SER (1)	4	/	/	20		MeOH
MIP5	SER (1)	4	/	/	20	1 wt % based on total monomers	CHCl$_3$
MIP6	SER (1)	4	/	/	20		ACN
MIP7	SER (1)	4	/	/	20		toluen
MIP8	SER (1)	4	8	/	12		CHCl$_3$
MIP9	SER (1)	4	8	/	12		ACN
MIP10	SER (1)	4	8	/	12		toluen
MIP11	SER (1)	4	/	8	12		CHCl$_3$
MIP12	SER (1)	4	/	8	12		ACN
MIP13	SER (1)	4	/	8	12		toluen

The mini-MIP library was synthesized by varying functional monomer, porogen, and the form of the template, as illustrated in the Table 1. The molar ratio between the template, functional monomer, and cross-linker was 1/4/20. In the case of mini-MIPs, 34.1 mg (0.1 mmol) of sertraline in HCl salt form (SER HCl) or 30.8 mg (0.1 mmol) free base sertraline (SER), 34 µL (0.4 mmol) methacrylic acid (MAA), and 380 µL (2 mmol) ethylene glycol dimethacrylate (EGDMA) was used. A total of 560 µL of porogen (either CHCl$_3$, methanol—MeOH, or acetonitrile—ACN) was added, with the exception of anhydrous toluene, where 580 μL was needed due to solubility issues. For polymers prepared using two functional monomers, we changed the ratio to 1/4/8/12 for the template (30.8 mg SER, 0.1 mmol), functional monomer (34 μL MAA, 0.4 mmol), co-monomer (860 μL of methyl methacrylate (mMA) or 970 μL of 2-hydroxyethyl methacrylate (HEMA), 0.8 mmol), and 227 μL of the EGDMA cross-linker (1.2 mmol). We used 1 wt % of the initiator 2,2′-azobis(2,4-dimethyl valeronitrile) (V-65) for synthesis of polymers on the basis of total monomers.

The synthetic procedure was identical for all MIPs. The monomers and the template were first mixed and dissolved in the porogen solvent. Then cross-linker EGDMA was added and the solution was mixed again. Finally, the initiator V-65 was added. The solution was mixed, purged with N_2 for 10 min, and polymerized at 50 °C for 24 h in an oven. After 24 h, the polymerization was carried out for another 2 hours at 70 °C. The corresponding non-imprinted polymers (NIPs) were prepared following the identical procedures in the absence of the template.

Best-performing MIPs and their corresponding NIPs were later prepared in a 10 times larger quantity, maintaining the same polymer compositions and ingredients. The polymers were then crushed and sieved into 25–50 μm particle size. Both MIPs and NIPs underwent Soxhlet extraction in 10% of acetic acid in methanol for 96 h until no SER was detected by a high-performance liquid chromatograph coupled with a diode array detector (HPLC-DAD). The polymers were further washed with water and MeOH to remove the acetic acid, before drying them in the oven at 50 °C for 24 h. The dried polymers were used for further physical and analytical characterization.

2.2. Selection of the Material: Batch Rebinding

Batch rebinding tests were performed in both water and acetonitrile (ACN). A total of 5 mg of each MIP and the corresponding NIP was weighed and placed in 1.5 mL Eppendorf tubes containing 500 μL of the SER solution with increasing concentrations: 0.1, 0.4, 1.0, 2.0, 3.0, and 4.0 mM. We used SER × HCl for rebinding in water, and SER in the free base form for the rebinding in ACN. All the experiments were performed after the equilibrium had been reached, i.e., after 20 h (see Section 2.5). The suspension was centrifuged at 10,000 rpm for 15 min. The supernatant was diluted 10 times with the mixture of 50% ACN and 50% 20 mM phosphate buffer at pH 3.70 (mobile phase) and subsequently quantified by HPLC-DAD analysis. The levels of bound compounds to the MIP/NIP for each solvent mixture were estimated from plotted calibration curves. We plotted the data in the form of rebinding isotherms using the bi-Langmuir isotherm as the best fit ($R^2 > 0.90$). The capacity, affinity, and selectivity were calculated for each polymer. Capacity was reported as the mass of bound compound per gram of polymer. Affinity was determined as the distribution ratio (D), the ratio between the amount of SER bound to the polymer (B), and the remaining SER in the supernatant (F). The selectivity was calculated as the imprinting factor (IF), comparing the D of MIP to the D of its corresponding NIP. All the parameters were calculated at equilibrium at the highest added concentration of 4.0 mM. On the basis of the results in both ACN and water, we chose three best performing MIPs for further testing.

2.3. Reusability Experiments

Reusability of the chosen MIPs and NIPs was tested by repeating 4 times the batch rebinding of 0.1 mM SER in ultrapure water (UW) on the same material, while following any changes in the performance. Between the cycles, the polymers were washed with 1 mL 1% trifluoroacetic acid (TFA) in MeOH (30 min) and 1 mL of MeOH (15 min) in order to remove SER from polymers. Solvent-free polymers were obtained by drying in the oven for 1 h at 60 °C. The experiment was performed in 5 parallels.

2.4. Cross-Reactivity Experiments

The cross-reactivity of the 3 materials selected as described in Section 2.2 was evaluated by binding experiments for antidepressants and their structurally related compounds: NS,

SEK, FLU, ESC, PXT, BUP, and BUC. The cross-reactivity was assessed through selectivity factor (α), the capacity, and the difference in binding between MIP and NIP for each compound. A was calculated as the ratio between the D of SER and D of the tested compound.

The cross-reactivity experiments were performed separately for each compound in UW, applying the same conditions as for SER rebinding tests (see Section 2.2.). The experiments were performed at the concentration of 1 mM, which was selected on the basis of the maximal solubility of NS in UW. SEK binding was evaluated in ACN due to solubility limitation. The concentrations in the supernatant were again determined with the HPLC-DAD.

2.5. Time to Reach Equilibrium

The time to reach the equilibrium state was estimated in batch experiments in UW. A total of 5 mg of each chosen polymer and AC were shaken for 15 min, 30 min, 1 h, 4 h, 8 h and 20 h. The 0.5 mL solutions contained a mixture of SER and the compounds included in Section 2.4 (test mixture), each added at the final concentration of 0.1 mM. The removal percentage was determined by HPLC-DAD.

2.6. Binding in WW Matrix: Influence of pH, Salts, and Chemical Oxygen Demand

The behavior of the chosen polymers and AC was observed in WW matrix spiked with the test mixture, again at the final concentration of 0.1 mM. The binding experiments were performed in 3 different matrices: UW, artificial wastewater (WW1) [43], and actual wastewater (WW2) obtained from a Slovenian wastewater treatment plant (WWTP). The WW was filtered (see SM, Section 1.1) before spiking in order to remove particulates and microorganisms that could have influenced the removal. The pH of the WWs was measured using the pH electrode by Wissenschaftlich-Technische Werkstätten GmbH (Weilheim, Germany) and the chemical oxygen demand (COD) was determined on a spectrophotometer using Hach reagents for water analysis, LCK 314 and 514.

We researched the influence of 2 parameters most often reported to influence the binding: pH and the presence of salt ions [44–46]. Since the reported pH of WW is between 6 and 8, the performance of the polymers was tested by batch tests in 50 mM phosphate buffer solutions with pH adjusted to 6.0, 7.0, or 8.0 with either a 2 mM HCl or 1 mM NaOH solution. The influence of salt ions was observed by comparing the binding in UW and in NaCl solutions at the concentrations of 0.1 M and 1.0 M.

2.7. Upscale Experiment

In order to observe the performance of the materials on a larger scale and at lower concentration of substrate, we packed the material into SPE cartridges by separately weighing 50 mg of MIP, NIP, or AC. MIPs and NIPs were sedimented beforehand in a mixture of MeOH and water ($v/v = 80/20$) four-times for 1.5 h to avoid the loss of material through the frit. For the same reason, AC mesh size 100–400 was used.

The materials were first washed with 5 mL of MeOH and 5 mL of UW water. Then, the cartridges were stacked on top of Oasis HLB cartridges in order to bind the remainder of the unbound compounds. The method used for Oasis HLB conditioning, equilibration, loading, and elution was adapted from our article on photodegradation of SER [9].

A total of 50 mL of WW2 spiked with the mixture of compounds at concentrations of 0.4 µM was loaded at the flow rate of 2 mL min^{-1} on to each material. The solution then flowed directly onto the Oasis HLB cartridge. After loading, the Oasis HLB cartridges were dried for 30 min and then eluted with 3×0.6 mL of triethylamine in MeOH. The elution solvent was evaporated, and the extracts were redissolved in 0.5 mL the HPLC mobile phase and filtered through 0.45 µm syringe filters before the HPLC measurements.

2.8. Leaching Evaluation

To examine the applicability of developed MIPs as SPE extraction materials, we checked the potential leaching of the template from the MIP. As reported under the upscale experiment (Section 2.7), 50 mg of each MIP was packed in the SPE column, conditioned, loaded, and eluted with 5 mL 1% TFA in MeOH. The extract was dried under nitrogen at 40 °C and the amount of leaching was quantified with a Nexera X2 ultra high performance liquid chromatograph (UHPLC, Schimadzu, Kyoto, Japan) coupled to the hybrid quadrupole-linear ion trap mass spectrometry analyzer QTRAP 4500 (Sciex, Framingham, MA, USA) following the method developed by Gornik et al., (2020a) [9].

2.9. Chemical and Morphological Characterization

Fourier transform infrared (FTIR) spectroscopy was performed on IRAffinity-1S (Schimadzu, Kyoto, Japan).

Elemental analysis was performed on a 2400, Series II, CHNS/O Analyzer (Perkin-Elmer, Waltham, MA, USA).

BET surface area analysis was performed with Porozimeter TriStar II (Micromeritics, Norcross, GA, USA).

The morphological characteristics were observed using a scanning electron microscope (SEM). The images were recorded with JSM-7600F (JEOL Ltd., Tokyo, Japan).

2.10. HPLC Measurements

For the determination of SER, NS, SEK, FLU, ESC, PXT, BUP, and BUC in the solutions, we utilized an HPLC-DAD (1260 Infinity Agilent Technologies, Santa Clara, CA, USA). For separation, we applied the column Zorbax Eclipse C-18 column (150 mm × 4.6 mm, 5 μm) (Agilent Technologies, Santa Clara, CA, USA). The injection volume was 10 μL or 20 μL, depending on the tested concentration range. The mobile phases were (A) ACN and (B) 20 mM phosphate buffer at pH 3.70. The gradient started with 70% B for 2 min, decreased to 61% in 13 min, then increased back to 70% B in 0.1 min and was kept as so for 1.5 min. The flow rate was 1 mL·min^{-1}. The retention times of the compounds were 3.27 min for BUP, 4.04 min for BUC, 6.20 min for ESC, 8.52 min for PXT, 11.95 min for NS, 12.65 min for FLU, and 13.04 min for SER. SEK was determined with a separate method at flow 2 mL·min^{-1}, isocratic elution at 70% A and 30% B. Other parameters coincided with the previous method. SEK eluted at 3.80 min.

3. Results and Discussion

All experiments with the exception of the reusability experiments ($n = 5$) were performed in duplicate. The inter-day repeatability reported as the relative standard deviation (RSD) for experiments performed in UW was <5% and in WW < 6%.

3.1. MIP Synthesis, Selection, and Reusability

We optimized the polymer composition to tune recognition properties of the material. The initiator (V-65) and cross-linker (EGDMA) were kept constant for all polymerization experiments, while different porogens and co-monomers were added in order to obtain water compatibility, increase capacity, and improve selectivity. The behavior of MIPs compared to their corresponding NIPs was evaluated in batch rebinding experiments performed in water and ACN at different concentrations to generate binding isotherms and calculate the capacity, affinity, and IF. The data we obtained during ACN rebinding experiments enabled us to quantify the binding on the basis only of specific interactions, such as hydrogen bonding, with the minimal non-specific hydrophobic effect [47], while our prime goal was recognition of the investigated compounds in water.

EGDMA in combination with MAA in different porogen solvents is one of the most commonly reported compositions of MIPs to date [30,48,49], including those in MIPs imprinted with SER [50,51]. Unlike in the literature [50,51], we observed no imprinting in MIPs where SER was used in its salt form (MIPs 1–3 in Table 1). Adding the extracted free base form of SER, on the other hand, resulted in successful imprinting. As shown in Figures 2 and 3, we observed higher capacities and affinities in water compared to those in ACN for most tested MIPs, except for MIP11 and MIP12. However, compared to ACN, the Ifs in water were lower in all the cases, indicating loss of selectivity in water. This can be justified by the hydrophobic effect established in polar solvents such as water, and disrupting the formation of hydrogen bonds. In order to improve the recognition abilities in water, we tested the influence of adding co-monomers mMA or HEMA (see examples MIP8–MIP13 in Table 1). Here, the ratios between monomers and cross-linker we applied were based on the results from Dirion et al., (2003). HEMA was chosen on the basis of the reports on improved Ifs in water [47,48], while the mMA was selected as its more non-polar alternative. MIPs with the mMA added into the polymerization mixture (MIP8, MIP9, and MIP10) had a similar IF in ACN, as compared to MIPs 5–7, which were prepared by MAA only (Table 1). However, the Ifs in water were slightly higher for all three materials (Figure 3). The capacities and affinities of MIP8, MIP9, and MIP10 in ACN were higher, yet lower or comparable in water. Compared to MIPs 5–7, adding HEMA as a co-monomer (MIP11, MIP12, and MIP13) did not improve the capacity or affinity of the MIPs in ACN. Additionally, both parameters were noticeably lower in UW. The considerable improvement was, however, observed in IF; the highest was that of MIP13. This high IF is in agreement with the results of Dirion et al., (2013).

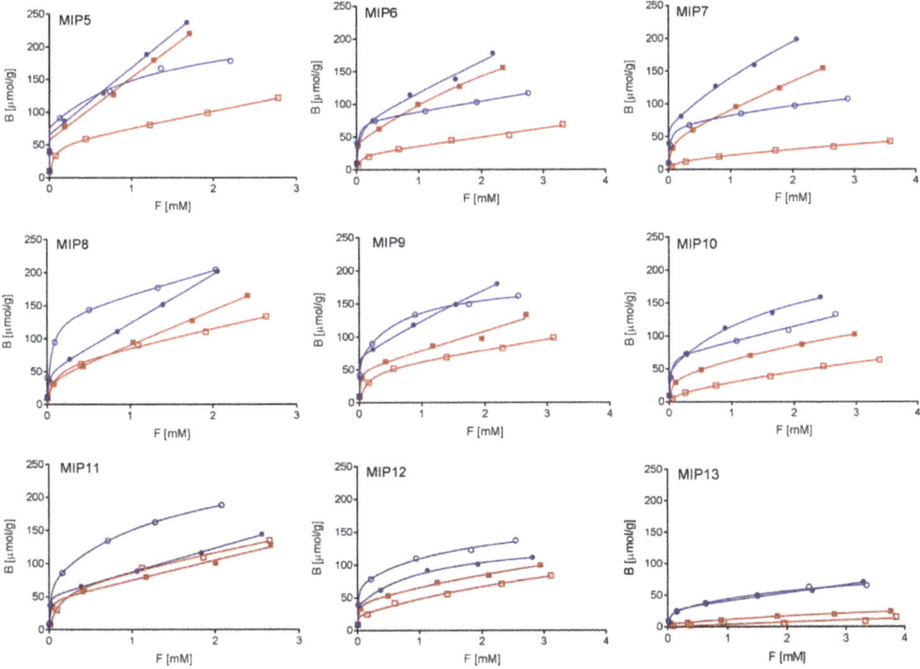

Figure 2. Rebinding isotherms for MIP (blue symbols) and non-imprinted polymer (NIP; red symbols) combinations 5–13 in ultrapure water (UW; full symbols) and acetonitrile (ACN; empty symbols).

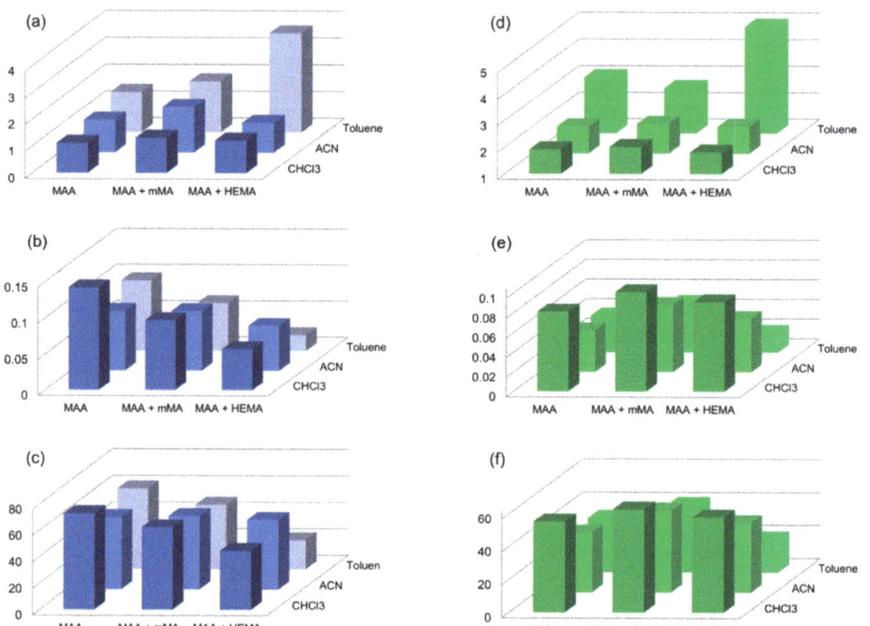

Figure 3. (**a**) The selectivity (imprinting factor, IF), (**b**) the affinity (L·g^{-1}), and (**c**) the capacity (mg·g^{-1}) of the MIPs in UW in blue. (**d**) The selectivity (IF), (**e**) the affinity (L·g^{-1}), and (**f**) the capacity (mg·g^{-1}) of the MIPs in ACN in green.

As seen in Table 1 and Figure 3, the porogen severely influenced the selectivity, capacity, and affinity of the MIPs. This happens as it affects the stability of the "pre-polymerization complex" (i.e., interactions between functional monomers and the template in the chosen porogen), which plays a crucial role in the imprinting effect. If the porogen disrupts hydrogen bonds between the template and monomers, no specific binding is observed, as can be seen in the case of MeOH (MIP4). On the contrary, using a more non-polar aprotic porogen, the pre-polymerization complex is stabilized, resulting in higher IF, which we showed in MIPs 7, 10, and 13 synthetized in toluene, as compared to those synthetized in ACN (MIPs 6, 9, 12) or CHCl$_3$ (MIPs 5, 8, 11) (Table 1, Figure 3a,d) [47].

Determining rebinding characteristics allowed us to select three most promising materials for further testing. In terms of capacity and affinity in water, the material MIP5 was chosen. MIP13 was chosen for its highest IF. Lastly, MIP9 was chosen because it combines satisfactory selectivity, capacity, and affinity in both water and ACN. The chosen polymers were reusable, with the maximum observed decrease in the capacity for SER in four consecutive rebinding experiments being only 2%.

3.2. Cross-Reactivity

We determined the cross-reactivity of three selected MIPs for the following antidepressants and structurally related compounds (Figure 1): BUC, BUP, ESC, PXT, NS, SEK, and FLU. While SEK, the metabolite of SER, was also initially included, it however showed very poor binding in ACN and no observed selectivity for any of the three MIPs. Its binding will therefore be based on non-specific interactions only. As for its poor solubility in water, it was thus excluded from further testing.

The results of cross-reactivity tests are selectivity factors (α) reported in Table 2. In general, α for each compound were comparable between the three selected MIPs, with the exception of BUP in MIP13. Here, the factor α 3.29, as compared to 9.60 and 9.76 for MIP5 and MIP9, respectively, indicated more cross-reactivity of MIP13 towards BUP. As reported in Table 2, for NS the selectivity factor was below 1, indicating better binding as compared to SER, which is reasoned by the absence of the methyl group in the chemical structure (Figure 1). Among the SSRI compounds, FLU and PXT had the factors slightly above 1, meaning comparable binding, while the factor for ESC varied between 2.7 and 2.9. The fact that ESC was the only SSRI with a tertiary amine in the structure, together with the favorable α for NS, suggests the impact of steric hindrance of the hydrogen bond-forming amino group on cross-reactivity. The size of the binding site seemed to be of lesser importance, considering that PXT and FLU are larger molecules as compared to SER and NS. BUC and BUP showed higher α in all three MIPs, which is justified by them being less structurally related to the SSRI group. Additionally, their amino groups are also sterically more hindered (tert-butyl group and tertiary amine).

Table 2. The capacity and selectivity factor of MIP5, MIP9, and MIP13 and the difference in binding of each compound between MIP and the corresponding NIP at 1 mM concentration of each tested analyte.

	MIP5			MIP9			MIP13		
Compound	Capacity (mg·g^{-1})	Selectivity Factor (α)	% (MIP-NIP)	Capacity (mg·g^{-1})	Selectivity Factor (α)	% (MIP-NIP)	Capacity (mg·g^{-1})	Selectivity Factor (α)	% (MIP-NIP)
SER	26.6 ± 0.6	1.0	8.4	24.4 ± 0.4	1.0	17.4	11.0 ± 0.3	1.0	25.6
NS	26.3 ± 0.3	0.8	7.2	23.1 ± 0.4	1.0	17.1	13.0 ± 0.1	0.7	33.5
FLU	25.3 ± 0.6	1.3	4.9	22.4 ± 0.4	1.2	16.2	10.3 ± 0.2	1.1	25.3
ESC	20.3 ± 0.5	2.9	4.7	18.8 ± 0.6	2.7	14.8	5.5 ± 0.2	2.8	10.8
PXT	27.3 ± 0.1	1.0	3.4	26.3 ± 0.4	1.0	16.8	11.3 ± 0.1	1.1	25.0
BUP	7.9 ± 0.1	9.6	2.2	6.5 ± 0.3	9.8	8.0	3.5 ± 0.3	3.3	12.3
BUC	10.3 ± 0.3	8.7	9.4	9.7 ± 0.3	7.0	9.4	2.2 ± 0.4	7.0	6.1

In general, the capacities of the three MIPs for SSRIs followed the same pattern as in SER binding. The highest capacity was observed in MIP5, closely followed by MIP9, and with more than half-lower capacities observed in MIP13 (Figure S1). Furthermore, we compared the binding to the corresponding NIPs. The difference between MIP and NIP was the largest in the case of MIP/NIP13 and the lowest in MIP/NIP5 (Table 2).

3.3. Time to Reach the Equilibrium

The time to reach equilibrium was tested for the chosen polymers and AC. A 0.1 mM test mixture was added. Figure S2 illustrates that for AC, the equilibrium was reached within 1 h; in cases of MIP5 and MIP9, the equilibrium was reached in 4 h; and for MIP13, in 20 h. For the NIPs, similar times to reach the equilibrium were shown as for their corresponding MIPs. As also depicted from Figure 4, AC non-selectively bound all the available compounds until their concentrations in the solvent reached below the limit of quantification (LOQ $\approx$ 0.001 mM).

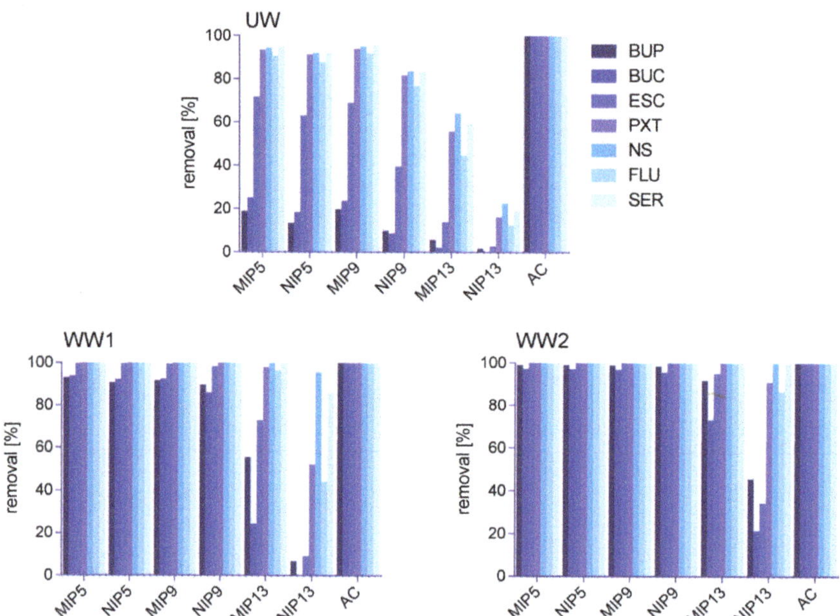

Figure 4. The impact of matrices (UW, artificial wastewater (WW1), and actual wastewater (WW2)) on the performance of MIPs, NIPs, and AC in the batch rebinding test.

3.4. Effect of WW Matrix

With the underlying objective to remove pharmaceuticals from WWs, we tested the capacity of MIPs to bind them. This way, we evaluated the ability of MIPs to be applied as sorbents in WW treatment systems. Aiming to get closer to the conditions during WW treatment, we employed the pH adjusted to 6–8 and simulated actual WW matrix composition. By comparing the results between the binding of BUP, BUC, ESC, PXT, NS, FLU and SER in UW, WW1, and WW2, we observed large differences in the removals of the test compounds (Figure 4), whereas AC removed all the tested compounds in any matrix to below LOQ concentrations (0.001 mM). In contrast with our expectations, as shown in Figure 4, the removal efficiencies of MIPs were lowest in UW and highest in the most complex matrix, WW2. In line with the trends shown in the capacity experiments, MIP5 and MIP9 showed best performance, closely followed by NIP5, NIP9, MIP13, and finally NIP13. By investigating the reason for such behavior, we determined the pH and COD of each inspected matrix. The pH values of UW, WW1, and WW2 were approximately 7, 7.2, and 8.2, respectively, whereas we measured COD at <15 mg·L^{-1} for WW1 (LOQ of the test) and 379 mg·L^{-1} for WW2. On the contrary, the literature reports either no change (up to 690 mg·L^{-1} COD) or a slight decrease in adsorption of their chosen templates to their MIPs at high COD values (over 800 mg·L^{-1}) [52–55]. The MIPs in these cases used similar reagents to those in our synthesis, i.e., MAA and EGDMA, albeit in different ratios, and employed DCM or ACN as porogens and 2,2′-azobisisobutyronitrile (AIBN) as the initiator [52–55]. Hence, we did not expect the higher COD values to be the cause behind the increased removal.

In order to deeper investigate the reasons behind the positive impact of matrix complexity on the removal of pharmaceuticals, we performed the rebinding experiments at different pH values and salt concentrations. Here, the imprinted and non-imprinted polymers showed similar trends, with the most notable differences for MIP and NIP13, as portrayed in Figure 5. The pH in the range of 6 to 8 had almost no influence on the binding with differences below 1%. The only exception was NIP13, with differences between pH 6 and pH 8 ranging up to 7.8%. On the other hand, the increasing salt concentration improved the removal of pharmaceuticals. This finding was further supported by the improved binding found during the pH tests, which were performed in phosphate buffer, as compared to the binding in UW. Our results are consistent with the findings of Kempe and Kempe (2010), where elevated concentrations of salts had a significant influence on the removal of penicillin G from solution and followed the Hofmeister series. As seen in Kempe and Kempe (2010), the higher removal was of non-specific nature, observed in both MIP and NIP [46]. The kosmotropic ions seem to promote the formation of stable interactions between the polymers and tested compounds. Since phosphate ions are more kosmotropic than chloride ions, this would also explain the larger effect in the buffer solutions, despite their lower concentrations [56].

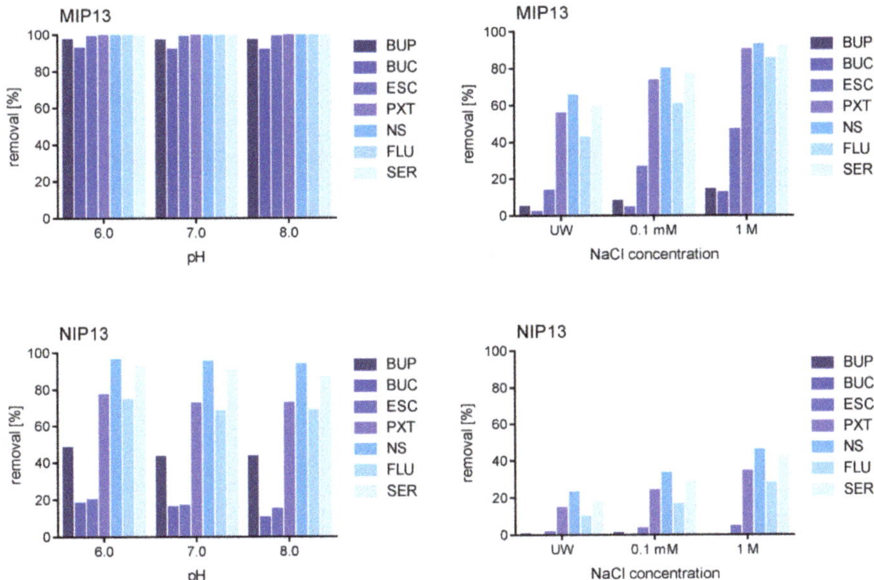

Figure 5. The effect of pH and presence of salt ions in MIP13 and NIP13.

3.5. Upscale Experiment

The performance of the materials was evaluated as the difference between the initial concentration and the remainder extracted by Oasis HLB SPE. This way, we avoided underestimating the performance of AC, since completely eluting compounds off the AC is a known difficulty [23]. The results on the performance of selected materials in the upscale experiment are shown in Figure 6. The main difference from the batch (mini-MIP) experiments is the less efficient binding to AC (Figure 6). The two main reasons behind this may involve the shorter contact time between the material and WW, or lower capacity of the material due to the non-specific binding of other matrix components. Since in the batch experiment AC showed shortest time to reach equilibrium, the latter is more probable. Furthermore, several reports showed AC performance deteriorating with an increase of matrix complexity (e.g., COD, total dissolved solids) [23,55].

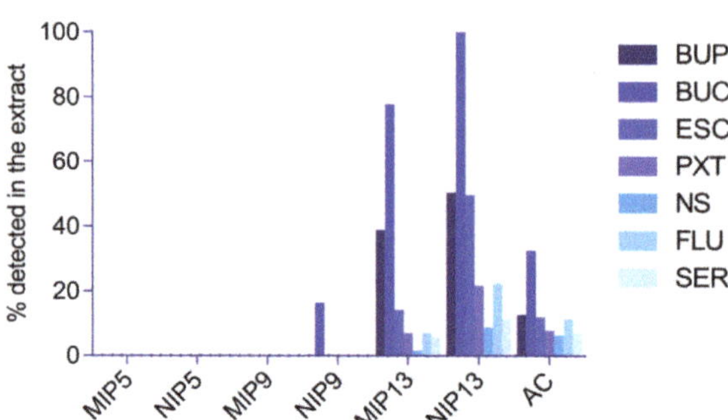

Figure 6. The remainder of compounds detected in the Oasis HLB extracts in the upscale experiment.

In the Oasis HLB extracts from MIP5, NIP5, and MIP9, none of the investigated pharmaceuticals were detected. On the contrary, as expected, their highest remainder was determined in the Oasis HLB extracts from NIP13, again implying its lowest binding capacity.

While non-specific rebinding is not desired in MIPs that are used, for example, in sample preparation or chromatography, we show here that this phenomenon is favorable in WW treatment. As Le Noir et al., (2007) pointed out, it only becomes a problem if it causes lower capacity and affinity of the selective binding [39]. MIP5 and MIP9 both showed higher capacities compared to MIP13, and even NIP5 and NIP9 performed better under tested conditions. This means that a larger amount of MIP13 would have to be used to achieve the competitive removal efficiencies. However, specific interactions of MIPs will likely play a more important role at higher volumes and more complex matrices. At the same time, we show that the NIPs, which are based on non-specific binding only, are less negatively affected by matrix, as compared to AC, and along with their easy recyclability they could therefore pose a less expensive alternative for the removal of pharmaceuticals.

3.6. Leaching

As an alternative to sorption in WW treatment, we also considered the developed MIPs for SPE extraction of environmental samples. As for our hypothesis, MIP could be employed as an SPE sorbent in order to selectively extract targeted compounds, thus reducing the suppressing effect of matrix interferences in further liquid chromatography coupled to mass spectrometry (LC–MS) analysis. Such sorbents may potentially be employed in a highly sensitive analytical method for an ultra-trace level determination of contaminants in WW [57]. MIPs have previously been used for SPE several times [58–61]. However, given the fact that the template in polymerization (SER) is also the analyte in the LC–MS method, the MIP sorbent would have to pass the "leaching test", which means that it would have to show a negligible leaching and thus avoid interfering with the assessment of trace-level analytes in the subsequent LC–MS analysis. Leaching of SER from the material was tested on UHPLC-QTRAP, applying the instrumental method developed by Gornik et al., (2020a) [9]. By using 5 mL of 1% TFA in MeOH, we eluted up to 3.5 μg of SER from the MIPs. Alternative methods for template removal, such as microwave or ultrasound-assisted extraction, heating under pressure, or even the use of another acid during Soxhlet extraction, could have lessened the leaching from the MIPs. On the other hand, the more extreme conditions could also have damaged or distorted the imprinted cavities and thus decreased the selectivity, affinity, and capacity of the MIPs [62,63]. Furthermore, the synthesis of MIPs and the subsequent washing procedures triggered the formation of SER transformation products (NS, SEK, hydroxyl-SER) [9], which in turn leached off the materials, thus interfering the environmental analysis. Unfortunately, this makes the material inappropriate for the determination of SER residues including its metabolites and transformation products at trace levels. Finding an appropriate dummy template that would substitute SER and produce a MIP cross-reactive towards SSRI could be a viable solution to such a problem [35]. Nonetheless, the synthetized material can still be applied to SPE of the remaining tested pharmaceuticals (Figure 1).

3.7. Characterization

The FTIR spectra for the chosen MIP/NIP pairs 5, 9, and 13 can be found in Figure 7. The broad band visible at approximately 3500 cm^{-1} corresponds with the stretching vibration of the hydroxyl group from MAAs COOH group. The stretch bands around 2950 cm^{-1} in all the spectra are part of the C–H vibration present in MAA, mMA, HEMA, and EGDMA. The band around 1720 cm^{-1} represents the vibration from the carboxylic C=O group that can be associated with the C=O groups from MAA, mMA, and EGDMA. The 1250 and 1140 cm^{-1} stretch bands contributed to the stretching of C–O also present in all three compounds. The stretch bands corresponded with the polymerized material. Since the composition of the synthetized materials did not vary strongly, the resulting FTIR spectra were accordingly similar.

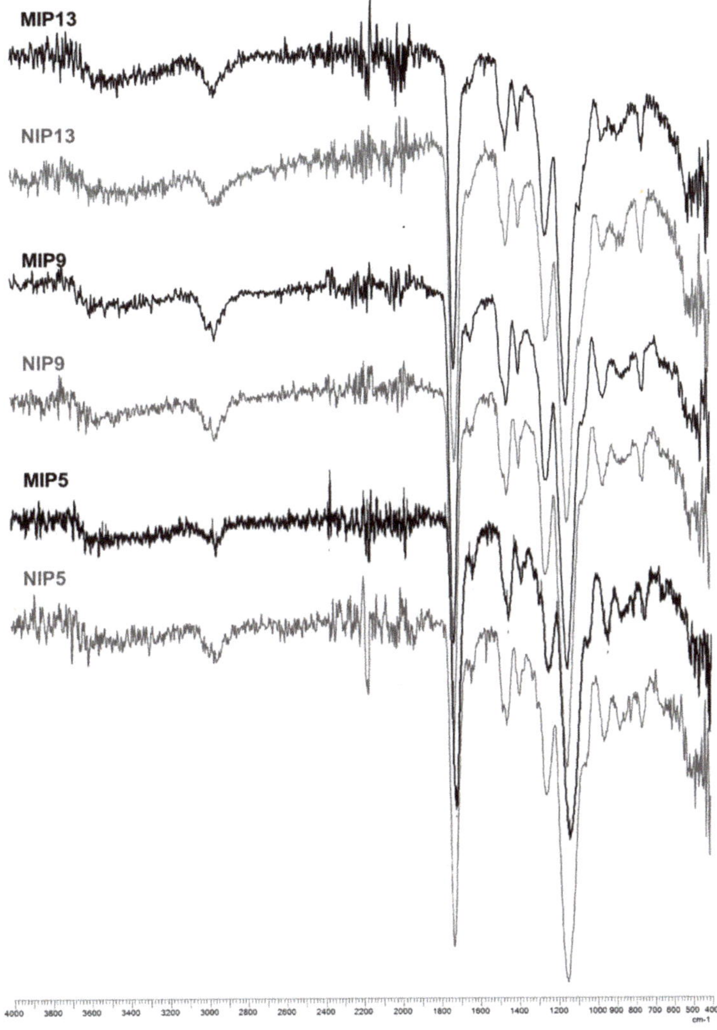

Figure 7. The FTIR spectra of MIPs and NIPs 5, 9, and 13.

The results of the elemental analysis of the MIP and NIP pairs 5, 9, and 13 are reported in Table 3. The results are in accordance with the expected values of the synthetized material. With this measurement, we confirmed that the added reagents reacted in the expected ratio.

Table 3. Results of the elemental analysis for MIPs and NIPs 5, 9, and 13.

MIP 5	**% C**	**% H**	**NIP 5**	**% C**	**% H**
Theoretical	60.21	7.11	Theoretical	60.21	7.11
Actual	59.56	7.81	Actual	59.68	8
Deviation	0.65	−0.7	Deviation	0.53	−0.89
MIP 9	**% C**	**% H**	**NIP 9**	**% C**	**% H**
Theoretical	59.99	7.28	Theoretical	59.99	7.28
Actual	59.55	8.18	Actual	60.2	8.25
Deviation	0.44	−0.9	Deviation	−0.21	−0.97
MIP 13	**% C**	**% H**	**NIP 13**	**% C**	**% H**
Theoretical	59.27	7.23	Theoretical	59.27	7.23
Actual	57.00	7.60	Actual	58.03	8.17
Deviation	2.27	−0.37	Deviation	1.24	−0.94

The BET surface area, pore size, and pore volume of the MIPs and NIPs are reported in Table 4. As expected, the larger the surface area and pore volume of the tested polymers, the higher the reported capacity and affinity. All three parameters were comparable between MIP and NIP pairs 5 and 9, with BET surface areas for MIP/NIP 5 in the $200\,\mathrm{m^2 \cdot g^{-1}}$ range and MIP/NIP 9 at the $100\,\mathrm{m^2 \cdot g^{-1}}$ range. However, NIP13 exhibited a more than five times lower BET surface area and pore volume compared to its corresponding MIP (Table 4). A similar difference was observed in MIP and NIP pairs using HEMA as the copolymer in toluene in the research by Dirion et al., (2003). They reported that stronger swelling was observed for the NIPs and similar elution times measured for void markers (acetone or MeOH) in their chromatographic evaluations of the polymers. This indicated a smaller difference between the MIP and NIP in their swollen state.

Table 4. Brunauer–Emmett–Teller (BET) surface area of MIPs and NIPs 5, 9, and 13.

Material	BET Area ($\mathrm{m^2 \cdot g^{-1}}$)	Pore Size (nm)	Pore Volume ($\mathrm{cm^3 \cdot g^{-1}}$)
MIP5	193.8	7.7	0.374061
NIP5	262.1	7.2	0.470623
MIP9	136.0	10.3	0.349167
NIP9	125.7	9.6	0.300946
MIP13	27.4	7.6	0.051835
NIP13	5.5	6.6	0.009074

The SEM images of the surface of our polymers in Figure 8 support the surface area and pore volume measurements. While the morphology of MIP5/NIP5 and MIP9/NIP9 were comparable, the surfaces of MIP13 and NIP13 were dissimilar. These differences in the morphology between MIP and NIP 13 indicate that care should be taken when NIPs are used for the evaluation of MIP selectivity. Comparing a material imprinted with a completely different compound or the determination of α between the template and other compounds can offer more information [35,48].

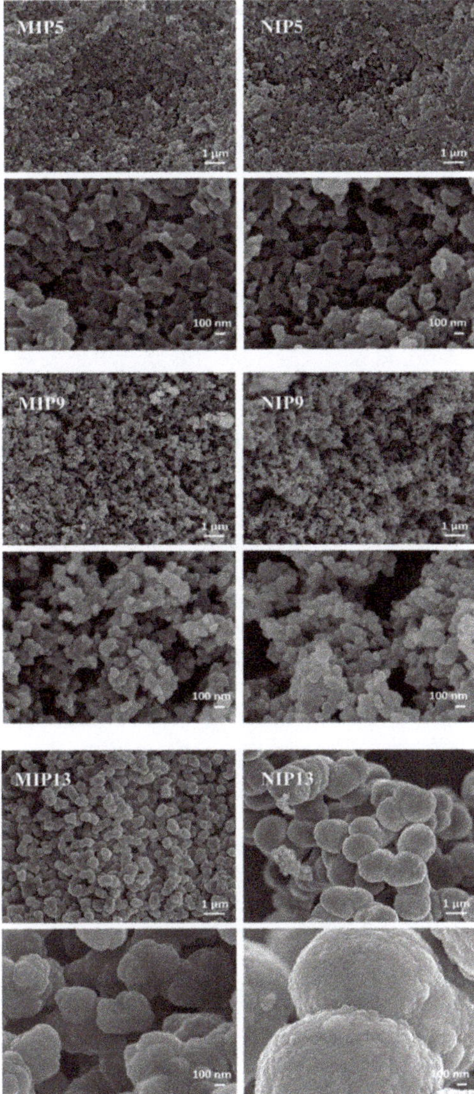

Figure 8. SEM images of MIPs and NIPs 5, 9, and 13.

Compared to AC with a surface area of 1400 m^2·g^{-1} [64], the surface areas of MIPs and NIPs were 5 to 253 times lower. Nevertheless, some of them showed superior binding characteristics in WW.

4. Conclusions

This study investigated the ability of MIPs imprinted with the free base form of SER to remove SSRIs and their metabolites. The functional monomers and porogens revealed a strong impact on the capacity, affinity, and selectivity of the synthetized MIPs. The three selected MIPs showed cross-reactivity towards the SSRIs and the metabolite norsertraline, whereas they bound a lesser amount of the competitors BUP and BUC. Further, the loss of selectivity towards the metabolite SEK was probably due to the loss of the amino group, which was thus found crucial for selective binding to the MIP. The performance of both the imprinted and non-imprinted materials was strongly influenced by the presence of salt ions, which improved their performance in WW. The performance of MIPs was stable throughout WW-relevant pH range 6–8. Compared to AC, the synthetized polymers had at least five times lower surface area and required a longer equilibration time. This slower mass transfer was particularly evident when selective binding was the main driving force behind the removal, as observed in MIP13. However, the capacity in WW for two out of the three tested MIPs surpassed that of AC, and thus both the non-specific and specific interactions showed an important role for the removal from WW. The surface area calculated from the BET isotherm for the MIPs correlated with a higher removal and more non-specific interactions. The advantage of the MIPs is also their reusability that, together with the lower number of regeneration cycles needed due to slower fouling, will cut the costs of the treatment. Unfortunately, the MIPs were found inappropriate for SPE of samples containing trace levels of SER due to continuous leaching of the template and its degradation products. Future work should include a large-scale experiment confirming the advantages of the synthetized material for the removal of SSRIs from WW.

Supplementary Materials: The following are available online at https://www.mdpi.com/2073-4360/13/1/120/s1: List of standards, chemicals, and materials; Standard solution preparation; Pre-preparation of the ingredients. Figure S1: Cross-reactivity of MIP5, MIP9, and MIP13 in UW. Figure S2: Time to reach the equilibrium for MIP5, MIP9, MIP13, and activated carbon (AC).

Author Contributions: Conceptualization, T.G., S.S., B.S. and T.K.; methodology, T.G., S.S., B.S. and T.K.; analysis and investigation, T.G., L.L., M.K. and S.S.; writing—original draft preparation, T.G.; writing—review and editing, S.S., B.S., T.K. and E.H.; supervision, B.S. and T.K.; project administration, B.S. and T.K.; funding acquisition, B.S., E.H. and T.K. All authors have read and agreed to the published version of the manuscript.

Funding: The authors acknowledge the financial support from the Slovenian Research Agency (research core funding no. P1-0143) and project J1-6744 (Development of Molecularly Imprinted Polymers and their application in environmental and bio-analysis). This work was supported by a STSM Grant from the NEREUS COST Action ES1403 and the Erasmus plus program.

Acknowledgments: Special thanks go to Amadeja Koler from the University of Maribor, Faculty of Chemistry and Chemical Engineering, PolyOrgLab, Maribor, Slovenia, for the work on chemical characterization of the polymers.

Conflicts of Interest: The authors declare no conflict of interest.

References

1. Radjenović, J.; Petrović, M.; Barceló, D. Fate and distribution of pharmaceuticals in wastewater and sewage sludge of the conventional activated sludge (CAS) and advanced membrane bioreactor (MBR) treatment. *Water Res.* **2009**, *43*, 831–841. [CrossRef] [PubMed]
2. Sarpong, K.A.; Xu, W.; Huang, W.; Yang, W. The Development of Molecularly Imprinted Polymers in the Clean-Up of Water Pollutants: A Review. *Am. J. Anal. Chem.* **2019**, *10*, 202–226. [CrossRef]
3. Gómez-Oliván, L.M. *The Handbook of Environmental Chemistry*; Springer International Publishing: Cham, Switzerland, 2019; Volume 66.
4. Bauer, M.; Monz, B.U.; Montejo, A.L.; Quail, D.; Dantchev, N.; Demyttenaere, K.; Garcia-Cebrian, A.; Grassi, L.; Perahia, D.G.; Reed, C.; et al. Prescribing patterns of antidepressants in Europe: Results from the Factors Influencing Depression Endpoints Research (FINDER) study. *Eur. Psychiatry* **2008**, *23*, 66–73. [CrossRef] [PubMed]
5. Fuentes, A.V.; Pineda, M.D.; Venkata, K.C.N. Comprehension of Top 200 Prescribed Drugs in the US as a Resource for Pharmacy Teaching, Training and Practice. *Pharmacy* **2018**, *6*, 43. [CrossRef]
6. NIJZ Poraba Ambulantno Predpisanih Zdravil v Sloveniji v Letu 2019. Available online: https://www.nijz.si/sites/www.nijz.si/files/publikacije-datoteke/publikacija_220520_koncno_0.pdf (accessed on 22 November 2020).
7. Arnnok, P.; Singh, R.R.; Burakham, R.; Pérez-Fuentetaja, A.; Aga, D.S. Selective Uptake and Bioaccumulation of Antidepressants in Fish from Effluent-Impacted Niagara River. *Environ. Sci. Technol.* **2017**, *51*, 10652–10662. [CrossRef]
8. Bergersen, O.; Hanssen, K.Ø.; Vasskog, T. Anaerobic treatment of sewage sludge containing selective serotonin reuptake inhibitors. *Bioresour. Technol.* **2012**, *117*, 325–332. [CrossRef]

9.   Gornik, T.; Vozic, A.; Heath, E.; Trontelj, J.; Roskar, R.; Zigon, D.; Vione, D.; Kosjek, T. Determination and photodegradation of sertraline residues in aqueous environment. *Environ. Pollut.* **2020**, *256*, 113431. [CrossRef]
10.   Gornik, T.; Kovacic, A.; Heath, E.; Hollender, J.; Kosjek, T. Biotransformation study of antidepressant sertraline and its removal during biological wastewater treatment. *Water Res.* **2020**, *181*, 115864. [CrossRef]
11.   Mole, R.A.; Brooks, B. Global scanning of selective serotonin reuptake inhibitors: Occurrence, wastewater treatment and hazards in aquatic systems. *Environ. Pollut.* **2019**, *250*, 1019–1031. [CrossRef]
12.   Schultz, M.M.; Furlong, E.T.; Kolpin, D.W.; Werner, S.L.; Schoenfuss, H.L.; Barber, L.B.; Blazer, V.S.; Norris, D.O.; Vajda, A.M. Antidepressant Pharmaceuticals in Two U.S. Effluent-Impacted Streams: Occurrence and Fate in Water and Sediment, and Selective Uptake in Fish Neural Tissue. *Environ. Sci. Technol.* **2010**, *44*, 1918–1925. [CrossRef]
13.   OECD. Pharmaceutical Residues in Freshwater: Hazards and Policy Responses-Policy Highlights. In *OECD Studies on Water*; OECD Publishing: Paris, France, 2019; Volume 2019.
14.   Park, J.-W.; Heah, T.P.; Gouffon, J.S.; Henry, T.; Sayler, G.S. Global gene expression in larval zebrafish (Danio rerio) exposed to selective serotonin reuptake inhibitors (fluoxetine and sertraline) reveals unique expression profiles and potential biomarkers of exposure. *Environ. Pollut.* **2012**, *167*, 163–170. [CrossRef] [PubMed]
15.   Vaclavik, J.; Sehonova, P.; Hodkovicova, N.; Vecerkova, L.; Blahova, J.; Franc, A.; Marsalek, P.; Mares, J.; Tichy, F.; Svobodova, Z.; et al. The effect of foodborne sertraline on rainbow trout (Oncorhynchus mykiss). *Sci. Total. Environ.* **2020**, *708*, 135082. [CrossRef]
16.   Xie, Z.; Lu, G.; Li, S.; Nie, Y.; Ma, B.; Liu, J. Behavioral and biochemical responses in freshwater fish Carassius auratus exposed to sertraline. *Chemosphere* **2015**, *135*, 146–155. [CrossRef] [PubMed]
17.   Silva, L.J.; Pereira, A.M.; Meisel, L.M.; Lino, C.; Pena, A. Reviewing the serotonin reuptake inhibitors (SSRIs) footprint in the aquatic biota: Uptake, bioaccumulation and ecotoxicology. *Environ. Pollut.* **2015**, *197*, 127–143. [CrossRef] [PubMed]
18.   Salgado, R.; Marques, R.; Noronha, J.; Mexia, J.; Carvalho, G.; Oehmen, A.; Reis, M. Assessing the diurnal variability of pharmaceutical and personal care products in a full-scale activated sludge plant. *Environ. Pollut.* **2011**, *159*, 2359–2367. [CrossRef] [PubMed]
19.   Cao, J.; Fu, B.; Zhang, T.; Wu, Y.; Zhou, Z.; Zhao, J.; Yang, E.; Qian, T.; Luo, J. Fate of typical endocrine active compounds in full-scale wastewater treatment plants: Distribution, removal efficiency and potential risks. *Bioresour. Technol.* **2020**, *310*, 123436. [CrossRef]
20.   Ek, M.; Baresel, C.; Magnér, J.; Bergström, R.; Harding, M. Activated carbon for the removal of pharmaceutical residues from treated wastewater. *Water Sci. Technol.* **2014**, *69*, 2372–2380. [CrossRef]
21.   Guillossou, R.; Le Roux, J.; Mailler, R.; Vulliet, E.; Morlay, C.; Nauleau, F.; Gasperi, J.; Rocher, V. Organic micropollutants in a large wastewater treatment plant: What are the benefits of an advanced treatment by activated carbon adsorption in comparison to conventional treatment? *Chemosphere* **2019**, *218*, 1050–1060. [CrossRef]
22.   Kårelid, V.; Larsson, G.; Björlenius, B. Pilot-scale removal of pharmaceuticals in municipal wastewater: Comparison of granular and powdered activated carbon treatment at three wastewater treatment plants. *J. Environ. Manag.* **2017**, *193*, 491–502. [CrossRef]
23.   Crini, G.; Lichtfouse, E. Advantages and disadvantages of techniques used for wastewater treatment. *Environ. Chem. Lett.* **2019**, *17*, 145–155. [CrossRef]
24.   Silva, B.; Martins, M.; Rosca, M.; Rocha, V.; Lago, A.; Neves, I.C.; Tavares, T. Waste-based biosorbents as cost-effective alternatives to commercial adsorbents for the retention of fluoxetine from water. *Sep. Purif. Technol.* **2020**, *235*, 116139. [CrossRef]
25.   Jaria, G.; Calisto, V.; Gil, M.V.; Otero, M.; Esteves, V.I. Removal of fluoxetine from water by adsorbent materials produced from paper mill sludge. *J. Colloid Interface Sci.* **2015**, *448*, 32–40. [CrossRef] [PubMed]
26.   Ebrahimi, F.; Orooji, Y.; Alizadeh, A. Applying Membrane Distillation for the Recovery of Nitrate from Saline Water Using PVDF Membranes Modified as Superhydrophobic Membranes. *Polymers* **2020**, *12*, 2774. [CrossRef] [PubMed]
27.   Orooji, Y.; Jaleh, B.; Homayouni, F.; Fakhri, P.; Kashfi, M.; Torkamany, M.J.; Yousefi, A.A. Laser Ablation-Assisted Synthesis of Poly (Vinylidene Fluoride)/Au Nanocomposites: Crystalline Phase and Micromechanical Finite Element Analysis. *Polymers* **2020**, *12*, 2630. [CrossRef] [PubMed]
28.   Mashile, G.P.; Dimpe, K.M.; Nomngongo, P.N. A Biodegradable Magnetic Nanocomposite as a Superabsorbent for the Simultaneous Removal of Selected Fluoroquinolones from Environmental Water Matrices: Isotherm, Kinetics, Thermodynamic Studies and Cost Analysis. *Polymers* **2020**, *12*, 1102. [CrossRef] [PubMed]
29.   Awual, R. A novel facial composite adsorbent for enhanced copper(II) detection and removal from wastewater. *Chem. Eng. J.* **2015**, *266*, 368–375. [CrossRef]
30.   Cantarella, M.; Carroccio, S.C.; Dattilo, S.; Avolio, R.; Castaldo, R.; Puglisi, C.; Privitera, V. Molecularly imprinted polymer for selective adsorption of diclofenac from contaminated water. *Chem. Eng. J.* **2019**, *367*, 180–188. [CrossRef]
31.   Qiu, L.; Jaria, G.; Gil, M.; Feng, J.; Dai, Y.; Esteves, V.I.; Otero, M.A.; Calisto, V. Core−Shell Molecularly Imprinted Polymers on Magnetic Yeast for the Removal of Sulfamethoxazole from Water. *Polymers* **2020**, *12*, 1385. [CrossRef]
32.   Sellergren, B. *Molecularly Imprinted Polymers: Man-Made Mimics of Antibodies and Their Application in Analytical Chemistry*; Elsevier: Amsterdam, The Netherlands, 2000.
33.   Affinisep Affinimip. Available online: https://www.affinisep.com/spe-kits-applications/spe-kit-for-sample-preparation/affinimip-spe-selectives-mip-spe-cartridges (accessed on 3 August 2020).

34. SupelMIP® SPE, Supelco. Available online: https://www.sigmaaldrich.com/analytical-chromatography/sample-preparation/spe/supelmip.html (accessed on 3 August 2020).

35. Mattiasson, B.; Ye, L. *Molecularly Imprinted Polymers in Biotechnology (Advances in Biochemical Engineering/Biotechnology)*; Springer International Publishing: Cham, Switzerland, 2015; Volume 150.

36. Sellergren, B.; Allender, C.J. Molecularly imprinted polymers: A bridge to advanced drug delivery. *Adv. Drug Deliv. Rev.* **2005**, *57*, 1733–1741. [CrossRef]

37. Altintas, Z.; Chianella, I.; Da Ponte, G.; Paulussen, S.; Gaeta, S.; Tothill, I. Development of functionalized nanostructured polymeric membranes for water purification. *Chem. Eng. J.* **2016**, *300*, 358–366. [CrossRef]

38. An, F.-Q.; Li, H.-F.; Guo, X.-D.; Hu, T.-P.; Gao, B.; Gao, J.-F. Design of novel "imprinting synchronized with crosslinking" surface imprinted technique and its application for selectively removing phenols from aqueous solution. *Eur. Polym. J.* **2019**, *112*, 273–282. [CrossRef]

39. Le Noir, M.; Lepeuple, A.-S.; Guieysse, B.; Mattiasson, B. Selective removal of 17β-estradiol at trace concentration using a molecularly imprinted polymer. *Water Res.* **2007**, *41*, 2825–2831. [CrossRef] [PubMed]

40. Lu, Y.C.; Mao, J.H.; Zhang, W.; Wang, C.; Cao, M.; Wang, X.D.; Wang, K.Y.; Xiong, X. A novel strategy for selective removal and rapid collection of triclosan from aquatic environment using magnetic molecularly imprinted nano−polymers. *Chemosphere* **2020**, *238*, 124640. [CrossRef] [PubMed]

41. Tan, F.; Sun, D.; Gao, J.; Zhao, Q.; Wang, X.; Teng, F.; Quan, X.; Chen, J. Preparation of molecularly imprinted polymer nanoparticles for selective removal of fluoroquinolone antibiotics in aqueous solution. *J. Hazard. Mater.* **2013**, *750*–757. [CrossRef]

42. DeVane, C.L.; Liston, H.L.; Markowitz, J.S.; De Vane, C.L. Clinical Pharmacokinetics of Sertraline. *Clin. Pharmacokinet.* **2002**, *41*, 1247–1266. [CrossRef]

43. Kosjek, T.; Heath, E.; Kompare, B. Removal of pharmaceutical residues in a pilot wastewater treatment plant. *Anal. Bioanal. Chem.* **2007**, *387*, 1379–1387. [CrossRef]

44. Abdouss, M.; Asadi, E.; Azodi-Deilami, S.; Beik-Mohammadi, N.; Aslanzadeh, S.A. Development and characterization of molecularly imprinted polymers for controlled release of citalopram. *J. Mater. Sci. Mater. Electron.* **2011**, *22*, 2273–2281. [CrossRef]

45. Chapuis, F.; Mullot, J.-U.; Pichon, V.; Tuffal, G.; Hennion, M.-C. Molecularly imprinted polymers for the clean-up of a basic drug from environmental and biological samples. *J. Chromatogr. A* **2006**, *1135*, 127–134. [CrossRef]

46. Kempe, H.; Kempe, M. Influence of salt ions on binding to molecularly imprinted polymers. *Anal. Bioanal. Chem.* **2009**, *396*, 1599–1606. [CrossRef]

47. Horemans, F.; Weustenraed, A.; Spivak, D.A.; Cleij, T.J. Towards water compatible MIPs for sensing in aqueous media. *J. Mol. Recognit.* **2012**, *25*, 344–351. [CrossRef]

48. Dirion, B.; Cobb, Z.; Schillinger, E.; Andersson, A.L.I.; Sellergren, B. Water-Compatible Molecularly Imprinted Polymers Obtained via High-Throughput Synthesis and Experimental Design. *J. Am. Chem. Soc.* **2003**, *125*, 15101–15109. [CrossRef] [PubMed]

49. Yan, H.; Row, K.H. Characteristic and Synthetic Approach of Molecularly Imprinted Polymer. *Int. J. Mol. Sci.* **2006**, *7*, 155–178. [CrossRef]

50. Arvand, M.; Hashemi, M. Synthesis by precipitation polymerization of a molecularly imprinted polymer membrane for the potentiometric determination of sertraline in tablets and biological fluids. *J. Braz. Chem. Soc.* **2012**, *23*, 392–402. [CrossRef]

51. Khalilian, F.; Kermani, F.K. Selective Dispersive Solid Phase Extraction of Ser-Traline Using Surface Molecu-larly Imprinted Polymer Grafted on SiO$_2$/Graphene Oxide. *J. Chem. Health Risks* **2017**, *7*, 49–60.

52. Krupadam, R.J.; Patel, G.P.; Balasubramanian, R. Removal of cyanotoxins from surface water resources using reusable molecularly imprinted polymer adsorbents. *Environ. Sci. Pollut. Res.* **2011**, *19*, 1841–1851. [CrossRef]

53. Krupadam, R.J.; Khan, M.S.; Wate, S.R. Removal of probable human carcinogenic polycyclic aromatic hydrocarbons from contaminated water using molecularly imprinted polymer. *Water Res.* **2010**, *44*, 681–688. [CrossRef]

54. Krupadam, R.J.; Bhagat, B.; Wate, S.R.; Bodhe, G.L.; Sellergren, B.; Anjaneyulu, Y. Fluorescence Spectrophotometer Analysis of Polycyclic Aromatic Hydrocarbons in Environmental Samples Based on Solid Phase Extraction Using Molecularly Imprinted Polymer. *Environ. Sci. Technol.* **2009**, *43*, 2871–2877. [CrossRef]

55. Murray, A.; Örmeci, B. Application of molecularly imprinted and non-imprinted polymers for removal of emerging contaminants in water and wastewater treatment: A review. *Environ. Sci. Pollut. Res.* **2012**, *19*, 3820–3830. [CrossRef]

56. Tadeo, X.; López-Méndez, B.; Castaño, D.; Trigueros, T.; Millet, O. Protein Stabilization and the Hofmeister Effect: The Role of Hydrophobic Solvation. *Biophys. J.* **2009**, *97*, 2595–2603. [CrossRef]

57. Zhou, W.; Yang, S.; Wang, P.G. Matrix effects and application of matrix effect factor. *Bioanalysis* **2017**, *9*, 1839–1844. [CrossRef]

58. Azizi, A.; Bottaro, C.S. A critical review of molecularly imprinted polymers for the analysis of organic pollutants in environmental water samples. *J. Chromatogr. A* **2020**, *1614*, 460603. [CrossRef] [PubMed]

59. Martín-Esteban, A.; Sellergren, B. 2.17-Molecularly Imprinted Polymers. In *Comprehensive Sampling and Sample Preparation*; Pawliszyn, J., Ed.; Academic Press: Oxford, UK, 2012; pp. 331–344.

60. Sellergren, B.; Lanza, F. *Techniques and Instrumentation in Analytical Chemistry*; Elsevier: Amsterdam, The Netherlands, 2001; Volume 23, pp. 355–375.

61. Turiel, E.; Esteban, A.M. 8-Molecularly imprinted polymers. In *Solid-Phase Extraction*; Poole, C.F., Ed.; Elsevier: Amsterdam, The Netherlands, 2020; pp. 215–233.

62. Ellwanger, A.; Karlsson, L.; Owens, P.K.; Berggren, C.; Crecenzi, C.; Ensing, K.; Bayoudh, S.; Cormack, P.A.; Sherrington, D.; Sellergren, B. Evaluation of methods aimed at complete removal of template from molecularly imprinted polymers. *Analyst* **2001**, *126*, 784–792. [CrossRef] [PubMed]
63. Lorenzo, R.A.; Carro, A.; Alvarez-Lorenzo, C.; Concheiro, A. To Remove or Not to Remove? The Challenge of Extracting the Template to Make the Cavities Available in Molecularly Imprinted Polymers (MIPs). *Int. J. Mol. Sci.* **2011**, *12*, 4327–4347. [CrossRef] [PubMed]
64. Activated Charcoal C3345. Available online: https://www.sigmaaldrich.com/catalog/product/sigald/c3345 (accessed on 8 September 2020).

 *polymers*

Article

Chain Entanglement of 2-Ethylhexyl Hydrogen-2-Ethylhexylphosphonate into Methacrylate-Grafted Nonwoven Fabrics for Applications in Separation and Recovery of Dy (III) and Nd (III) from Aqueous Solution

Hiroyuki Hoshina *, Jinhua Chen *, Haruyo Amada and Noriaki Seko

Department of Advanced Functional Materials Research, Takasaki Advanced Radiation Research Institute, Quantum Beam Science Research Directorate, National Institutes for Quantum and Radiological Science and Technology, 1233 Watanuki-machi, Takasaki, Gunma 370-1292, Japan; amada.haruyo@qst.go.jp (H.A.); seko.noriaki@qst.go.jp (N.S.)
* Correspondence: hoshina.hiroyuki@qst.go.jp (H.H.); chen.jinhua@qst.go.jp (J.C.); Tel.: +81-27-346-9125 (J.C.)

Received: 27 October 2020; Accepted: 9 November 2020; Published: 11 November 2020

Abstract: A nonwoven fabric adsorbent loaded with 2-ethylhexyl hydrogen-2-ethylhexylphosphonate (EHEP) was developed for the separation and recovery of dysprosium (Dy) and neodymium (Nd) from an aqueous solution. The adsorbent was prepared by the radiation-induced graft polymerization of a methacrylate monomer with a long alkyl chain onto a nonwoven fabric and the subsequent loading of EHEP by hydrophobic interaction and chain entanglement between the alkyl chains. The adsorbent was evaluated by batch and column tests with a Dy (III) and Nd (III) aqueous solution. In the batch tests, the adsorbent showed high Dy (III) adsorptivity close to 25.0 mg/g but low Nd (III) adsorptivity below 1.0 mg/g, indicating that the adsorbent had high selective adsorption. In particular, the octadecyl methacrylate (OMA)-adsorbent showed adsorption stability in repeated tests. In the column tests, the OMA-adsorbent was also stable and showed high Dy (III) adsorptivity and high selectivity in repeated adsorption–elution circle tests. This result suggested that the OMA-adsorbent may be a promising adsorbent for the separation and recovery of Dy (III) and Nd (III) ions.

Keywords: selective adsorption; dysprosium; neodymium; fabric adsorbent; radiation; graft polymerization

---

1. Introduction

Rare earths including scandium, yttrium, and 15 lanthanoid elements, have recently become indispensable materials for the high-tech industry. Due to the uneven distribution of rare-earth sources in the world, almost all rare earths are supplied by limited countries [1]. Therefore, it is necessary to recycle used rare earths to ensure a stable supply of these materials in many countries [2–7]. Among the rare earths, dysprosium (Dy) and neodymium (Nd) are listed as "critical materials" by the United States due to supply issues and their importance to electronics and electrical technology [8,9]. For example, neodymium and dysprosium are key components of permanent magnets, such as NdFeB magnets. The demand for the separation and recovery of dysprosium and neodymium from used permanent magnets and scraps generated during manufacturing is increasing [10–20].

The technology for the separation and recovery of dysprosium and neodymium from used permanent magnets has been extensively studied [21,22]. The most common method of recovering dysprosium and neodymium from waste materials involves leaching them in an acid solution and purifying the leached ions by solvent extraction [11–13,23]. Organophosphorus compounds such as

2-ethylhexyl hydrogen-2-ethylhexylphosphonate and di(2-ethylhexyl)phosphoric acid, carboxylic acid such as neodecanoic acid and naphthenic acid, and methyltrioctylamine chloride are usually used as extractants for rare-earth ions due to their good separation and recovery performance [22–26]. However, solvent extraction requires a large number of separation steps, a long processing time, and a large space for all necessary equipment. On the other hand, other methods such as chemical precipitation and ionic liquids extraction are also used for the separation and recovery of rare earths. Although the chemical precipitation process is simple and low in cost, the purity and recovery ratio of the resulting product are usually low, while the ionic liquid extraction cost is high for actual application [21]. Currently, the effective separation and recovery of rare earths from an aqueous solution requires relatively simple processes [27,28]. Adsorption techniques using adsorbents, such as inorganic particles, ion-exchange resins, and polymer ligands, are attractive for the separation and recovery of rare-earth ions [29–36]. This is because the adsorption process does not require much energy and water and can be easily operated anywhere by batch or column methods [37].

Inorganic particles, such as clay minerals, activated carbon, and magnetite nanoparticles, are highly suitable for removing heavy metals from water and wastewater. In many cases, these inorganic materials show high adsorption but low selectivity [37–39]. On the other hand, adsorbents with special ligands or chelating functional groups can be designed to selectively separate and recover target metal ions in water. These adsorbents, including ion-exchange resins and polymer ligands, can be prepared by introducing functional groups onto polymer materials by the radiation-induced graft polymerization method. This method can introduce new functional properties while maintaining the properties of the trunk polymers [40–48]. Various vinyl monomers have been radiation-grafted onto trunk polymers, such as polyethylene [41,42], polypropylene [43,44], fluoropolymers [45], and cellulose [46,47]. Furthermore, graft polymerization can be applied to various types of materials, such as films [45], fabrics [30,45–47], fibers [46], and particles [48]. Various adsorbents have been developed using this technology for the recovery and removal of metal ions from environmental water and industrial wastewater [46–51]. In the design of these adsorbents, it is important to select the most suitable functional groups based on the metal ion that needs to be adsorbed.

We noticed that 2-ethylhexyl hydrogen-2-ethylhexylphosphonate (EHEP), used as an extractant in the solvent extraction process, has two alkyl chains on each molecule [10,24,52]. In this study, we attempted to load EHEP onto polyethylene-coated polypropylene (PE/PP) nonwoven fabrics to develop a novel adsorbent for rare-earth ions. For this purpose, we grafted a polymerized methacrylate monomer with a long alkyl chain onto the fabrics. The EHEP was then loaded onto the grafted fabrics by hydrophobic interaction and chain entanglement between the alkyl chains. Here, since the EHEP is only physically bonded on the fabrics by hydrophobic interaction and chain entanglement, the loss of EHEP is a concern in practical applications. Therefore, the stability of EHEP-loaded adsorbents needs to be confirmed for practical use.

Four methacrylate monomers with different alkyl chain lengths—butyl methacrylate (BMA), hexyl methacrylate (HMA), dodecyl methacrylate (DMA), octadecyl methacrylate (OMA)—were radiation-grafted onto the PE/PP nonwoven fabrics in this study. The grafted fabrics were then loaded with EHEP to prepare the adsorbents. The adsorbents were tested in batch and column modes using Dy (III) and Nd (III) ion solutions [18]. The effects of the alkyl chain length of the monomers on the stability and adsorption performance of the EHEP-loaded absorbents were studied and evaluated.

2. Experimental

2.1. Materials

The trunk material used for graft polymerization was a nonwoven fabric composed of polyethylene-coated polypropylene (PE/PP) fibers, provided by Kurashiki Textile Manufacturing Co., Ltd., Kurashiki, Japan. The PE on the fiber surface is easy to be radiation-grafted, and the PP core makes the fiber mechanically stronger. Furthermore, the PE/PP nonwoven fabric is relatively cheap among artificial fabrics and has a large specific surface. The four methacrylate monomers—butyl methacrylate (BMA), hexyl methacrylate (HMA), dodecyl methacrylate (DMA), and octadecyl methacrylate (OMA)—are of chemical reagent grade and were purchased from Fujifilm Wako Pure Chemical Corporation, Tokyo, Japan. 2-Ethylhexyl hydrogen-2-ethylhexylphosphonate (EHEP) was provided by Daihachi Chemical Industry Co., Ltd., Tokyo, Japan. The other reagents, such as Tween 20 surfactant, methanol, ammonia water, HCl solution, Dy (III) (Dy_2O_3 in 5 wt.% HNO_3) standard solution, and Nd (III) (Nd_2O_3 in 5 wt.% HNO_3) solution, were purchased from Kanto Chemical Co., Inc., Tokyo, Japan. All chemicals were used without further purification. In this study, the deionized Mili-Q water with a high resistivity of 18 MΩ cm was used.

2.2. Graft Polymerization of Methacrylate Monomers

Figure 1 shows the process of preparing the fabric adsorbents. Graft polymerization was performed using a preirradiation method. In this study, either PE nonwoven fabric or PP nonwoven fabric could be used as trunk polymers. However, the mechanical strength of common PE nonwoven fabric is significantly lower than that of PP nonwoven fabric, while the PP nonwoven fabric deteriorates faster than PE nonwoven fabric. Therefore, we chose the PE-coated PP nonwoven fabric as the polymer trunk for radiation grafting. The PE/PP nonwoven fabric with a size of 5 cm × 8 cm was placed in a polyethylene bag, purged with nitrogen gas to create an oxygen-free environment, and electron beam preirradiated at −80 °C (dry ice) with a beam energy of 2 MeV at a current of 3 mA to generate radicals on the fabric. The preirradiated fabric was removed and filled into a glass ampoule, which was evacuated and filled with a nitrogen-bubbled monomer solution to immerse the fabric completely. The ampoule was placed in a temperature-controlled oven. Under these conditions, the radicals initiated graft polymerization. The monomer structures and grafting conditions are shown in Table 1. After graft polymerization, the fabric was washed with methanol to remove residual monomers and homopolymers and dried in an oven at 60 °C for more than 24 h.

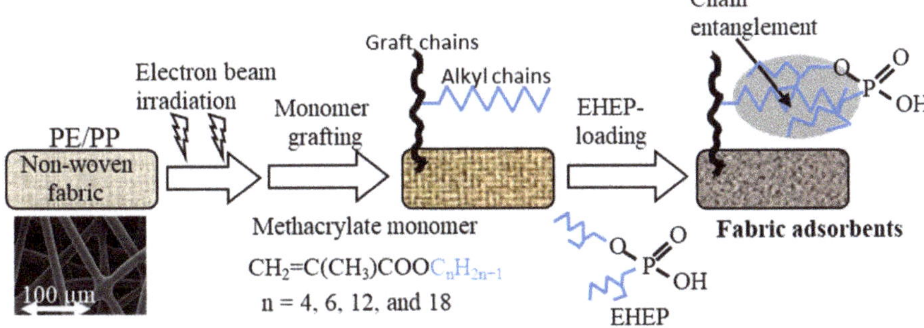

Figure 1. Preparation of the fabric adsorbents by graft polymerization of methacrylate monomers and the subsequent 2-ethylhexyl hydrogen-2-ethylhexylphosphonate (EHEP) loading.

The degree of grafting and the density of alkyl chains of the grafted fabrics were calculated using the following equations.

$$\text{Degree of grafting (\%)} = (W_g - W_0)/W_0 \times 100 \tag{1}$$

$$\text{Density of alkyl chains (mmol/g)} = 1000 \times (W_g - W_0)/M/W_g \tag{2}$$

where W_0 and W_g are the dry weights (mg) of the fabrics before and after graft polymerization, and M is the molecular weight of the monomers as shown in Table 1.

Table 1. Methacrylate monomers and grafting conditions used in this study.

Name	Molecular Structures	M *	Grafting Conditions **		
			Dose *** (kGy)	Temp. (°C)	Time (min)
Butyl methacrylate (BMA)	$CH_2=CCH_3COOC_4H_9$	142	10	40	15
Hexyl methacrylate (HMA)	$CH_2=CCH_3COOC_6H_{13}$	170	10	60	30
Dodecyl methacrylate (DMA)	$CH_2=CCH_3COOC_{12}H_{25}$	254	300	60	180
Octadecyl methacrylate (OMA)	$CH_2=CCH_3COOC_{18}H_{37}$	338	100	60	120

* M is the molecular weight of the monomers; ** For the monomer solutions, monomer concentrations were fixed at 5.0 wt.% in water for BMA, HMA, and DMA, and in a water/methanol mixture solvent (1:1 in weight) for OMA; 0.5 wt.% of Tween 20 surfactant was added to the monomer solutions. *** Preirradiation was performed at −80 °C (dry ice) in an oxygen-free environment.

2.3. Loading of EHEP onto the Grafted Fabrics

A 50 wt.% EHEP solution of ethanol was uniformly dropped onto the grafted fabric for EHEP loading. The EHEP-loaded fabric was dried in a vacuum oven at 40 °C to remove the ethanol solvent. EHEP loading of the resulting fabric adsorbent was calculated by the following equation.

$$\text{EHEP loading (mmol/g)} = 1000 \times (W_a - W_g)/306/W_a \tag{3}$$

where W_a is the dry weights (mg) of EHEP-loaded fabric, and 306 is the molecular weight of EHEP. The prepared fabric adsorbents with different monomers were named BMA-, HMA-, DMA-, and OMA-adsorbent, respectively.

2.4. Characterization

Fourier transform infrared (FTIR) spectroscopic analysis was performed with an FTIR spectrophotometer in the attenuated total reflectance (ATR) mode (Spectrum One, PerkinElmer, Inc., Tokyo, Japan). The scanning range and resolution were 500–2500 cm^{-1} and 1 cm^{-1}, respectively.

The hydrophobicity of the grafted fabric was examined by measuring the contact angle with a contact angle meter (CA-X, Kyowa Interface Science Co., Ltd., Tokyo, Japan).

2.5. Batch Adsorption Tests

The prepared fabric adsorbent was evaluated by batch adsorption tests. The test solution contained 100 ppm Dy (III) and 100 ppm Nd (III). The pH of the test solution was adjusted to 2.0 by ammonia water. The fabric adsorbent with a size of 2 cm × 2 cm was immersed in 50 mL of test solution in a glass bottle. The bottle was placed on a shaker and shaken at a rate of 150 rpm at 25 °C for 3.0 h. After the adsorption test, the adsorbent was washed with deionized water to remove the unadsorbed ions on them.

To elute the adsorbed ions, the fabric adsorbent was immersed in 50 mL of 1.0 M HCl solution in a glass bottle, and the bottle was shaken at a rate of 150 rpm at 25 °C for 1.0 h. After elution, the fabric adsorbent was washed with deionized water and adsorption was repeated under the same conditions as the first adsorption test.

The ion concentrations in the adsorption and elution solutions were analyzed before and after each test with an inductively coupled plasma optical emission spectrometer (ICP-OES, Optima 8300, PerkinElmer, Inc., Tokyo, Japan). The adsorptivity (mg/g) of the fabric adsorbent was calculated as follows.

$$\text{Adsorptivity (mg/g)} = 1000 \times (C_0 - C_i) \times V/W_a \tag{4}$$

where C_0 (mg/mL) and C_i (mg/mL) are the metal ion concentrations in the solution before and after the adsorption, respectively, and V (mL) is the volume of the solution.

2.6. Column Adsorption Tests

For the column adsorption tests, the fabric adsorbent with a diameter of 7.0 mm was packed into a column with an inner diameter of 7.0 mm. The volume of the adsorbent packed in the column was 0.2 mL. The test solution (100 ppm Dy (III) and 100 ppm Nd (III), pH 2) was passed through the column at a space velocity (SV) of 100 h^{-1} at 25 °C. The SV is calculated by dividing the solution flow rate (mL/h) by the volume of adsorbent in the column (fixed at 0.2 mL in this study). A fraction collector was used to continuously collect the effluent from the column, and the ion concentrations were detected by ICP-OES. By plotting the relationship between C_i and bed volume (BV), the ion concentration curve of the effluent was obtained. Here, C_i is the ion concentration of the effluent at BV, and BV is calculated by dividing the total effluent volume from the column by the adsorbent volume (0.2 mL).

The adsorptivity (mg/g) of the adsorbent packed in the column was calculated by the following equation

$$\text{Adsorptivity (mg/g)} = 1000 \times \sum (C_0 - C_i)\, \Delta V_i / W_a \tag{5}$$

where ΔV_i (mL) and C_i (mg/mL) are the volume and concentration of each collected effluent during the adsorption, respectively.

After the adsorption test, the adsorbent was thoroughly washed by passing deionized water through the column. Then, 1.0 M HCl solution of the eluent was passed through the column with a space velocity of 100 h^{-1} at 25 °C until no metal ions were detected in the effluent. The eluted amount (mg/g) and recovery ratio were calculated by the following equations.

$$\text{Eluted amount (mg/g)} = 1000 \times \sum C_i\, \Delta V_i / W_a \tag{6}$$

$$\text{Recovery ratio (\%)} = \text{Eluted amount} / \text{Adsorptivity} \times 100 \tag{7}$$

where ΔV_i (mL) and C_i (mg/mL) are the volume and concentration of each collected effluent during the elution, respectively.

After the elution test, the adsorbent in the column was thoroughly washed with deionized water and used for the adsorption test again to evaluate its stability.

3. Results and Discussion

3.1. Synthesis of EHEP-Loaded Adsorbent

The adsorbent was prepared by the radiation-induced graft polymerization of methacrylate with a long alkyl chain onto PE/PP nonwoven fabric and the subsequent loading of EHEP by hydrophobic interaction and chain entanglement between the alkyl chains. Here, the EHEP organophosphorus compound has a special affinity for Dy (III) ions. The grafting results and the density of EHEP loading are summarized in Table 2.

Table 2. Degree of grafting and alkyl group density of the grafted fabrics, and the EHEP loading of the corresponding adsorbents.

Grafted Monomers	Degree of Grafting (%)	Alkyl Group Density * (mmol/g)	EHEP Loading ** (mmol/g)
Butyl methacrylate (BMA)	51	2.39	1.24
Hexyl methacrylate (HMA)	62	2.24	1.26
Dodecyl methacrylate (DMA)	102	1.99	1.24
Octadecyl methacrylate (OMA)	219	2.03	1.22

* Alkyl group density of the monomer-grafted fabric was calculated using Equation (2); ** EHEP loading was calculated by the weight increase of the grafted fabric before and after EHEP loading using Equation (3).

As shown in Table 2, four monomers with different alkyl chain lengths—BMA, HMA, DMA, and OMA—were radiation-grafted onto the fabrics. For comparison, the alkyl chain density in the grafted fabric was adjusted to be close to 2.0 mmol/g. For this reason, the degree of grafting was significantly different for each monomer and increased in proportion to the molecular weight of the grafted monomer. For example, to obtain a similar alkyl chain density of 2.0 mmol/g, the degree of grafting for the BMA is 51%, while it is 219% for the OMA. The latter is approximately four times higher than that of the former.

To obtain similar alkyl chain densities of the grafted fabrics, BMA grafting was carried out by immersing the 10 kGy preirradiated fabric into a 5.0 wt.% BMA emulsion at 40 °C for 15 min, while for HMA grafting, a higher temperature of 60 °C and longer grafting time of 30 min were needed. We also carried out BMA grafting at 60 °C. However, the grafting rate was too fast to control the graft yielding. For monomers with longer alkyl chains, preirradiation doses higher than 100 kGy were used to generate more radicals in the fabrics. This is because the steric hindrance effects of the monomers inhibited the graft polymerization from reaching a high degree of grafting. Furthermore, a mixture solvent of methanol and water in the ratio of 1:1 was used for OMA grafting. Here, the addition of methanol to the monomer solution increased the affinity between the fabric and the monomer, thereby enhancing the radiation grafting [53].

The loading of EHEP onto the grafted fabric was achieved by dropping the EHEP solution of ethanol onto the grafted fabric to reach a loading density of approximately 1.2 mmol/g. After removing ethanol by evaporation, the adsorbent was obtained.

3.2. Materials Characterization

The FTIR results shown in Figure 2 confirmed that the BMA, HMA, DMA, and OMA monomers were graft polymerized onto the PE/PP nonwoven fabrics and EHEP was loaded onto the OMA-grafted fabric. The peaks of the PE/PP nonwoven fabric only appeared at 1472, 1462, 1375, 731, and 718 cm^{-1}, corresponding to the characteristic absorptions of PE [54], indicating that the PP fiber was completely coated by PE. After grafting, new peaks at 1730 and 1155 cm^{-1}, attributed to the C=O and C–O stretching of methacrylate, respectively, were observed (Figure 2b–e) [55,56]. After loading EHEP onto the OMA-grafted fabric, new peaks at 1250 (P–O–C), 1050 (P–O–C), and 980 (P=O) cm^{-1} were observed, as shown in Figure 2f [25]. These results indicated that the methacrylate monomers were grafted onto the fabrics and EHEP was loaded onto the OMA-grafted fabric.

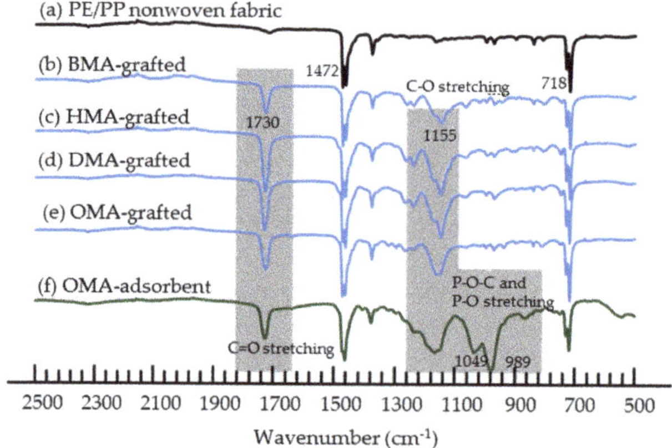

Figure 2. FTIR spectra of (**a**) PE/PP nonwoven fabric, (**b**) BMA-grafted PE/PP nonwoven fabric, (**c**) HMA-grafted PE/PP nonwoven fabric, (**d**) DMA-grafted PE/PP nonwoven fabric, (**e**) OMA-grafted PE/PP nonwoven fabric, and (**f**) OMA-adsorbent prepared by loading of EHEP onto the OMA-grafted PE/PP nonwoven fabric.

The surface properties of the BMA-, HMA-, DMA-, and OMA-grafted fabrics were evaluated by a contact angle meter. A high contact angle indicates the high hydrophobicity of the sample. Pictures of water droplets on the surface with the smallest and largest contact angles are shown in Figure 3a, b, respectively. The contact angle of the BMA-grafted fabric was 97° (Figure 3a), while that of the OMA-grafted fabric was 112° (Figure 3b). For comparison, the contact angles of the grafted fabrics are summarized in Figure 3c. The contact angle increased with the increase of the alkyl chain length of the grafted monomers. The OMA-grafted fabric had the highest hydrophobicity due to the longest alkyl chains of the grafted monomers as well as the highest degree of grafting (see Table 2). It was expected that the grafted fabric with high hydrophobicity was more conducive to the physical bonding of the alkyl chain of EAEH for loading.

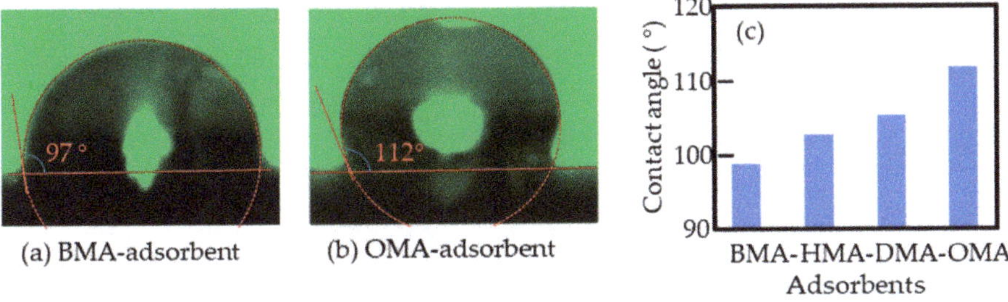

Figure 3. Water droplets on the BMA-adsorbent (**a**) and OMA-adsorbent (**b**). Contact angle of the water droplets on the four fabric adsorbents (**c**).

3.3. Batch Adsorption Tests

The adsorbent performance was evaluated in advance by a batch adsorption test. The aqueous solution of 100 ppm Dy (III) and 100 ppm Nd (III) at pH 2 was used as the adsorption solution. After adsorption, the adsorbent was immersed in 1.0 M HCl solution to completely elute the adsorbed ions and washed with adequate water to conduct the adsorption test again.

The results of the batch adsorption test are shown in Table 3 and Figure 4. In the first adsorption test, all adsorbents had similar Dy (III) adsorptivity around 25.0 mg/g. The EHEP loaded in the fabric is a cationic extractant, which is known to extract metal ions from aqueous solution and can be labeled HA. The adsorption is an ion-exchange process in which one Dy (III) ion combines three EHEPs to form a DyA_3 structure in the adsorbent [57,58]. The similar adsorptivity was due to the similar EHEP loading (1.20 mmol/g) of the four adsorbents. However, the Nd (III) adsorptivity for each adsorbent was considerably small (less than 1.0 mg/g). Therefore, the EHEP-loaded adsorbents had a high adsorption selectivity for Dy (III) and could be used for separation and recovery.

Table 3. Summary of the first and repeated batch adsorption tests.

Adsorbents	1st Adsorption *			Repeated Adsorption **			W_a/W_b
	W_b (mg)	C_{Dy-1} (mg/g)	C_{Nd-1} (mg/g)	C_{Dy-r} (mg/g)	C_{Nd-r} (mg/g)	W_a (mg)	
BMA	54	28.6	0.1	11.4	0.1	40	0.74
HMA	66	26.2	0.6	15.0	0.1	51	0.77
DMA	80	24.9	0.4	22.7	0	74	0.93
OMA	86	26.0	1.0	25.3	0.8	82	0.95

* First adsorption was performed using the new adsorbent, and W_b is the dry weight of the new adsorbent, C_{Dy-1} and C_{Nd-1} are the Dy(III) and Nd(III) adsorptivities of the first adsorption, respectively; ** Repeated adsorption was performed after the adsorbent diluted and adequate water-washed, C_{Dy-r} and C_{Nd-r} are the Dy(III) and Nd(III) adsorptivities of the repeated adsorption, respectively, and W_a is the dry weight of the used adsorbent after the repeated adsorption and dilution.

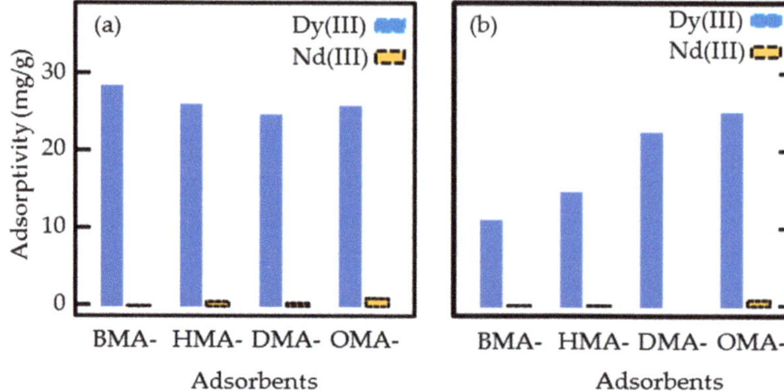

Figure 4. The Dy (III) and Nd (III) adsorptivity of the four nonwoven fabric adsorbents. (**a**) First adsorption test. (**b**) Second adsorption test using the refreshed adsorbents after elution and washing. Initial adsorption solution: 100 ppm Dy (III) and 100 ppm Nd (III) at a pH of 2.0 at 25 °C.

In the repeated adsorption tests, the OMA-adsorbent retained a high Dy (III) adsorptivity of 25.3 mg/g. In contrast, the Dy (III) adsorptivities of BMA-, HMA-, and DMA-adsorbents were significantly reduced to were 11.4, 15.0, and 22.7 mg/g, respectively. The decrease of Dy (III) adsorptivity might be due to the loss of EHEP loaded in the fabric during the repeated tests. As shown in Table 3, after repeated adsorption tests, the weight of the OMA-adsorbent was almost unchanged, while the weight of the BMA-adsorbent was reduced by 26%. The shorter the alkyl chain length of the grafted monomer, the more the weight of the adsorbent decreased due to the loss of EHEP. According to these results, the OMA-adsorbent with the longest alkyl chain was chosen for the column adsorption test.

3.4. Column Adsorption Tests

Column adsorption and elution were carried out using the same adsorption and elution solutions as the above batch tests. The solution was passed through the column at a space velocity of 100 h^{-1}. As shown in Figure 5, the Dy (III) was completely adsorbed up to a higher bed volume (BV) of 80. After that, the concentration of Dy (III) in the effluent gradually increased, reaching 98 ppm at a BV of 400 (similar to the concentration of the fed solution, 100 ppm). The total Dy (III) adsorbed from the solution was calculated using Equation (5) to be 43.6 mg/g. The adsorption is an ion-exchange process between the metal ions and the proton of EHEP loaded in the fabrics; that is, one Dy (III) ion can bond with three phosphate groups. Therefore, for a 1.2 mmol/g EHEP-loaded adsorbent, the calculated adsorption capacity is close to 64.8 mg/g. The detected value of 43.6 mg/g is lower than the calculated value, which is due to the adsorption equilibria at the low Dy (III) concentration of the feed solution. Even then, it is still much higher than in the case of using hybrid silica nanoparticles, as reported by Topel et al., where the Dy (III) adsorption is 0.019 mmol/g or 30.9 mg/g [57]. In contrast, the Nd (III) was completely adsorbed up to a lower BV of 40, and the Nd (III) concentration rapidly increased up to 130 ppm at a BV of 144, which was higher than that of the fed solution (100 ppm). This is because the adsorbed Nd (III) was replaced by Dy (III), indicating that the OMA-adsorbent was favorable for Dy (III) adsorption. The Nd (III) adsorptivity of the adsorbents in the column was also calculated using Equation (5) to be 4.2 mg/g, which was one-tenth of the Dy (III) adsorption.

The adsorbed Dy (III) and Nd (III) were eluted by passing 1.0 M HCl solution through the column. The maximum concentrations of Dy (III) and Nd (III) in the effluent were 373 and 38 ppm, respectively. The recovery ratios of Dy (III) and Nd (III) calculated using Equations (6) and (7) were 99% and 98%, respectively, indicating that almost all metal ions were eluted by the 1.0 M HCl solution within a BV of 160 (from 550 to 710 BV in Figure 5).

After the first adsorption, water washing, HCl elution, and water washing, the repeated column test was continued (Figure 5). The concentration curves of Dy (III) and Nd (III) for the repeated adsorption test show similar behavior as the first adsorption test. These results indicate that the OMA-adsorbent was stable for repeated use in the separation and recovery of Dy (III) and Nd (III) ions from an aqueous solution.

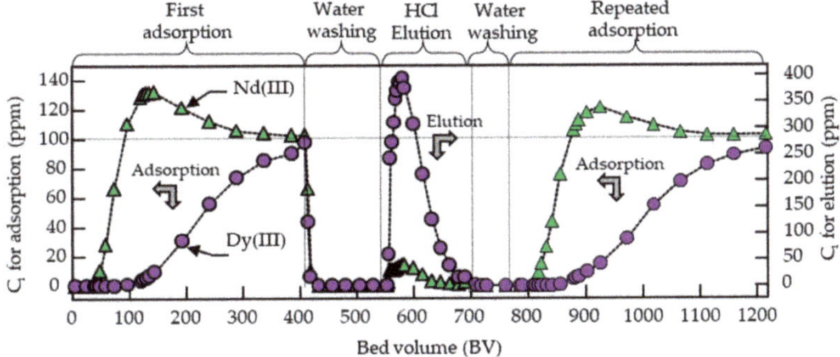

Figure 5. Profile of the adsorption and elution of Dy (III) and Nd (III) ions with an OMA-adsorbent. Adsorption solution: 100 ppm Dy (III) and 100 ppm Nd (III), pH 2.0; elution solution: 1.0 M HCl; space velocity (SV) = 100 h^{-1}; temperature = 25 °C; the total 1200 BV means that the adsorption–elution–adsorption process was operated for 12 h under the fixed space velocity of 100 h^{-1}.

4. Conclusions

A fabric adsorbent for the separation and recovery of Dy (III) and Nd (III) from an aqueous solution was successfully prepared by graft polymerization of methacrylate with a long alkyl chain onto the nonwoven fabric and loading EHEP by hydrophobic interaction and chain entanglement between the alkyl chains.

In the batch adsorption tests, the adsorbents showed a high Dy (III) adsorptivity above 25.0 mg/g and a low Nd (III) adsorptivity below 1.0 mg/g, indicating that the adsorbents had a high Dy (III) selective adsorption. However, only the OMA-adsorbent with the longest alkyl chain was stable and retained its high Dy (III) adsorption performance in repeated adsorption tests.

In the column adsorption test with the OMA-adsorbent, the adsorptivities of Dy (III) and Nd (III) were 43.6 and 4.2 mg/g, respectively. The Dy (III) adsorptivity was approximately ten times higher than that of the Nd (III) adsorptivity. Similar adsorption performance of the adsorbents was observed in the repeated tests. These results demonstrate that the OMA-adsorbent was stable for repeated use. The high stability of the OMA-adsorbents due to the loss of EHEP was suppressed by the strong hydrophobic interaction and chain entanglement between the long alkyl chains.

The OMA-adsorbent can be synthesized easily and economically by immersing the irradiated nonwoven fabric in the monomer solution and EHEP solution in sequence. The obtained adsorbent can be used in batch mode or column mode without any other separation process. Even if the adsorbent is operated in a strong acid, it is stable without any weight loss. Furthermore, the adsorbent has a high selectivity to Dy (III) ions. Therefore, the OMA-adsorbent developed in this study can effectively separate and recover Dy (III) and Nd (III) from an aqueous solution and is expected to contribute to the recovery of rare-earth metals from NdFeB permanent magnet scraps in the future.

Author Contributions: Conceptualization, H.H. and N.S.; methodology, formal analysis and investigation H.H., J.C. and H.A.; writing—original draft preparation, H.H.; writing—review and editing, J.C.; project administration, N.S.; All authors have read and agreed to the published version of the manuscript.

Funding: This research was partly funded by the Japan Society for the Promotion of Science (JSPS) KAKENHI Grant Numbers JP17K00632 (Grant-in-Aid for Scientific Research (C)).

Conflicts of Interest: The authors declare no conflict of interest.

References

1. Schreiber, A.; Marx, J.; Zapp, P.; Hake, J.F.; Voßenkaul, D.; Friedrich, B. Environmental impacts of rare earth mining and separation based on eudialyte: A new European way. *Resources* **2016**, *5*, 32. [CrossRef]
2. Akcil, A.; Akhmadiyeva, N.; Abdulvaliyev, R.; Abhilash; Meshram, P. Overview on extraction and separation of rare earth elements from red mud: Focus on scandium. *Min. Process. Extr. Metall. Rev.* **2018**, *39*, 145–151. [CrossRef]
3. Grandell, L.; Lehtilä, A.; Kivinen, M.; Koljonen, T.; Kihlman, S.; Lauri, L.S. Role of critical metals in the future markets of clean energy technologies. *Renew. Energy* **2016**, *95*, 53–62. [CrossRef]
4. Golev, A.; Scott, M.; Erskine, P.D.; Ali, S.H.; Ballantyne, G.R. Rare earths supply chains: Current status, constraints and opportunities. *Resour. Policy* **2014**, *41*, 52–59. [CrossRef]
5. Dushyantha, N.; Batapola, N.; Ilankoon, I.M.S.K.; Rohitha, S.; Premasiri, R.; Abeysinghe, B.; Ratnayake, N.; Dissanayake, K. The story of rare earth elements (REEs): Occurrences, global distribution, genesis, geology, mineralogy and global production. *Ore Geol. Rev.* **2020**, *122*, 103521. [CrossRef]
6. Binnemans, K.; Jones, P.T.; Blanpain, B.; Gerven, T.V.; Yang, Y.; Walton, A.; Buchert, M. Recycling of rare earths: A critical review. *J. Clean. Prod.* **2013**, *51*, 1–22. [CrossRef]
7. Machacek, E.; Richter, J.L.; Habib, K.; Klossek, P. Recycling of rare earths from fluorescent lamps: Value analysis of closing-the-loop under demand and supply uncertainties. *Resour. Conserv. Recycl.* **2015**, *104*, 76–93. [CrossRef]
8. U.S. Department of Energy. *Critical Materials Strategy*; U.S. Department of Energy: Washington, DC, USA, 2011; pp. 1–191.
9. Rademaker, J.H.; Kleijn, R.; Yang, Y. Recycling as a strategy against rare earth element criticality: A systemic evaluation of the potential yield of NdFeB magnet recycling. *Environ. Sci. Technol.* **2013**, *47*, 10129–10136. [CrossRef]
10. Thakur, N.V.; Jayawant, D.V.; Iyer, N.S.; Koppiker, K.S. Separation of neodymium from lighter rare earths using alkyl phosphonic acid, PC 88A. *Hydrometallurgy* **1993**, *34*, 99–108. [CrossRef]
11. Yoon, H.S.; Kim, C.J.; Chung, K.W.; Kim, S.D.; Lee, J.Y.; Kumar, J.R. Solvent extraction, separation and recovery of dysprosium (Dy) and neodymium (Nd) from aqueous solutions: Waste recycling strategies for permanent magnet processing. *Hydrometallurgy* **2016**, *165*, 27–43. [CrossRef]
12. Riaño, S.; Binnemans, K. Extraction and separation of neodymium and dysprosium from used NdFeB magnets: An application of ionic liquids in solvent extraction towards the recycling of magnets. *Green Chem.* **2015**, *17*, 2931–2942. [CrossRef]
13. Riaño, S.; Foltova, S.S.; Binnemans, K. Separation of neodymium and dysprosium by solvent extraction using ionic liquids combined with neutral extractants: Batch and mixer-settler experiments. *RSC Adv.* **2020**, *10*, 307–316. [CrossRef]
14. Ding, Y.; Harvey, D.; Wang, N.L. Two-zone ligand-assisted displacement chromatography for producing high-purity praseodymium, neodymium, and dysprosium with high yield and high productivity from crude mixtures derived from waste magnets. *Green Chem.* **2020**, *22*, 3769–3783. [CrossRef]
15. Alcaraz, L.; Escudero, M.E.; Alguacil, F.J.; Llorente, I.; Urbieta, A.; Fernández, P.; López, F.A. Dysprosium removal from water using active carbons obtained from spent coffee ground. *Nanomaterials* **2019**, *9*, 1372. [CrossRef] [PubMed]
16. Gergoric, M.; Ravaux, C.; Steenari, B.M.; Espegren, F.; Retegan, T. Leaching and recovery of rare-earth elements from neodymium magnet waste using organic acids. *Metals* **2018**, *8*, 721. [CrossRef]
17. Kim, J.; Azimi, G. Recovery of scandium and neodymium from blast furnace slag using acid baking–water leaching. *RSC Adv.* **2020**, *10*, 31936–31946. [CrossRef]
18. Demey, H.; Lapo, B.; Ruiz, M.; Fortuny, A.; Marchand, M.; Sastre, A.M. Neodymium recovery by chitosan/iron (III) hydroxide [ChiFer (III)] sorbent material: Batch and column systems. *Polymers* **2018**, *10*, 204. [CrossRef]
19. Matsumiya, M.; Yamada, T.; Kikuchi, Y.; Kawakami, S. Removal of iron and boron by solvent extraction with ionic liquids and recovery of neodymium metal by direct electrodeposition. *Solvent Extr. Ion Exch.* **2016**, *34*, 522–534. [CrossRef]
20. Riaño, S.; Petranikova, M.; Onghena, B.; Vander Hoogerstraete, T.; Banerjee, D.; Foreman, M.R.S.; Binnemans, K. Separation of rare earths and other valuable metals from deep-eutectic solvents: A new alternative for the recycling of used NdFeB magnets. *RSC Adv.* **2017**, *7*, 32100–32113. [CrossRef]

21. Zhang, Y.; Gu, F.; Su, Z.; Liu, S.; Anderson, C.; Jiang, T. Hydrometallurgical recovery of rare earth elements from NdFeB permanent magnet scrap: A review. *Metals* **2020**, *10*, 841. [CrossRef]

22. Ismail, N.A.; Aziz, M.A.A.; Yunus, M.Y.M.; Hisyam, A. Selection of extractant in rare earth solvent extraction system: A review. *Int. J. Recent Technol. Eng.* **2019**, *8*, 728–742.

23. Xie, F.; Zhang, T.A.; Dreisinger, D.; Doyle, F. A critical review on solvent extraction of rare earths from aqueous solutions. *Miner. Eng.* **2014**, *56*, 10–28. [CrossRef]

24. Xia, Y.; Xiao, L.; Xiao, C.; Zeng, L. Direct solvent extraction of molybdenum(VI) from sulfuric acid leach solutions using PC-88A. *Hydrometallurgy* **2015**, *158*, 114–118. [CrossRef]

25. Zhang, F.; Dai, J.; Wang, A.; Wu, W. Investigation of the synergistic extraction behavior between cerium (III) and two acidic organophosphorus extractants using FT-IR, NMR and mass spectrometry. *Inorg. Chim. Acta* **2017**, *466*, 333–342. [CrossRef]

26. Tasaki-Handa, Y.; Abe, Y.; Ooi, K.; Narita, H.; Tanaka, M.; Wakisaka, A. Selective crystallization of phosphoester coordination polymer for the separation of neodymium and dysprosium: A thermodynamic approach. *J. Phys. Chem. B* **2016**, *120*, 12730–12735. [CrossRef] [PubMed]

27. Bogart, J.A.; Lippincott, C.A.; Carroll, P.J.; Schelter, E.J. An operationally simple method for separating the rare-earth elements neodymium and dysprosium. *Angew. Chem. Int. Ed.* **2015**, *54*, 8222–8225. [CrossRef] [PubMed]

28. Swain, N.; Mishra, S. A review on the recovery and separation of rare earths and transition metals from secondary resources. *J. Clean. Prod.* **2019**, *220*, 884–898. [CrossRef]

29. Dong, Z.; Liu, J.; Yuan, W.; Yi, Y.; Zhao, L. Recovery of Au (III) by radiation synthesized aminomethyl pyridine functionalized adsorbents based on cellulose. *Chem. Eng. J.* **2016**, *283*, 504–513. [CrossRef]

30. Hayashi, N.; Chen, J.; Seko, N. Nitrogen-containing fabric adsorbents prepared by radiation grafting for removal of chromium from wastewater. *Polymers* **2018**, *10*, 744. [CrossRef]

31. Abney, C.W.; Mayes, R.T.; Saito, T.; Dai, S. Materials for the recovery of uranium from seawater. *Chem. Rev.* **2017**, *117*, 13935–14013. [CrossRef]

32. Liu, S.; Xu, M.; Yu, T.; Han, D.; Peng, J.; Li, J.; Zhai, M. Radiation synthesis and performance of novel cellulose-based microsphere adsorbents for efficient removal of boron (III). *Carbohyd. Polym.* **2017**, *174*, 273–281. [CrossRef] [PubMed]

33. Chi, F.; Zhang, S.; Wen, J.; Xiong, J.; Hu, S. Functional polymer brushes for highly efficient extraction of uranium from seawater. *J. Mater. Sci.* **2019**, *54*, 3572–3585. [CrossRef]

34. Li, R.; Li, Y.; Zhang, M.; Xing, Z.; Ma, H.; Wu, G. Phosphate-based ultrahigh molecular weight polyethylene fibers for efficient removal of uranium from carbonate solution containing fluoride ions. *Molecules* **2018**, *23*, 1245. [CrossRef]

35. Tran, T.H.; Okabe, H.; Hidaka, Y.; Hara, K. Removal of metal ions from aqueous solutions using carboxymethyl cellulose/sodium styrene sulfonate gels prepared by radiation grafting. *Carbohyd. Polym.* **2017**, *157*, 335–343. [CrossRef]

36. Li, C.; Zhang, Y.; Peng, J.; Wu, H.; Li, J.; Zhai, M. Adsorption of Cr (VI) using cellulose microsphere-based adsorbent prepared by radiation-induced grafting. *Radiat. Phys. Chem.* **2012**, *81*, 967–970. [CrossRef]

37. Ray, P.Z.; Shipley, H.J. Inorganic nano-adsorbents for the removal of heavy metals and arsenic: A review. *RSC Adv.* **2015**, *5*, 29885–29907. [CrossRef]

38. Topel, S.D.; Legaria, E.P.; Tiseanu, C.; Rocha, J.; Nedelec, J.M.; Kessler, V.G.; Seisenbaeva, G.A. Hybrid silica nanoparticles for sequestration and luminescence detection of trivalent rare-earth ions (Dy^{3+} and Nd^{3+}) in solution. *J. Nanopart. Res.* **2014**, *16*, 2783. [CrossRef]

39. Gu, S.; Kang, X.; Wang, L.; Lichtfouse, E.; Wang, C. Clay mineral adsorbents for heavy metal removal from wastewater: A review. *Environ. Chem. Lett.* **2019**, *17*, 629–654. [CrossRef]

40. Nasef, M.M.; Gürsel, S.A.; Karabelli, D.; Güven, O. Radiation-grafted materials for energy conversion and energy storage applications. *Progr. Polym. Sci.* **2016**, *63*, 1–41. [CrossRef]

41. Zhilyaeva, N.A.; Lytkina, A.A.; Mironova, E.Y.; Ermilova, M.M.; Orekhova, N.V.; Shevlyakova, N.V.; Tverskoy, V.A.; Yaroslavtsev, A.B. Polyethylene with radiation-grafted sulfonated polystyrene membranes for butane and butenes separation. *Chem. Eng. Res. Des.* **2020**, *161*, 253–259. [CrossRef]

42. Sherazi, T.A.; Guiver, M.D.; Kingston, D.; Ahmad, S.; Kashmiri, M.A.; Xue, X. Radiation-grafted membranes based on polyethylene for direct methanol fuel cells. *J. Power Sour.* **2010**, *195*, 21–29. [CrossRef]

43. Mandal, D.V.; Bhunia, H.; Bajpai, P.K.; Bhalla, V.K. Thermal degradation kinetics and estimation of lifetime of radiation grafted polypropylene films. *Radiat. Phys. Chem.* **2017**, *136*, 1–8. [CrossRef]

44. Bondar, Y.; Kim, H.J.; Yoon, S.H.; Lim, Y.J. Synthesis of cation-exchange adsorbent for anchoring metal ions by modification of poly (glycidyl methacrylate) chains grafted onto polypropylene fabric. *React. Funct. Polym.* **2004**, *58*, 43–51. [CrossRef]

45. Chen, J.; Seko, N. Cleavage of the graft bonds in PVDF-g-St films by boiling xylene extraction and the determination of the molecular weight of the graft chains. *Polymers* **2019**, *11*, 1098. [CrossRef] [PubMed]

46. Seko, N.; Basuki, F.; Tamada, M.; Yosaii, F. Rapid removal of arsenic(V) by zirconium(IV) loaded phosphoric chelate adsorbent synthesized by radiation induced graft polymerization. *React. Funtc. Polym.* **2004**, *59*, 235–241. [CrossRef]

47. Ha, H.; Wu, L.; Tai, H.; Zhang, Z.; Wei, J.; Wu, J. Study of radiation grafting of styrene on cotton cellulose. *Radiat. Phys. Chem.* **1995**, *46*, 823–827.

48. Gad, H.M.H.; Hamed, M.M.; Eldahab, H.A.; Moustafa, M.E.; El-Reefy, S.A. Radiation-induced grafting copolymerization of resin onto the surface of silica extracted from rice husk ash for adsorption of gadolinium. *J. Mol. Liq.* **2017**, *231*, 45–55. [CrossRef]

49. Ueki, Y.; Seko, N. Synthesis of Fibrous metal adsorbent with a piperazinyl—Dithiocarbamate group by radiation-induced grafting and its performance. *ACS OMEGA* **2020**, *5*, 2947–2956. [CrossRef]

50. Shibata, T.; Seko, N.; Amada, H.; Kasai, N.; Saiki, S.; Hoshina, H.; Ueki, Y. Evaluation of a cesium adsorbent grafted with ammonium 12-molybdophosphate. *Radiat. Phys. Chem.* **2016**, *119*, 247–252. [CrossRef]

51. Othman, N.A.F.; Selambakkannu, S.; Yamanobe, T.; Hoshina, H.; Seko, N.; Abdullah, T.A.T. Radiation grafting of DMAEMA and DEAEMA-based adsorbents for thorium adsorption. *J. Radioanal. Nucl. Chem.* **2020**, *324*, 429–440. [CrossRef]

52. Zhang, W.; Avdibegović, D.; Koivula, R.; Hatanpää, T.; Hietala, S.; Regadío, M.; Harjula, R. Titanium alkylphosphate functionalised mesoporous silica for enhanced uptake of rare-earth ions. *J. Mater. Chem. A* **2017**, *5*, 23805–23814. [CrossRef]

53. Ueki, Y.; Saiki, S.; Hoshina, H.; Seko, N. Biodiesel fuel production from waste cooking oil using radiation-grafted fibrous catalysts. *Radiat. Phys. Chem.* **2018**, *143*, 41–46. [CrossRef]

54. Fonseca, J.L.C.; Apperley, D.C.; Plasma, J.P.S. Plasma polymerization of tetramethylsilane. *Chem. Mater.* **1993**, *5*, 1676–1682. [CrossRef]

55. Nava-Ortiz, C.A.B.; Burillo, G.; Bucio, E.; Alvarez-Lorenzo, C. Modification of polyethylene films by radiation grafting of glycidyl methacrylate and immobilization of β-cyclodextrin. *Radiat. Phys. Chem.* **2009**, *78*, 19–24. [CrossRef]

56. Ormond-Prout, J.; Dupin, D.; Armes, S.P.; Foster, N.J.; Burchell, M.J. Synthesis and characterization of polypyrrole-coated poly (methyl methacrylate) latex particles. *J. Mater. Chem.* **2009**, *19*, 1433–1442. [CrossRef]

57. Yin, S.H.; Li, S.W.; Xie, F.; Zhang, L.B.; Peng, J.H. Study on the aqueous solution behavior and extraction mechanism of Nd (III) in the presence of the complexing agent lactic acid with di-(2-ethylhexyl) phosphoric acid. *RSC Adv.* **2015**, *5*, 64550–64556. [CrossRef]

58. Pathak, S.K.; Tripathi, S.C.; Singh, K.K.; Mahtele, A.K.; Dwivedi, C.; Juby, K.A.; Bajaj, P.N. PC-88A impregnated polymeric beads: Preparation, characterization and application for extraction of Pu (IV) from nitric acid medium. *Radiochim. Acta* **2013**, *101*, 761–771. [CrossRef]

Publisher's Note: MDPI stays neutral with regard to jurisdictional claims in published maps and institutional affiliations.

 *polymers*

Article

Removal of Rhodamine B from Water Using a Solvent Impregnated Polymeric Dowex 5WX8 Resin: Statistical Optimization and Batch Adsorption Studies

Moonis Ali Khan [1],*, **Momina** [2], **Masoom Raza Siddiqui** [1], **Marta Otero** [3], **Shareefa Ahmed Alshareef** [1] and **Mohd Rafatullah** [4],*

[1] Chemistry Department, College of Science, King Saud University, Riyadh 11451, Saudi Arabia; mrsiddiqui@ksu.edu.sa (M.R.S.); 438203872@student.ksu.edu.sa (S.A.A.)

[2] School of Chemical Engineering, Universiti Sains Malaysia, Engineering Campus, Nibong Tebal 14300, Penang, Malaysia; momina.anees@gmail.com

[3] CESAM - Centre for Environmental and Marine Studies, Department of Environment and Planning, University of Aveiro, Campus de Santiago, Aveiro 3810-193, Portugal; marta.otero@ua.pt

[4] School of Industrial Technology, Universiti Sains Malaysia, Main campus 11800, Penang, Malaysia

* Correspondence: mokhan@ksu.edu.sa or moonisalikhan@gmail.com (M.A.K.); mrafatullah@usm.my (M.R.)

Received: 29 January 2020; Accepted: 19 February 2020; Published: 24 February 2020

Abstract: Herein, commercially available Dowex 5WX8, a cation exchange polymeric resin, was modified through solvent impregnation with t-butyl phosphate (TBP) to produce a solvent impregnated resin (SIR), which was tested for the removal of rhodamine B (RhB) from water in batch adsorption experiments. The effect of SIR dosage, contact time, and pH on RhB adsorption was studied and optimized by response surface methodology (RSM), interaction, Pareto, and surface plots. Scanning electron microscopy (SEM) and Fourier transform infrared spectroscopy (FTIR) were respectively used for characterizing SIR surface morphology and identifying active binding sites before and after RhB adsorption. SEM showed that the pristine SIR surface was covered with irregular size and shape spots with some pores, while RhB saturated SIR surface was non-porous. FTIR revealed the involvement of electrostatic and π–π interactions during RhB adsorption on SIR. Dosage of SIR, contact time, and their interaction significantly affected RhB adsorption on SIR, while pH and its interaction with dosage and contact time did not. The optimum identified experimental conditions were 0.16 g of SIR dose and 27.66 min of contact time, which allowed for 98.45% color removal. Moreover, RhB adsorption equilibrium results fitted the Langmuir isotherm with a maximum monolayer capacity (q_{max}) of 43.47 mg/g.

Keywords: modified polymeric resin; t-butyl phosphate impregnation; polymer based adsorbents; dye adsorption; response surface methodology

---

1. Introduction

Textile industries are among the largest consumers of water, dyes, and different types of chemicals, resulting in the generation of large volumes of highly toxic effluents. The discharge of these effluents without prior treatment can be lethal to the environment. Usually, the textile effluents are rich in color, pH, chemical oxygen demand, inorganic salts, turbidity, and temperature [1]. According to the United States Environmental Protection Agency (USEPA), textile waste is mainly divided into four principal classifications, namely hard-to-treat, high volume, dispersible, and hazardous and toxic wastes [2]. Rhodamine B (RhB) is one of the most widely used cationic water-soluble organic dyes, and it is toxic to aquatic environments. It reduces sunlight penetration into water bodies, which can be lethal for aquatic life due to limited availability of oxygen for respiration [3–5]. Therefore, before effluent discharge, it is necessary to apply intensive treatment processes to minimize its concentration in water bodies.

Membrane filtration, flocculation, biological treatments, photocatalytic oxidation, and adsorption [6–10] are some of the commonly used textile effluent treatment processes. On the other hand, the application of polymer materials in water treatment and selective sequestration has impressively developed in the last decades, with the production of novel materials and composites, post-polymerization modifications, introduction of functional groups, and development of supramolecular assemblies and nanomaterials [11–16]. However, these methods have their own limitations and efficiencies in terms of cost effectiveness. Recently, functionalized polymeric resins have become an alternative to commercial activated carbon and other adsorbents due to economic concerns and regeneration properties. Polymeric resins are characterized by their high surface area, moderate swelling, and narrow pore size distribution. In order to improve the adsorption characteristics of such resins, their surface properties can be modified using the advantage of adsorbate and adsorbent interaction. Mostly, ion-exchange resins are prepared by styrene divinylobenzene cross-linked co-polymer, which is comparatively lower in cost than activated carbon and serve several advantages as a matrix [17]. Additionally, the polystyrene based matrix has the potential to provide excellent chemical and physical stability together with resistance to degradation by oxidation or hydrolysis. Apart from ion-exchange resins, solvent extraction and liquid–liquid extraction, which work on the principle that solute distributes itself in a certain ratio with immiscible solvents, have attracted considerable attention in recent years [18]. The merits of solvent extraction include rapid and very selective separations that are usually highly efficient [18]. Solvent impregnated resins (SIRs) pose synergic merits of both ion-exchange and solvent extraction. A SIR is described as a liquid complexing agent dispersed homogeneously in a solid polymeric medium. In SIR removal processes, a specific solute is extracted from the aqueous phase to the organic phase inside the pores of the resin. The resin acts as carrier of the solvent and reduces the entrainment and irreversible emulsification that occur during solvent extraction [19]. Previously, a non-functional macroporous polymeric resin was used as a polymeric support for the removal of dyes from wastewater [20]. However, macroporous resins have lower retention capacity and slower kinetic diffusion compared to gel-type resins. Therefore, gel-type resins have higher removal efficiency than conventional macroporous polymeric resins. Moreover, the optimization of operational conditions for dye adsorption using SIR can improve the removal efficiency of dye. Thus, in this study, a gel-type Dowex cation exchange resin was used to remove RhB dye from aqueous solution. Moreover, the effect of interaction of operational conditions on the removal of RhB dye was also studied. Response surface methodology (RSM) is effective, reliable, and very comprehensive as compared to other conventional optimization processes [21]. It is a statistical tool that is very effective for designing, analyzing, and optimizing the effect of independent factors for the prediction of response output [21]. Therefore, the aim of this work was to investigate the operating conditions for removal of RhB dye using a 2^3 full factorial design. Moreover, an equilibrium batch study was performed at optimum conditions to study the mechanism of dye removal using SIR.

2. Experimental

2.1. Chemicals, Reagents, and Adsorbent

Dowex 5WX8 gel-type cation exchange resin (BDH, England, UK) with particle size 0.39–1.00 mm and 50%–58% moisture content was used as an adsorbent. The resin contained a styrene divinyl benzene matrix, having sulfonic acid as a matrix active functional group. Rhodamine B (RhB: $C_{28}H_{31}ClN_2O_3$) (Sigma-Aldrich, Darmstadt, Germany) with respective color index and molecular weight 45170 and 479.2 g/mol, synonymously known as basic violet 10, was used as an adsorbate. Tributyl phosphate (TBP: $C_{12}H_{27}O_4P$) was obtained from Sigma-Aldrich, Darmstadt, Germany. All the chemicals and reagents used during the study were of analytical reagent (A.R) grade or as itemized. Ultra-pure deionized (D.I: Millipore, Burlington, MA, USA) water was used throughout the study.

2.2. Synthesis of Solvent Impregnated Dowex 5WX8 Resin

Initially, Dowex 5WX8 resin was washed with D.I water in order to remove inorganic impurities and monomeric material. Thereafter, the resin was overnight dried in an oven at 70 °C. The resin was impregnated with TBP (hydrophobic in nature) through the wet impregnation method reported elsewhere [22]. Briefly, undiluted TBP and resin in a volume to weight ratio of 6.0 was used to impregnate resin in a conical flask. Resin was aged for 24 h in TBP to achieve highest impregnation efficiency [20]. Further, the impregnated resin was separated from TBP through filtration and thoroughly rinsed with D.I water to remove unimpregnated traces of TBP. Then, solvent impregnated Dowex 5WX8 resin (SIR) was ready to use for adsorption studies.

2.3. Characterization of Solvent Impregnated Dowex 5WX8 Resin

Fourier transform infrared (FT-IR: Is10 Nicolet Thermo Scientific, Waltham, MA, USA) analysis was carried out to determine the available functional groups on Dowex 5WX8 resin and SIR (both pristine and RhB saturated) surfaces. The surface morphology of Dowex 5WX8 resin and SIR (both pristine and RhB saturated) was analyzed by scanning electron microscopy (SEM: Zeiss, model EVO Ma10, Oberkochen, Germany).

2.4. Batch Scale Adsorption

The RhB adsorption studies over SIR were performed at room temperature by varying operation parameters viz. initial pH (pH_i: 2–8), SIR dose (m: 0.1–0.5 g), and contact time (t: 5–30 min). A series of 10 mL RhB solutions of initial concentration C_o: 100 mg/L were equilibration with 0.1–0.5 g SIR in 25 mL conical flasks over a shaker at 230 rpm. At predetermined contact times, solid/solution phases were separated, and residual RhB concentrations in solutions were determined by using a Shimadzu UV-Visible Spectrophotometer at λ_{max}: 554 nm. The amount of RhB adsorbed at any time t onto SIR was calculated as:

$$Adsorbed\ concentration\ at\ time\ t\ (q_t, mg/g) = (C_o - C_t) \times \frac{V}{m} \tag{1}$$

where V (L) is the volume of RhB solution, C_o (mg/L) is the initial RhB concentration, C_t (mg/L) is the remaining RhB concentration in solution at any time t, and m (g) is the mass of SIR.

The amount of RhB adsorbed on SIR at equilibrium, which was attained in 30 min under shaking, was calculated as:

$$Adsorbed\ concentration\ at\ equilibrium\ (q_e, mg/g) = (C_o - C_e) \times \frac{V}{m} \tag{2}$$

where C_e (mg/L) is the equilibrium concentration of RhB in solution.

The decolorization efficiency (D.E, %) was calculated as follows:

$$Decolourization\ efficiency\ (D.E.\ \%) = \frac{C_o - C_e}{C_o} \times 100 \tag{3}$$

2.5. Design of Experiments and Optimization of Parameters

Two level (low level −1 and high level +1) factorial design (2^3) of response surface methodology (RSM) was applied for three independent variables (factors), namely the operational parameters dosage of SIR (A), contact time (B), and initial pH (C), to predict D.E. (%) (response factor, y) from RhB dye solution using solvent impregnated resin (Table 1). A total 12 runs including 4 times of replication for center point was carried out (2^3 = 8 runs; 8 runs + 4 replications for center point = 12 runs) using Design Expert (6.0.10) (Stat Ease, Minneapolis, MN, USA). Suitable approximation can be determined for the true functional relationship between the process response, y, and the set of factors by first order model or second order model. When the response linearly varies with the independent variable, then the first order model, which is given by Equation (4), is satisfied.

$$y = \beta_0 + \beta_1 x_1 + \ldots \beta_k x_k + \varepsilon \tag{4}$$

where, y is the response, β_0 is the offset term, $\beta_1, \ldots, \beta_k$ are the effect term, $x_1, \ldots, x_k$ are the independent variables, and ε is the random error term. When a curvature is detected in the system, second order model is selected and expressed by the following equation (Equation (5)):

$$y = \beta_0 + \sum_{i=1}^{k} \beta_1 x_1 + \sum_{i=1}^{k} \beta_{ii} x_i^2 + \sum_{i=1}^{n} \sum_{i<j}^{n} \beta_{ij} x_i x_j \tag{5}$$

where y is the predicted response, β_0 is the constant, β_1 is the linear effect, β_{ii} is the square effect, and β_{ij} is the interaction effect.

Table 1. Factors, levels, and ranges of the parameters considered for the factorial design.

Terms	Factors	Levels		
		−1	0	+1
A	SIR dose (g)	0.1	0.3	0.5
B	Contact time (min)	5	17.5	30
C	Initial pH of RhB solution	2	5	8

For the optimization of experimental design, the statistical software Minitab 16 was used. The results were analyzed by estimating the response of the dependent response variable to obtain the effects, coefficient, and other statistical parameters. The conditions for optimization of adsorption process were obtained from Minitab 16 (Minitat LLC., Penn State University, PA, USA) as well. By using the analysis of variance (ANOVA), the determination coefficient (r^2) and statistical significance were determined.

2.6. Isotherm Modeling

Langmuir and Freundlich isotherm models in linearized forms were fitted to data on RhB adsorption onto SIR. The Langmuir isotherm model, which assumes monolayer adsorption over homogenous sites on the adsorbent surface and equal activation energy for each molecule, is given by Equation (6) in its linearized form:

$$\frac{1}{q_e} = \frac{1}{q_{max} K_L C_e} + \frac{1}{q_{max}} \tag{6}$$

where q_e (mg/g) is the amount of RhB adsorbed on SIR, C_e (mg/L) is the saturated amount of RhB adsorption at equilibrium concentration, and q_{max} (mg/g) is the maximum monolayer adsorption capacity of RhB on SIR. The constants K_L and q_{max} can be calculated from a linear plot of $1/q_e$ vs. $1/C_e$. The characteristics of the fitting to the Langmuir equation are given by a dimensionless number, R_L (Equation (7)), which indicates the type of isotherm to be irreversible ($R_L = 0$), favorable ($0 < R_L < 1$), linear ($R_L = 1$), or unfavorable ($R_L > 1$) [23].

$$R_L = \frac{1}{1 + K_L C_o} \tag{7}$$

The Freundlich isotherm, which is an empirical model, is usually associated with multilayer adsorption of RhB molecules over heterogenous adsorption sites and can be expressed in linearized form as:

$$\log q_e = \log K_F + \frac{1}{n} \log C_e \tag{8}$$

where K_F is a Freundlich constant, and n is a parameter related to the binding strength changes with the adsorption density. If $1/n = 0$, it indicates that the extent of adsorption is independent between two phase concentration; $1/n < 1$ indicates favorable chemical adsorption; $1/n > 1$ indicates a cooperative adsorption [23].

3. Results and Discussion

3.1. Pre and Post-Adsorption Characterization

The surface morphologies of Dowex 5WX8 resin and SIR (both pristine and RhB saturated) were analyzed using SEM with 300X magnification, as illustrated in Figure 1a–d. The raw Dowex 5WX8 resin has a smooth surface with some pores (Figure 1a). After TBP solvent impregnation over raw Dowex 5WX8 resin, the whole SIR surface was covered with spots of irregular size and shape (Figure 1b). This confirms successful impregnation of raw Dowex 5WX8 resin [24]. The structural pores after impregnation remained unchanged, as shown in Figure 1c. These pores were well occupied by RhB molecules during adsorption, displayed by protruding occupation of pores (Figure 1d). The FT-IR spectrum of raw Dowex 5WX8 resin (Figure 1e) showed a strong band centered at 3420 cm^{-1}, ascribed to hydroxyl (–OH) group stretching, due to the presence of internal moisture in raw Dowex 5WX8 resin. The conjoint bands at 2927 and 2852 cm^{-1} were associated with C–H stretching vibrations for saturated aliphatic species. The bands between 1483 and 1510 cm^{-1} were due to CH$_3$ deformation in amino acid or hydrochloride compounds in raw Dowex 5WX8 resin. Moreover, the bands at 1010 and 1033 cm^{-1} represented C–O stretching of cyclic alcohol in raw resin. However, after impregnation of raw Dowex 5WX8 resin with TBP, the band at 1033 cm^{-1} became sharp and intense due to P–O–C stretching, thus confirming the attachment of phosphorous with C–O. A band at 1226–1237 cm^{-1} was due to P=O stretching in the phosphate group, and a band at 1383–1388 cm^{-1} showed CH$_3$ deformation in the t-butyl group. The presence of P=O and P–O–C groups, and CH$_3$ deformity indicate successful impregnation of raw Dowex 5WX8 resin to form SIR. After RhB adsorption on SIR, the band at 1033 cm^{-1} was displaced by a low intensity band at 1034 cm^{-1}, confirming its involvement in binding dye molecules during adsorption.

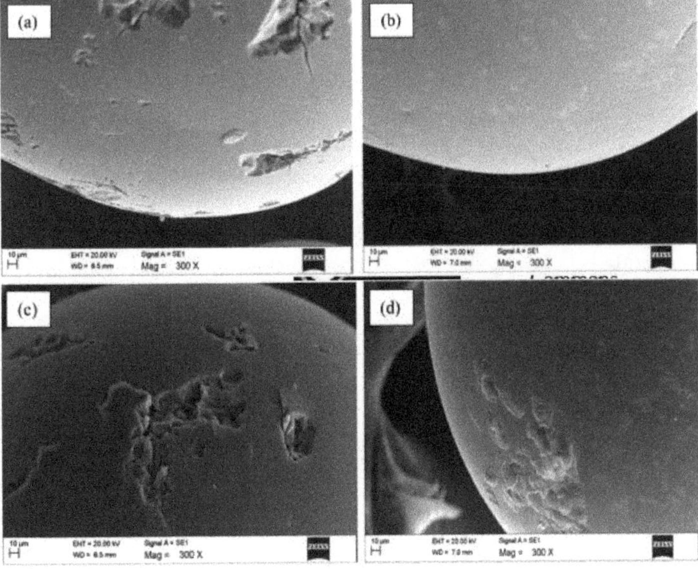

Figure 1. *Cont.*

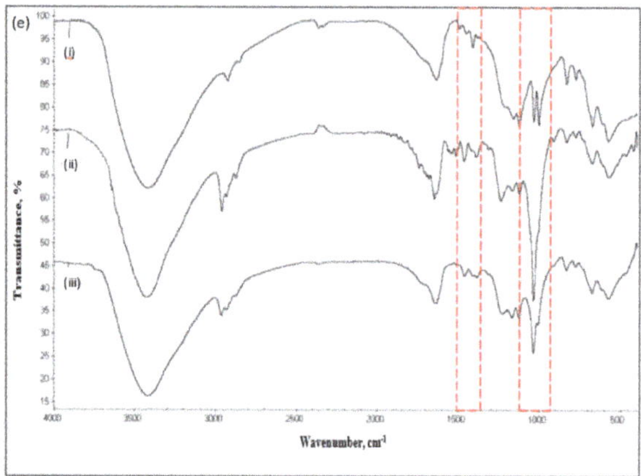

Figure 1. Scanning electron microscopic (SEM) images of raw Dowex 5WX8 polymeric resin (**a**), pristine SIR (**b, c**), RhB saturated SIR (**d**), and Fourier transform infrared (FT-IR) spectra of raw Dowex 5WX8 polymeric resin (**i**), pristine SIR (**ii**), and RhB saturated SIR (**iii**) (**e**).

3.2. Screening of Process Independent Variables

In this batch adsorption study, the interaction of operational parameters for the removal of RhB using SIR was examined. A total of 12 experimental runs were optimized using three dominant parameters for the removal of RhB from aqueous solution, which was calculated using Equation (3). It was used to achieve improved adsorption capacity of SIR by possible interaction of operational parameters (Table S1, Supplementary Materials). The results showed that two of the considered operational parameters, specifically dosage of SIR and contact time, influenced the removal of RhB from aqueous solution. Additionally, the surface modification of Dowex 5WX8 resin by impregnation to SIR led to improved surface characteristics for the removal of RhB.

The effect of variable interaction during the adsorption process was carried out using analysis of variance (ANOVA). Then, 2^2 fractional factorial designs were used to study the selected factors. The purpose of carrying out the fractional factorial design was to determine the factors that had a significant effect on D.E. (%). ANOVA for the fractional factorial design is given in Table 2. The ANOVA and response surface regression of D.E. (%) is tabulated in Table 3. The effect of pH for aqueous phase was found to be insignificant due to the p-value of 0.314, which was greater than 0.05. Moreover, the negative coefficient (–1.08) of pH pointed towards a decrease in adsorption efficiency as the pH increased. The overall prediction of the output model in terms of operational parameters showed that the model was suitable for predicting the adsorption of RhB on SIR (p < 0.05). The respective p-values of SIR dosage and contact time were 0.063 (nearer to 0.05) and 0.012, which showed that both parameters significantly influenced D.E. (%). Furthermore, the linear effects of both SIR dosage and contact time indicated that they were suitable for improving adsorption efficiency.

However, quadratic coefficients of SIR dosage and contact time inhibited the performance of RhB adsorption from aqueous solution (p > 0.05). Moreover, the negative coefficient of interaction of dosage and contact time was helpful in increasing adsorption efficiency. The polynomial first order and interactive regression model equation was developed using Minitab software. Therefore, the model equation is given as:

$$y = 91.374 + 11.215A + 17.214B - 20.055AB \tag{9}$$

where y is the D.E. (%) of RhB. In the aforementioned equation, a synergistic effect was indicated by the positive sign, while an antagonistic effect was indicated by the negative sign [25]. The ANOVA for the model is given in Table 4. It was deduced that the color removal was significant at 95% (p < 0.05) confidence level, which shows the validity of the model for RhB adsorption onto SIR.

Table 2. ANOVA analysis for the fractional factorial design carried out to determine the factors that have significant effect on the decolorization efficiency (D.E. (%)).

Term	Coefficient	SE Coefficient	T-value	P-value
Constant	71.78	0.9575	74.97	0.000
A	18.90	0.9575	19.74	0.000
B	21.53	0.9575	22.48	0.000
C	−1.08	0.9575	−1.13	0.341
AB	−20.93	0.9575	−21.86	0.000
BC	0.07	0.9575	0.08	0.943
AC	0.24	0.9575	0.25	0.817
ABC	0.99	0.9575	1.03	0.378

SE Coefficient = standard error of the coefficient.

Table 3. Estimated regression coefficients and ANOVA for optimization of decolorization efficiency (D.E. (%)).

Term	Coefficient	SE Coefficient	T-value	P-value
Constant	91.374	5.156	17.722	0.000
A	11.215	5.069	2.212	0.063
B	17.214	5.069	3.396	0.012
A^2	−12.117	7.472	−1.622	0.149
B^2	−6.863	7.472	−0.918	0.389
AB	−20.055	6.209	−3.230	0.014

SE Coefficient = standard error of the coefficient.

Table 4. ANOVA analysis of the model for the removal of RhB using the produced SIR.

Source	Degree of Freedom	Sum of Squares	Mean Squares	F-value	P-value
Regression	5	4972.66	994.53	6.45	0.015
A	1	754.69	754.69	4.89	0.063
B	1	1777.96	1777.96	11.53	0.012
A^2	1	701.09	405.48	2.63	0.149
B^2	1	130.08	130.08	0.84	0.389
AB	1	1608.84	1608.84	10.43	0.014
Residual error	7	1079.36	154.19		
Lack of fit	3	1052.53	350.84	52.30	0.001
Pure error	4	26.83	6.71		
Total	12	6052.02			

$$r^2 = 82.17\%$$
$$r^2 \text{ (adjusted)} = 69.43\%$$

The determination of coefficient (r^2) was used to evaluate the quality of the developed model [7,26]. The r^2 value of the color removal was 82.17%, which means that 0.8217 of total variation was explained by the model, while 17.83% of the variation was left unexplained. The importance of the effects of the operation variables and their interaction can be best described by the Pareto chart, as shown in Figure 2a. A student's t-test was performed to determine whether the calculated effects were significantly different from zero; these values for each effect are shown in the Pareto chart by horizontal columns [7]. The t-value for 95% confidence level was 2.013. The values exceeding the reference line were considered as significant for 95% confidence level, whereas values below the reference line were considered as insignificant. As shown in Figure 2a, the two parameters, dosage of SIR (A) and contact time (B), as well as their interaction (AB) were found to be significant at the 0.05 level. However, the effect of pH (C) and its interaction with dosage (AC) and contact time (BC) were below the reference line, which points to their insignificance for D.E. (%). According to previous studies, the effect of increasing pH was considered as favorable for the removal percentage of RhB [27,28]. However, in this study the effect of pH was not found to be significant. Therefore, the effects of dosage, contact time, and their interaction, which resulted in 97.45% of dye adsorption efficiency, were studied.

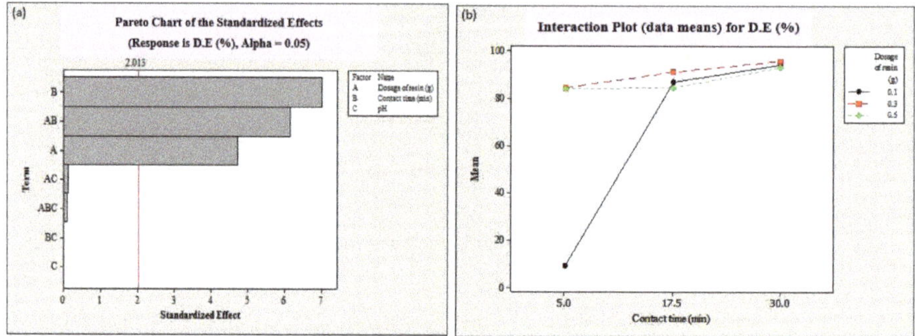

Figure 2. Pareto chart showing the effects and interactions of operational variables, namely dosage of resin (g), contact time (min) and pH, on the decolorization efficiency (D.E. (%)) by SIR (**a**), and interaction plot of decolorization efficiency (D.E. (%)) by SIR versus contact time (min) for the different dosages of resin, namely 0.1, 0.2, and 0.3 g of SIR (**b**).

The interaction of contact time and dosage of resin were described by interaction plots, illustrated in Figure 2b. The interaction plots show the D.E. (%) versus the contact time (min) for each dosage of SIR. It was found that as the contact time increased, the D.E. (%) increased and reached its maximum for a 0.3 g dosage of SIR. The interaction between contact time and SIR dosage improved the adsorption efficiency of RhB, as shown in Figure 3a. The response surface plots show the estimated value of D.E. (%) (the height of the surface represents the value of D.E. (%)) as a function of the independent variable. It must be highlighted that the surface plots represent the same results as observed in interaction plots.

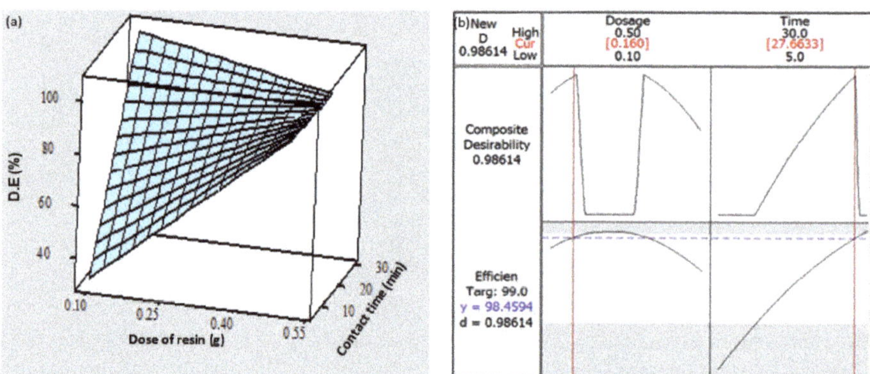

Figure 3. Surface plot showing the decolorization efficiency (D.E. (%)) as a function of the dosage of resin (g) and contact time (min) under a shaking speed of 230 rpm and pH 3.6 (**a**), and optimization plot for the determination of the optimum conditions, namely dosage of resin (g) and contact time (min), for a maximum decolorization efficiency (D.E. (%)) by SIR (**b**).

3.3. Optimization of Experiment

A standard RSM design called central composite design (CCD) was used to optimize the operating parameters. Optimum conditions of effective parameters with minimum number of experiments can be determined by this statistical technique. This method is also suitable to analyze the interaction and relationship between each parameter. The optimum conditions for the removal of RhB dye were obtained from the screening of operational parameters (Table S1, Supplementary Materials).

The optimum operational parameters for D.E. (97.45%) were achieved with 0.3 g SIR dosage and 30 min contact time. Since contact time played a major role in the extraction process [20], therefore extraction efficiency of TBP impregnated SIR increased steadily with time until it reached equilibrium. Moreover, an increasing amount of SIR increases the D.E. (%) because low amounts of SIR contain low amounts of extractant [20]. However, high amounts of SIR and TBP cause an increase of the dye solution acidity. Therefore, concentrated acidic medium causes back-extraction of extractant–dye complex, thus resulting in higher dye concentration [29].

The optimization plot, which is displayed in Figure 3b, was used to find optimum conditions for RhB removal by SIR. From the analysis of experimental data obtained, the optimum identified conditions were 0.16 g of SIR dose and 27.66 min of contact time. With the application of such optimum conditions, the predicted value of D.E. was 98.45%, which was experimentally verified to be fulfilled with a deviation of ± 0.1%.

3.4. Adsorption Isotherm

The adsorption isotherm is used to study the mechanism and pattern of adsorption at liquid-phase equilibrium [21,30,31]. Fittings of equilibrium data for the adsorption of RhB on SIR by Langmuir and Freundlich models were determined in this work. Linear plots of Langmuir and Freundlich isotherms and the respective parameters are shown in Figure S1 (Supplementary Materials). As can be seen, equilibrium results fitted the Langmuir isotherm ($r^2 = 0.99$) but not the Freundlich model ($r^2 = 0.087$). Therefore, it may be assumed that the adsorption of RhB on SIR was the monolayer on the surface of SIR, where the active sites and energies were homogenously distributed [32,33]. The fitted maximum monolayer adsorption capacity (q_{max}) of RhB on SIR was found to be 43.47 mg/g, and K_L was 0.0126 L/mg. Therefore, adsorption of RhB on SIR was found to be favorable because R_L was calculated (Equation (7)) to be 0.284, which is greater than 0 and smaller than 1.

3.5. Adsorption Mechanism

Dowex 50WX8, a cation exchange polymeric resin, was used in this study. It contains cross-linked styrene divinyl benzene co-polymer with sodium sulfonate groups as ion-exchange sites. The resin was impregnated with TBP, which contains certain functional groups that have significant influence on the adsorption of RhB dye (as revealed by FT-IR results in Figure 1e). Therefore, an increased RhB adsorption can be achieved using the interaction between the adsorbate (RhB) and modified resin (SIR). The structure and functional groups present on modified SIR resin were the main factors responsible for the adsorption of RhB dye on this resin. Rhodamine B dye has amino and carboxylic functional groups, which can be involved on its adsorption on modified SIR. According to FTIR results, the peaks for P–O–C disappeared after adsorption of dye on impregnated resin and were replaced by the C–O group of the RhB dye. On the other hand, phosphate was not detected in solution after the adsorptive removal of RhB. Therefore, phosphate groups might be responsible for binding the positively charged dye ions by modified SIR resin. Furthermore, the possible interaction that might be occurring between modified SIR resin and RhB dye can be electrostatic and π–π bonding, as shown in Figure 4.

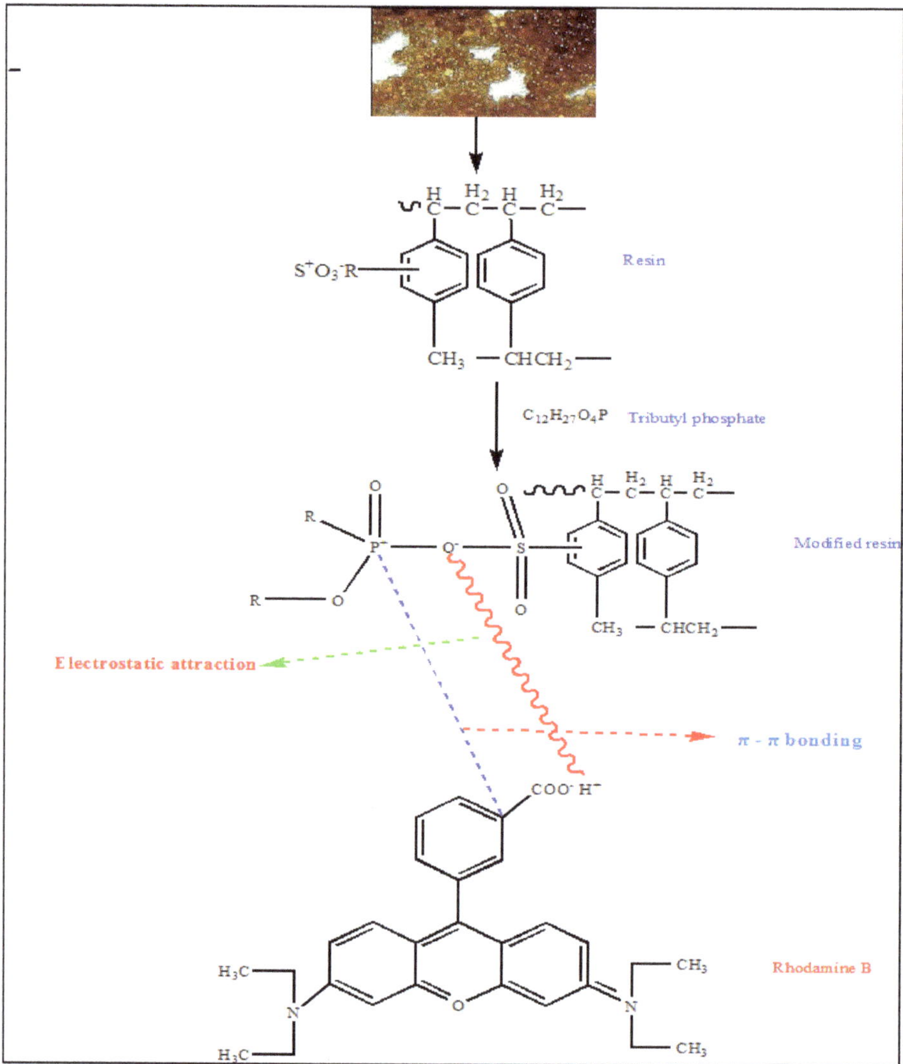

Figure 4. Schematic representation of SIR production and RhB adsorption mechanism on SIR.

4. Conclusions

A TBP impregnated polymeric resin was produced in this work and tested for the removal of RhB from water. FT-IR analysis showed the presence of a phosphate functional group on the surface of the solvent impregnated resin (SIR), which was indicative of successful impregnation of TBP over the resin. The effect of operational conditions, namely pH, adsorbent dosage, and contact time, on the adsorption of RhB onto SIR was studied and optimized. It was found that pH does not have a significant effect on the D.E. (%). The maximum color removal obtained was 97.45% at 100 mg/L of initial dye concentration, 230 rpm of shaking speed, pH 3.6, 0.3 g of resin dosage, and 30 min of contact time. The optimum conditions for the adsorption of RhB by SIR were identified as 0.16 g of SIR resin and 27.66 min of contact time, which gave 98.45% of color removal. The adsorption data fitted well the Langmuir isotherm model, which pointed to monolayer adsorption on the SIR surface, with homogeneous distribution of active sites and energies. Furthermore, interaction of RhB and SIR was inferred to be electrostatic and π–π bonding.

Supplementary Materials: The following are available online at http://www.mdpi.com/2073-4360/12/2/500/s1, Figure S1: Langmuir (a) and Freundlich (b) plots for the adsorption of RhB on SIR; Table S1: Results of response surface methodology design.

Author Contributions: Data curation, S.A.A.; Supervision, M.R.; Validation, M.R.S.; Visualization, M.O.; Writing—original draft, M.A.K. and M.; Revision, M.A.K. and M.O. All authors have read and agreed to the published version of the manuscript.

Funding: Deanship of Scientific Research, King Saud University: Research Group No. RG-1437-031.

Acknowledgments: The authors would like to extend their sincere appreciation to the Deanship of Scientific Research at King Saud University for funding this work through Research Group No. RG-1437-031. Furthermore, Marta Otero would like to thank FCT/MCTES for the financial support to CESAM (UID/AMB/50017/2019), through national funds and support by the FCT Investigator Program (IF/00314/2015).

Conflicts of Interest: The authors declare no conflict of interest.

References

1. Verma, A.K.; Dash, R.R.; Bhunia, P. A review on chemical coagulation/flocculation technologies for removal of color from textile wastewaters. *J. Environ. Manag.* **2012**, *293*, 154–168. [CrossRef] [PubMed]

2. Dasgupta, J.; Sikder, J.; Chakraborty, S.; Curcio, S.; Drioli, E. Remediation of textile effluents by membrane based treatment techniques: A state of the art review. *J. Environ. Manag.* **2015**, *147*, 55–72. [CrossRef] [PubMed]

3. Wabaidur, S.M.; Khan, M.A.; Siddiqui, M.R.; Alothman, Z.A.; Al-Ghamdi, M.S.; Al-Sohami, H.I. Dodecyl sulfate chain anchored bio-char to sequester triaryl methane dyes: Equilibrium, kinetics, and adsorption mechanism. *Desal. Water Treat.* **2017**, *67*, 357–370. [CrossRef]

4. Ahmad, T.; Rafatullah, M.; Ghazali, A.; Sulaiman, O.; Hashim, R. Oil palm biomass-Based adsorbents for the removal of water pollutants—A review. *J. Environ. Sci. Health Part C* **2011**, *29*, 177–222. [CrossRef]

5. Rafatullah, M.; Sulaiman, O.; Hashim, R.; Ahmad, A. Adsorption of methylene blue on low-cost adsorbents: A review. *J. Hazard. Mater.* **2010**, *177*, 70–80. [CrossRef]

6. Momina, M.; Isamil, S. Regeneration performance of clay-based adsorbents for the removal of industrial dyes: A review. *RSC Adv.* **2018**, *8*, 24571–24587. [CrossRef]

7. Rafatullah, M.; Ismail, S.; Ahmad, A. Optimization study for the desorption of methylene blue dye from clay based adsorbent coating. *Water* **2019**, *11*, 1304.

8. Rafatullah, M.; Ahmad, T.; Ghazali, A.; Sulaiman, O.; Danish, M.; Hashim, R. Oil palm biomass as a precursor of activated carbons: A review. *Crit. Rev. Environ. Sci. Technol.* **2013**, *43*, 1117–1161. [CrossRef]

9. El-Refaie, K.; Ghfar, A.A.; Wabaidur, S.M.; Khan, M.A.; Siddiqui, M.R.; Alothman, Z.A.; Alqadami, A.A. Muhammad, Cetyltrimethylammonium bromide intercalated and branched polyhydroxystyrene functionalized montmorillonite clay to sequester cationic dyes. *J. Environ. Manag.* **2018**, *219*, 285–293.

10. Aldawsari, A.; Khan, M.A.; Hameed, B.H.; Yaseen, A.; Alothman, Z.A.; Siddiqui, M.R. Development of activated carbon from *Phoenix dactylifera* fruit pits: Process optimization, characterization, and methylene blue adsorption. *Desal. Water Treat.* **2017**, *62*, 273–281. [CrossRef]

11. Das, R. Introduction. In *Polymeric materials for clean water. Springer Series on Polymer and Composite Materials*; Das, R., Ed.; Springer Publishing: New York, NY, USA, 2019; pp. 1–5.

12. Karimi-Maleh, H.; Shafieizadeh, M.; Taher, M.A.; Rezapour, M.; Orooji, Y. The role of magnetite/graphene oxide nano-composite as a high-efficiency adsorbent for removal of phenazopyridine residues from water samples, an experimental/theoretical investigation. *J. Mol. Liq.* **2020**, *298*, 112040. [CrossRef]

13. Orooji, Y.; Liang, F.; Razmjou, A.; Liu, G.; Jin, W. Preparation of anti-adhesion and bacterial destructive polymeric ultrafiltration membranes using modified mesoporous carbon. *Sep. Purif. Technol.* **2018**, *205*, 273–283. [CrossRef]

14. Orooji, Y.; Ghasali, E.; Emami, N.; Noorisafa, F.; Razmjou., A. ANOVA design for the optimization of TiO$_2$ coating on polyether sulfone membranes. *Molecules* **2019**, *24*, 2924. [CrossRef] [PubMed]

15. Razmjou, A.; Eshaghi, G.; Orooji, Y.; Hosseini, E.; Korayem, A.H.; Mohagheghian, F.; Boroumand, Y.; Noorbakhsh, A.; Asadnia, M.; Chen, V. Lithium ion-selective membrane with 2D subnanometer channels. *Water Res.* **2019**, *159*, 313–323. [CrossRef] [PubMed]

16. Razmjou, A.; Asadnia, M.; Hosseini, E.; Habibnejad Korayem, A.; Chen, V. Design principles of ion selective nanostructured membranes for the extraction of lithium ions. *Nat. Commun.* **2019**, *10*, 5793. [CrossRef] [PubMed]

17. Greluk, M.; Hubicki, Z. Evaluation of polystyrene anion exchange resin for removal of reactive dyes from aqueous solutions. *Chem. Eng. Res. Des.* **2013**, *91*, 1343–1351. [CrossRef]

18. El-Ashtoukhy, E.S.; Fouad, Y. Liquid-liquid extraction of methylene blue dye from aqueous solutions using sodium dodecylbenzenesulfonate as an extractant. *Alex. Eng. J.* **2015**, *54*, 77–81. [CrossRef]

19. Bokhove, J.; Schuur, B.; DeHaan, A. Resin screening for the removal of pyridine-derivatives from waste-water by solvent impregnated resin technology. *React. Funct. Polym.* **2013**, *73*, 595–605. [CrossRef]

20. Helaly, O.S.; El-Ghany, M.S.A.; Moustafa, M.I.; Abuzaid, A.H.; El-monem, N.M.A.; Ismail, I.M. Extraction of cerium (IV) using tributyl phosphate impregnated resin from nitric acid medium. *Trans. Nonferr. Met. Soc. China* **2012**, *22*, 206–214. [CrossRef]

21. Oyekanmi, A.A.; Ahmad, A.; Hossain, K.; Rafatullah, M. Adsorption of Rhodamine B dye from aqueous solution onto acid treated banana peel: Response surface methodology, kinetics and isotherm studies. *PLoS ONE* **2019**, *14*, e0216878. [CrossRef]

22. Kabay, N.; Cortina, J.L.; Trochimczuk, A.; Streat, M. Solvent-impregnated resins (SIRs)-methods of preparation and their applications. *React. Funct. Polym.* **2010**, *70*, 484–496. [CrossRef]

23. Arami, M.; Limaee, N.Y.; Mahmoodi, N.M.; Tabrizi, N.S. Equilibrium and kinetics studies for the adsorption of direct and acid dyes from aqueous solution by soy meal hull. *J. Hazard. Mater.* **2006**, *135*, 171–179. [CrossRef] [PubMed]

24. Fitzpatrick, F. Solvent Impregnated Resins for the Recovery of Gold from Gold (I) Thiourea Solutions. MSc. Thesis, Dublin City University, Dublin, Ireland, 1997; p. 51.

25. Low, L.W.; Teng, T.T.; Alkarkhi, A.F.M.; Ahmad, A.; Morad, N. Optimization of the adsorption conditions for the decolorization and COD reduction of methylene blue aqueous solution using low-cost adsorbent. *Water Air Soil Pollut.* **2011**, *214*, 185–195. [CrossRef]

26. Anouzla, A.; Abrouki, Y.; Souabi, S.; Safi, M.; Rhbal, H. Colour and COD removal of disperse dye solution by a novel coagulant: Application of statistical design for the optimization and regression analysis. *J. Hazard. Mater.* **2009**, *166*, 1302–1306. [CrossRef] [PubMed]

27. Annadurai, G.; Juang, R.-S.; Lee, D.-J. Use of cellulose-based wastes for adsorption of dyes from aqueous solutions. *J. Hazard. Mater.* **2002**, *92*, 263–274. [CrossRef]

28. Namasivayam, C.; Muniaswamy, N.; Gayatri, K.; Rani, M.; Ranganathan, K. Removal of dyes from aqueous solutions by cellulosic waste orange peel. *Bioresour. Technol.* **1996**, *57*, 37–43. [CrossRef]

29. Belkhouche, N.E.; Didi, M.A. Extraction of Bi (III) from nitrate medium by D2EHPA impregnated onto Amberlite XAD-1180. *Hydrometallurgy* **2010**, *103*, 60–67. [CrossRef]

30. Şölener, M.; Tunali, S.; Ozcan, A.S.; Ozcan, A.; Gedikbey, T. Adsorption characteristics of lead (II) ions onto the clay/poly (methoxyethyl) acrylamide (PMEA) composite from aqueous solutions. *Desalination* **2008**, *223*, 308–322. [CrossRef]

31. Oyekanmi, A.A.; Ahmad, A.; Hossain, K.; Rafatullah, M. Statistical optimization for adsorption of Rhodamine B dye from aqueous solutions. *J. Mol. Liq.* **2019**, *281*, 48–58. [CrossRef]

32. Li, H.; Liu, J.; Gao, X.; Liu, C.; Guo, L.; Zhang, S.; Liu, X.; Liu, C. Adsorption behavior of indium (III) on modified solvent impregnated resins (MSIRs) containing sec-octylphenoxy acetic acid. *Hydrometallurgy* **2012**, *121*, 60–67. [CrossRef]

33. Liu, J.; Gao, X.; Liu, C.; Guo, L.; Zhang, S.; Liu, X.; Li, H.; Liu, C.; Jin, R. Adsorption properties and mechanism for Fe (III) with solvent impregnated resins containing HEHEHP. *Hydrometallurgy* **2013**, *137*, 140–147. [CrossRef]

Article

Investigation of Biocidal Effect of Microfiltration Membranes Impregnated with Silver Nanoparticles by Sputtering Technique

Aline M. F. Linhares [1],*, Cristiano P. Borges [2] and Fabiana V. Fonseca [1] 

[1] School of Chemistry, Federal University of Rio de Janeiro, Horacio Macedo Av, 2030, Technology Center, I-124, University City, Rio de Janeiro 21941-909, Brazil; fabiana@eq.ufrj.br

[2] Chemical Engineering Program, COPPE, Federal University of Rio de Janeiro, Horacio Macedo Av, 2030, Technology Center, G-115, University City, Rio de Janeiro 21941-450, Brazil; cristiano@peq.coppe.ufrj.br

* Correspondence: alinemarquesrj@hotmail.com

Received: 18 June 2020; Accepted: 24 July 2020; Published: 29 July 2020

Abstract: Silver nanoparticles were loaded in microfiltration membranes by sputtering technique for the development of biocidal properties and biofouling resistance. This technology allows good adhesion between silver nanoparticles and the membranes, and fast deposition rate. The microfiltration membranes (15 wt.% polyethersulfone and 7.5 wt.% polyvinylpyrrolidone in N,N-dimethylacetamide) were prepared by phase inversion method, and silver nanoparticles were deposited on their surface by the physical technique of vapor deposition in a sputtering chamber. The membranes were characterized by Field Emission Scanning Electron Microscopy, and the presence of silver was investigated by Energy-Dispersive Spectroscopy and X-ray Diffraction. Experiments of silver leaching were carried out through immersion and filtration tests. After 10 months of immersion in water, the membranes still presented ~90% of the initial silver, which confirms the efficiency of the sputtering technique. Moreover, convective experiments indicated that 98.8% of silver remained in the membrane after 24 h of operation. Biocidal analyses (disc diffusion method and biofouling resistance) were performed against *Pseudomonas aeruginosa* and confirmed the antibacterial activity of these membranes with 0.6 and 0.7 log reduction of viable planktonic and sessile cells, respectively. These results indicate the great potential of these new membranes to reduce biofouling effects.

Keywords: silver nanoparticles; microfiltration; membranes; biofouling; sputtering

---

1. Introduction

Microfiltration process (MF) has many consolidated advantages over conventional separation processes, mainly due to its ease of operation and low energy consumption, being widely used for disinfecting water. An evaluation of the global MF membrane market has indicated an expected annual growth rate of 9.0% from 2018 to 2023 [1]. However, membranes' fouling and biofouling are major drawbacks, reducing permeate flux and increasing operational costs. Usually, authors consider that biofouling can be one of the most difficult deposits on membranes to eliminate, highlighting the fast microbes growth, even at low nutrient concentrations [2–4].

To minimize such problems, one approach is to modify the properties of the surface of the membrane. For instance, silver nanoparticles (AgNps) are known for their bactericidal characteristics, which can work against microorganisms growth and, consequently, against biofouling on membranes [2,5–7]. Besides water disinfection, microfiltration membranes with biofouling resistance can be used for several applications such as membrane bioreactors and pretreatment for nanofiltration and reverse osmosis.

The biocidal mechanism of silver nanoparticles is not completely understood. However, the most accepted hypothesis is that silver ions interact with thiol groups in proteins, resulting in inactivation of enzymes and leading to the production of reactive oxygen species (ROS). Another important mechanism is the adhesion of AgNps to the surface of microorganisms, which alters the exchange of nutrients, salts, and water. DNA damage, resulting from AgNps penetrating the bacterial cell, can be highlighted as well [8–14].

On the other hand, the use of AgNps membranes for the disinfection of water should consider any possibility of risks to human health. It is known that levels of silver up to 0.1 mg L^{-1} can be tolerated. However, long-term exposures to silver at high concentrations can generate skin darkening (argyria) [15–17]. Although the existing standards protect consumers from nanoproducts, there are still gaps in the assessment of risks for humans and some aspects need to be optimized, such as limits of toxicity, dose, and concentration to aquatic organisms and humans [14,16,18,19].

Silver nanoparticles may be synthesized by many methods and chemical reduction is the most commonly used. However, these routes present several reaction steps, and their residual solvents depict environmental concerns. Furthermore, AgNps tend to aggregate during their preparation in solution, thus the use of coating agents is necessary, which may reduce their antibacterial activity [2,20].

Physical techniques of vapor deposition, such as sputtering, allow the modeling of the size and distribution of the particles, as well as the high deposition rate and the good adhesion between AgNps and membranes. This technique uses high-energy ions to carry atoms from a target, which acts as a cathode, and deposit them onto a substrate. The last one acts as the anode in a sputtering chamber filled with inert gas. The releasing of Ag plasma ions moves with high kinetic energy and condenses as nanoparticles on the supporting material [21,22].

Previous studies reported concerning the inhibition of microorganism growth after contact between bacteria suspension and polymeric membranes impregnated with nanoparticles have been published. Among them, many reported the biocidal effect of AgNps membranes, but for short periods of time and with a continuous release of AgNps. This issue reduces the bactericidal properties of the membrane over time and can cause overestimations of the overall efficiency of the membrane [7,20,23]. For example, Dong et al. (2017) [7] showed 100% of mortality of *Escherichia coli* and *Bacillus subtilis* suspensions. Furthermore, Dong et al. (2019) [24] observed a significant suppression of *Serratia marcescens* using membranes loaded with AgNps. However, it should be considered that these studies evaluated the silver loss, either through membrane immersion or filtration process, only in short-term periods. On the other hand, Park et al. (2016) [20] verified the strong antibacterial activity against *E. coli*, *Pseudomonas aeruginosa*, and *Staphylococcus aureus*, even though the membrane was expected to last no longer than 97 days, while the estimation of Liu et al. (2015) [23] was of 340 days. Thus, nanoparticles leaching still poses a challenge to overcome, which suggests the development of new techniques to solve this problem [25,26].

In this work, silver nanoparticles were loaded on the surface of polymeric membranes by sputtering technique, aiming at the development of membranes with biocidal properties that would be resistant to biological fouling and capable of being used in the disinfection of water.

Silver leaching was extensively investigated through immersion and convective experiments.

2. Materials and Methods

2.1. Materials

Polyethersulfone (PES, MW 58 kDa) and polyvinylpyrrolidone (PVP, MW 360 kDa) were purchased from Basf, Ludwigshafen am Rhein, Germany and Sigma-Aldrich, St. Louis, MI, USA, respectively. The common solvent for both polymers was *N*,*N*-dimethylacetamide (DMAc, 99.5%), purchased from Tedia, Fairfield, Ohio, EUA. For microbiological experiments, deionized water was supplied by a Milli-Q apparatus (Merck KGaA, Darmstadt, Germany). Yeast extract, meat peptone, and agar were purchased from Kasvi, São José do Pinhais, Brazil. Magnesium sulfate, potassium phosphate monobasic, potassium phosphate dibasic, and glycerol were purchased from Vetec, Duque de Caxias, Brazil. These reagents were used as culture medium. PES and PVP were dried at 60 °C overnight before being used and the other reagents were used as received.

2.2. Microfiltration Membrane

The microfiltration membranes were prepared by phase inversion method [27]. Briefly, a polymer solution with 15 wt.% PES and 7.5 wt.% PVP in DMAc was prepared by continuous stirring at room temperature. After degassing overnight, the solution was cast onto a glass plate with a 200 μm thick casting knife and exposed to the ambient atmosphere (60% RH) for 100 s prior to immersion in a precipitation bath composed of 80 wt.% DMAc and 20 wt.% deionized water. After complete precipitation (~15 min), the resulting membrane was immersed three times in a deionized water bath for 2 h to remove any residual solvent.

2.3. Silver Nanoparticles Deposition

Silver nanoparticles were directly deposited on the surface of the microfiltration membrane (4.7 cm in diameter) by sputtering (Quorum Q150R ES, Quorum Technologies, Laughton, UK) at room temperature and under a low-pressure argon atmosphere (0.1 Pa). A silver target (99.9% Ag, 57 mm in diameter and 0.1 mm thick, Sigma Aldrich, St. Louis, MI, USA) located in the center of the vacuum chamber acts as a cathode and the MF membranes are used as a substrate for deposition. The Ag plasma ions move with high kinetic energy to be condensate as nanoparticles on the surface of the membrane.

The MF-AgNps membranes were obtained with 15 mA (power = 6.0 W) and 50 mA (power = 26.0 W) of sputtering current and 15 and 120 s of deposition time, respectively, as described elsewhere [28]. These membranes are referred to as MF-15mA-15s and MF-50mA-120s, respectively. The chosen conditions aimed to investigate the biocidal properties of the membranes with different content of silver (8.22 and 317.78 mg m^{-2} on MF-15mA-15s and MF-50mA-120s, respectively). For comparison purposes, a membrane without sputtering treatment was used as a control (MF-membrane).

2.4. MF-AgNps Membrane Characterization

The surfaces of MF-AgNps membranes were verified by Field Emission Scanning Electron Microscopy (FESEM, ZEISS Auriga 40, Ulm, Germany) and the presence of silver was evaluated by Energy-Dispersive Spectroscopy (EDS). Samples were coated with carbon using a metallizer (Quorum Emitech K550, Quorum Technologies, Kent, UK) for FESEM and with gold for EDS.

X-ray Diffraction (XRD, Rigaku Miniflex II, Rigaku, The Woodlands, TX, USA) was also used to verify the silver nanoparticles loaded by the sputtering technique and to assess their crystallinity. The data were collected in the 2-theta range of 5° to 90° and scanning speed of 0.05 s^{-1}. The average size was calculated using the Debye-Scherrer formula, presented in Equation (1):

$$D_p = \frac{K\lambda}{\beta_{1/2}\cos\theta} \tag{1}$$

where D_p is the crystallite size, K is a numerical factor referred to as the crystallite-shape factor ($K = 0.9$ is a good approximation), λ is the wavelength of the X-rays ($\lambda = 1.5418$ Å), $\beta_{1/2}$ is the full-width at half-maximum of the X-ray diffraction peak in radians, and θ is the diffraction angle.

Fourier Transform Infrared Spectroscopy (FTIR) analysis was conducted on Agilent Cary 630, Santa Clara, Califórnia, EUA (wavenumber range 500 to 4000 cm^{-1}; 32 scans at a resolution of 4 cm^{-1}).

Silver leaching from MF-AgNps membranes was evaluated through their immersion in water and their performance in convective experiments.

The immersion tests were performed following an established protocol that has been extensively reported in the literature [2,20,23,28,29]. The MF-AgNps membrane coupons (4.7 cm in diameter) were soaked in 50 mL of deionized water at 25 °C and stirred at 60 rpm in a shaker. Water samples were collected after 1, 4, and 24 h of immersion. Long duration tests were also conducted over 1 and 10 months of immersion without intermediate sampling. All samples were acidified to pH 2.0 with HNO_3 (2% v/v), and then analyzed by inductively coupled plasma-atomic emission spectrometry (ICP-AES) to quantify the amount of dissolved Ag.

Silver content in the membranes and silver loss percentage were also quantified by the digestion of other coupons of the MF-AgNps membranes that were immersed into HNO_3 (10% v/v) and sonicated for 3 h to ensure that all the AgNps present in the membrane were leached to solution. The total amount of Ag deposited onto the MF-AgNps membranes was assessed by ICP-AES [5,20].

The convective experiments were performed in a cross-flow membrane system, as illustrated in Figure 1, which had an effective membrane area of 45 cm^2.

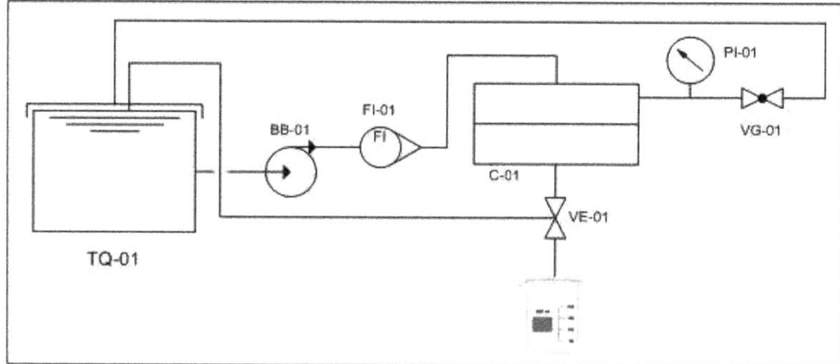

Figure 1. Cross-flow membrane system for silver leaching experiments.

The membrane system was designed to work with or without recirculation of permeate and retentate to the feed tank (TQ-01). The system consisted of a pump (BB-01) that transfers the solution from the feed tank to the permeation cell (C-01) with a flowrate meter (FI-01) between them. The permeate stream could be collected or recirculated by switching the valve (VE-01), and the retentate stream was continuously recirculated. The system's pressure was measured by a control pressure gauge (PI-01) and adjusted through a valve (VG-01).

The silver leaching experiments were conducted with the operational pressure set to 1.5 bar and a flow rate of 40 L h^{-1}.

Additionally, the water flux was calculated using Equation (2) at specific transmembrane pressure, and the permeability was acquired by the slope of the curve J_p × pressure (i.e., 0.5 to 1.5 bar).

$$J_p = \frac{V}{A \times t} \tag{2}$$

where J_p is the water flux (L h^{-1}m^{-2}), V is the permeate volume (L), A is the effective membrane area (m^{-2}), and t is the filtration time (h).

P. aeruginosa suspension (10^8 UFC mL^{-1}) was used as an organism probe to evaluate the rejection capacity (Equation (3)) of each membrane at 1.5 bar.

$$R\ (\%) = 1 - \frac{C_p}{C_f} \times 100 \tag{3}$$

where C_p and C_f is the final viable *P. aeruginosa* concentration of the permeate stream and feed solution, respectively.

2.5. Antibacterial Activity Tests

For the antibacterial activity tests, Gram-negative bacterium *P. aeruginosa* was selected as the model organism. *P. aeruginosa* is considered the paradigm organism for microbial biofilm studies due to its ability to quickly adhere to many different surfaces, its high reproduction rate, and its significance as a pathogen [25,30,31].

P. aeruginosa cells were inoculated into liquid culture medium and incubated with continuous stirring at 200 rpm overnight at 30 °C. This cell suspension served as a bacterial stock solution, which was further diluted to a specific concentration for each test.

The liquid culture medium was prepared with 5.0 g L^{-1} yeast extract, 5.0 g L^{-1} meat peptone, 0.2 g L^{-1} magnesium sulfate, 7.0 g L^{-1} potassium phosphate dibasic, 3.0 g L^{-1} potassium phosphate monobasic, and 30 g L^{-1} glycerol. The solid culture medium was produced from the same solution with the addition of 18 g L^{-1} agar.

2.5.1. The Disc Diffusion Method

The antibacterial activity of the MF-AgNps membranes was first investigated by a disc diffusion method against *P. aeruginosa*. Membrane coupons (17 mm in diameter) were previously sterilized by ultraviolet irradiation for 15 min. Then, the upper surface of the membranes, which holds the silver nanoparticles, was put in contact with the agar plates containing *P. aeruginosa* bacteria at a concentration of 10^6 colony forming units per mL (CFU mL^{-1}).

After incubation at 30 °C for 24, 48, and 72 h, the presence of inhibition zones was monitored and recorded by a digital camera. This inhibition ring, without microbial growth, served as an indicator of antibacterial activity. Furthermore, MF membrane (without AgNps) and an agar plate without membrane were also observed as control samples. All tests were made in triplicate.

2.5.2. The Biofouling Resistance Tests

The biofouling resistance test was performed to evaluate the activity of AgNps in the prevention of bacterial adhesion on the membrane surface. MF-AgNps membrane coupons (1 cm^2) were immersed into 10^7 CFU mL^{-1} *P. aeruginosa* suspensions and incubated for 24 h at 30 °C and 200 rpm stirring. Samples with MF membrane coupons and without membranes were also investigated as controls. All tests were made in triplicate.

After incubation, the planktonic cells in the supernatant and the sessile cells in the biofilm were counted. For total planktonic cells, the optical density at 600 nm (Shimatsu Mini 1240, Mumbai, India) was monitored and the bacterial concentration was determined. For viable planktonic cells, tenfold dilutions were spread onto agar plates and incubated overnight, and viable bacterial colonies were counted on the following day [5,31].

To measure the number of cells attached to the surface of the membrane (sessile cells), the coupon was rinsed with 20 mL of normal saline (0.9 wt.%) to ensure the removal of unattached cells. Then, the membrane coupon was placed in 15 mL of liquid culture medium and vortexed on the highest setting for 120 s in order to cause biofilm disruption. This supernatant was analyzed through optical density at 600 nm for total sessile cells. For the quantification of viable sessile cells, tenfold dilutions were spread onto agar plates, incubated overnight, and the CFU on the plates were counted on the following day.

The Log Reduction and the bacterial viability were calculated using Equations (4) and (5), respectively [20,32]:

$$Log\ Reduction = log_{10}\frac{N_0}{N} \tag{4}$$

$$Bacterial\ Viability\ (\%) = \frac{N}{N_0} \times 100 \tag{5}$$

where N is the number of viable cells in contact with MF-AgNps membranes and N_0 is the number of viable cells in contact with MF membrane (control—without AgNps).

In order to investigate the occurrence of biofouling in the MF membrane and MF-50mA-120s, a filtration experiment was carried out in a cross-flow membrane system, as indicated in Figure 1. The tests were conducted with recirculation of both permeate and retentate streams at 1.5 bar, with a flow rate of feed of 40 L h^{-1}. *P. aeruginosa* suspension of 10^8 CFU mL^{-1} was prepared and placed in the feed tank, and after 4.5 h of permeation, the viable sessile cells were quantified as described before.

3. Results and Discussion

3.1. Membrane Characterization

Figure 2 presents the FESEM photomicrographs of the upper surface and cross section of MF membrane (A and D) and the surfaces of MF-15mA-15s (B) and MF-50mA-120s (C). No significant difference in the surface pores of MF membrane and membranes loaded with silver nanoparticles was observed. The cross section of MF membrane showed a sponge-like morphology with interconnected pores.

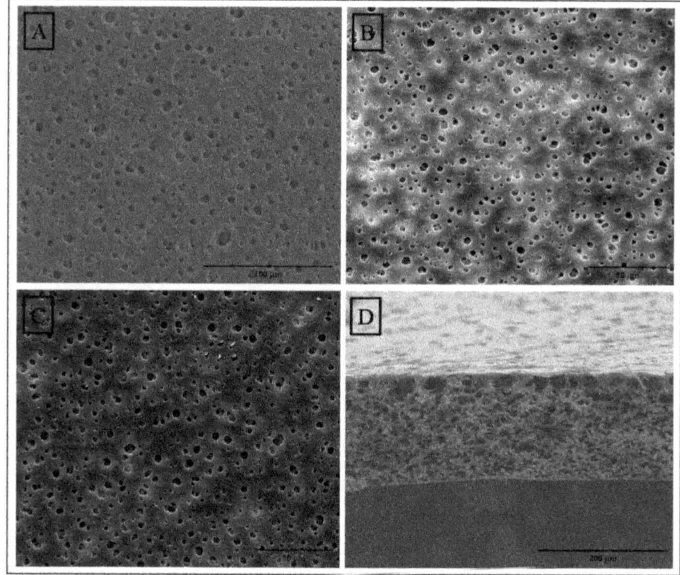

Figure 2. Field emission scanning electron microscopy (FESEM) photomicrographs of the surfaces (1000×) of (**A**) MF-membrane, (**B**) MF-15mA-15s, (**C**) MF-50mA-120s, and of (**D**) cross section of MF membrane (500×).

The EDS spectra of the membranes' surface are portrayed in Figure 3, where the presence of silver element is indicated by black arrows. In addition, Figure 4 shows their EDS mapping and FESEM images. FESEM images exhibit spherical particles on the surfaces of MF-15mA-15s (C) and MF-50mA-120s (D). These spherical particles present average diameters of 88 and 50 nm for MF-15mA-15s and MF-50mA-120s, respectively. The EDS mapping revealed uniform distribution of silver element and, as expected, a larger amount of AgNps in MF-50mA-120s, corroborating what was observed in FESEM images. These results suggest that the increases in sputtering time and sputtering current reduce the diameter of silver particles, which might be related to a higher nucleation rate for nanoparticle growth at higher sputtering current, and confirm that the silver nanoparticles were successfully impregnated on the surface of the MF membranes.

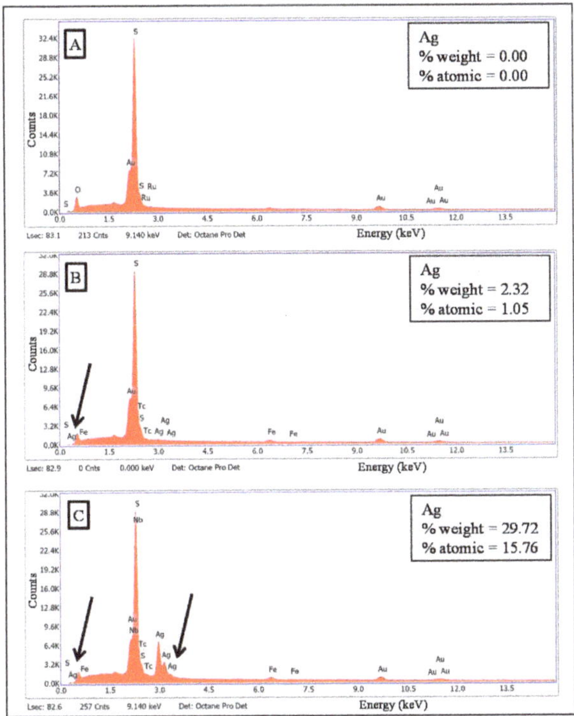

Figure 3. The energy dispersive X-ray (EDS) spectra of membranes (**A**) MF membrane, (**B**) MF-15mA-15s, and (**C**) MF-50mA-120s.

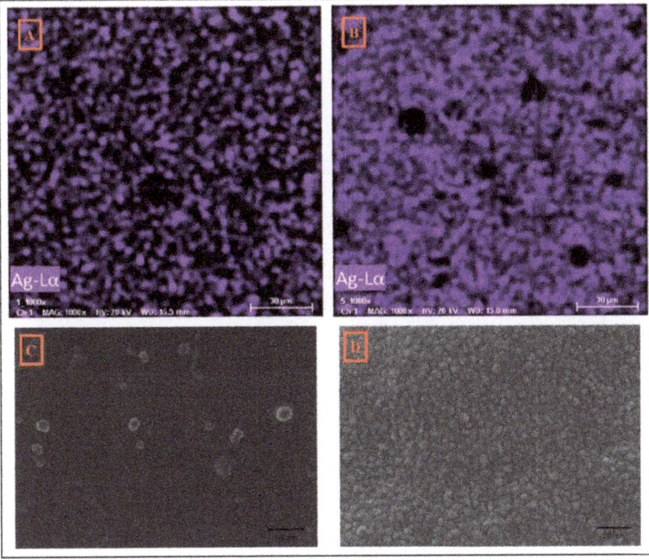

Figure 4. EDS mapping for silver element from (**A**) MF-15mA-15s and (**B**) MF-50mA-120s, and FESEM photomicrographs of the surface of the membranes (100,000×) of (**C**) MF-15mA-15s and (**D**) MF-50mA-120s.

X-ray diffraction patterns are shown in Figure 5 from 5° to 90°. The MF membrane exhibited a broad peak that corresponds to the amorphous structure of PES. For MF-50mA-120s membrane, a sharp peak is observed at 2 θ = 38°, which is attributed to the crystallinity of silver (black arrow) and represents (111) Bragg's reflections of face-centered cubic (fcc) structure [33]. These observations are in agreement with values of silver nanoparticles reported in the literature [2,6,33–35]. For the MF-15mA-15s membrane, there was no observed peak related to crystalline domains, which may be attributed to the low amount of silver deposited in these sputtering conditions.

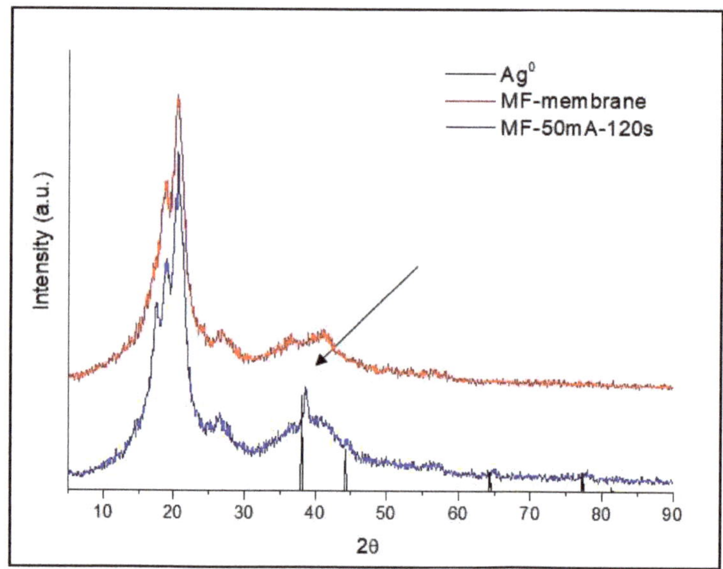

Figure 5. XRD patterns of Ag^0, MF membrane, and MF-50mA-120s.

The broadening of peaks in the X-ray diffraction pattern can be related to the particle size [33]; thus, the average size of 17.7 nm was estimated using Scherrer's equation for the silver nanoparticles in MF-50mA-120s.

The synthesis of nanoparticles by sputtering deposition techniques and their mechanisms of nucleation and growth were investigated by other studies for several metals such as silver [36,37], copper [38], cobalt [39], niobium [40], and palladium [41]. However, the detailed description of these mechanisms is very complex [39,42].

It has been also described that, after long periods (e.g., more than 6 min) of plasma treatment, the surface of PES membranes can be damaged, for instance, the molecular bonds C-C and C-H can be cleaved by argon plasma. On the other hand, the application of short periods of time seams not affect the polymer chains of the membranes [43,44]. Therefore, in this current work, the FTIR spectra (Figure S1 in Supplementary Materials) of the membranes, before and after the impregnation of the nanoparticles, revealed no significant effects of the sputtering technique. The aromatic C-H stretches at 3094 and 3062 cm^{-1} (Figure S1B), and aromatic C=C stretches at 1574 and 1481 cm^{-1} (Figure S1C) remained the same for both membranes.

The water permeability of MF membrane and membranes impregnated with silver nanoparticles are presented in Table 1. The results indicate that there was no significant difference in water permeability between MF membrane and modified membranes (MF-15mA-15s and MF-50mA-120s). This result is in agreement with reported works for different membranes characteristics and AgNps synthesis techniques [7,23,24,45].

Table 1. Water permeability of MF membrane and membranes loaded with AgNps.

Membrane	Water Permeability (L h^{-1} m^{-2} bar^{-1})	Rejection (%)
MF-membrane	6349.9 ± 475.1	26.4
MF-15mA-15s	6455.0 ± 519.9	-
MF-50mA-120s	6388.0 ± 564.8	78.3

These membranes present large pore sizes, and they are fabricated with hydrophilic polymers (PES and PVP). Such polymers can input high water flux to the membrane, especially on phase inversion technique, where part of the additive, i.e., PVP, can be entrapped in the membrane matrix, as reported in the literature [46–48].

The PES membrane contains sulfone and ether groups alternated between aromatic rings [44,49]. The FTIR spectra show characteristic bands of PES: (1) 3094 and 3062 cm^{-1} due to aromatic C–H stretch (Figure S1B) [46,49,50]; (2) 1574 and 1481 cm^{-1} due to aromatic C=C asymmetric stretch (Figure S1C) [44,50]; and (3) 1320 and 1296 cm^{-1} resulting from the anti-symmetric O=S=O stretch of the sulfone group (Figure S1D) [44,49,50]. Moreover, the FTIR spectra shows the presence of PVP with a characteristic band at 1650 cm^{-1} due to carbonyl group (Figure S1E) [46,47,50].

3.2. AgNps Releasing Test

In order to investigate the stability of silver nanoparticles loaded on MF-AgNps membranes, the concentration of silver leaching was also determined by immersion and filtration experiments (Figure 6). The percentage of the remaining Ag on the modified membranes was calculated based on the total amount of AgNps initially deposited on the surface of the membranes, corresponding to 8.22 and 317.78 mg m^{-2} on MF-15mA-15s and MF-50mA-120s, respectively.

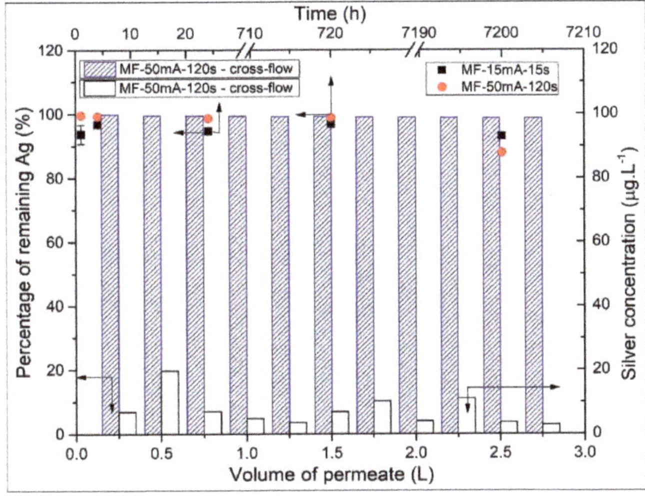

Figure 6. Percentage of the remaining Ag on MF membranes (**left axis**) over time (**upper axis**) in immersion test (**black square**) Ag loaded with 15 mA and 15 s of sputtering; (**red circle**) Ag loaded with 50 mA and 120 s of sputtering). Conditions: diameter of membrane = 4.7 cm, volume of deionized water = 50 mL, 60 rpm. Percentage of remaining Ag (**blue-dashed column, left axis**) and silver concentration leaching (**white column, right axis**) per volume of permeate (**lower axis**) in cross-flow experiment. Conditions: pressure = 1.5 bar, flow rate = 40 L h^{-1}, membrane area = 45 cm^2.

During the first hour of immersion, the MF-15mA-15s membrane lost 6.32 ± 2.98% silver content; however, considering the total period of 10 months (7200 h), a loss of 7.02% was observed, indicating that the biggest part of silver release occurs at the beginning of immersion in the water bath. This fast decrease in the release of silver is qualitatively similar to other studies [7,20,24,51] and may be explained by probable unattached AgNps on the surface of the membrane.

Besides the percentage of silver releasing from the MF-15mA-15s membrane, its concentration in water after 10 months was 20.0 μg L^{-1}, which is lower than the silver maximum contaminants limit of 100 μg L^{-1} described by the World Health Organization Guideline for Drinking Water [17] and the U.S. Environmental Protection Agency (USEPA).

Not only did the MF-15mA-15s membrane present a feasible characteristic on the entrapment of silver nanoparticles, but also the MF-50mA-120s showed 1.34 ± 0.13% of silver loss after 24 h of immersion. Furthermore, after 1 month of immersion, the silver concentration in water was similar to the one after 24 h of immersion (0.1 mg L^{-1}), which indicates the same trend of silver releasing.

After 10 months of immersion, there was still approximately 93.0 and 87.9% of impregnated silver on MF-15mA-15s (Figure 6, black square) and MF-50mA-120s (Figure 6, red circle), respectively.

Figure 6 also shows the results of silver leaching from MF-50mA-120s membrane during the cross-flow experiment. After 2.75 L of water permeation, there was still 98.5% of silver impregnated on the membrane surface (Figure 6, blue-dashed column), and the concentration of silver leached for each 0.25 L of permeate was lower than 100 μg L^{-1} (Figure 6, white column).

In addition, a test with recirculation of both permeate and retentate streams (full-recycle setup) was also performed during 24 h. The concentration of silver leached (in the feed tank) was 8.4 μg L^{-1} at the end of the experiment, which indicates that 98.8% of the initial silver impregnated on MF-50mA-120s remained on its surface after permeation. The water permeation test corroborates the silver loss after 24 h of immersion in the water bath and indicates that, even at a flowrate of 40 L h^{-1}, the MF-50mA-120s membrane showed a small loss of silver. The results of silver leaching in both experiments indicate that the sputtering technique is effective for impregnating and entrapping silver nanoparticles on the surface of membranes. This finding is important for the maintenance of the membrane's biocidal performance and for the minimization of silver leaching to the environment.

Evaluation of silver loss in previous studies of membranes impregnated with silver nanoparticles by chemical reduction method showed percentages of remaining silver of 99.38% and 98.75% after 24 h and six days of immersion, respectively [7]. Furthermore, a percentage of 97.0% was obtained after a cross-flow experiment [24]. However, in these studies, the immersion test was conducted for short periods and under reduced flowrate (cross-flow experiment) in comparison with this work. In addition, the chemical reduction method presents some disadvantages as several steps production and chemical reagents, the use of stabilizer agents and the residual solvents.

Another important fact is a continuous silver loss observed in other studies with physical methods and green synthesis [20,23]. In these cases, the authors indicated that the lifespans of their membranes were 97 and 340 days according to their silver leaching rate.

In fact, this current work highlights the evaluation of silver loss for long-term immersion and cross-flow experiments for membranes loaded with silver nanoparticles by a one-step production method with no residual reagents.

3.3. Antibacterial Activity Tests

3.3.1. The Disc Diffusion Method

As illustrated by the disk tests (see Figure 7), MF-membrane (A) had no significant effect on the growth of *P. aeruginosa*, while MF-AgNps membranes showed a clear area with no evidence of bacterial growth. An inhibition zone around the membranes of 0.5 and 0.8 mm was observed for MF-15mA-15s (B) and MF-50mA-120s (C), respectively. These results are similar to many reported works, which also found the inhibition zone around the substrate containing AgNps, and demonstrate that this antibacterial activity comes mainly from the silver and not from PES [23,51,52].

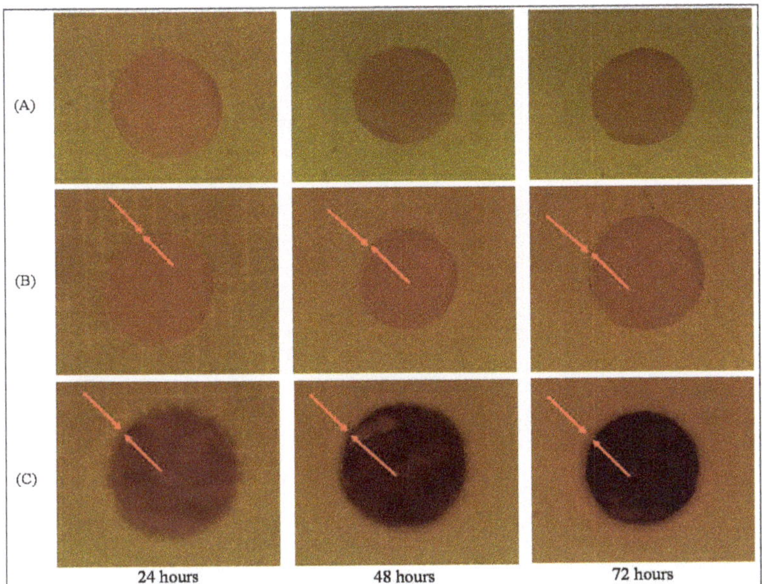

Figure 7. Diffusion disc method against *Pseudomonas aeruginosa* of (**A**) MF membrane (control sample), (**B**) MF-15mA-15s, and (**C**) MF-50mA-120s. Inhibition zone are indicated by red arrows.

In general, the inhibition zones reported by different authors are larger than the ones observed in this work, which can be explained to by the hypothesis that the release of silver nanoparticles is amplifying the biocidal zone. Once the nanoparticles are not well attached to the membranes, they can diffuse in the media and inhibit bacterial growth. However, in this current work, the silver release is reduced to a small region, corroborating this hypothesis. Thus, the sputtering technique was efficient at entrapping silver nanoparticles on the membrane, and their biocidal effect will be concentrated in this region, which may be noteworthy to inhibit biofouling formation.

3.3.2. The Biofouling Resistance Test

The quantification of total planktonic cells and total sessile cells, shown in Figure 8, did not indicate any significant difference between the membranes, with the CFU mL^{-1} being in the same order of magnitude for all membranes.

The quantification of viable planktonic cells grown in suspensions without membrane (control sample) and with MF membrane (without AgNps) did not show a significant difference (Figure 9). On the other hand, for MF-15mA-15s, there was a considerable reduction of ~0.6 log units in CFU mL^{-1} compared to MF-membrane. In this case, after the exposure to MF-AgNps membranes for 24 h, the viability of the planktonic cells was only 24.2%. This decrease demonstrated that the MF-15mA-15s membrane causes an expressive *P. aeruginosa* growth inhibition.

Figure 9 also depicts the viable sessile cells on the different membranes. A comparison between MF membrane and MF-15mA-15s indicates that the bacterial growth showed a 0.7 log unit reduction in CFU mL^{-1}, while the viability of the sessile cells was 19.8% after 24 h of exposure to AgNps. This result confirmed that the silver impregnated in MF-15mA-15s improved the bactericidal properties of the membranes, which is in agreement with other studies that investigated silver nanoparticles synthesized by chemical route [25,51].

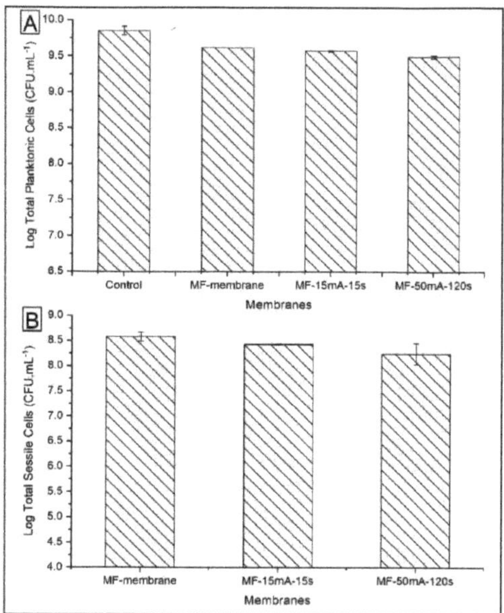

Figure 8. Total cells of *P. aeruginosa* after their exposure to MF membrane and MF-AgNps: (**A**) total planktonic cells and (**B**) total sessile cells. Number of bacteria initially inoculated on each sample: log CFU mL^{-1} = 7.0 at 30 °C, 200 rpm for 24 h of incubation.

The adhesion of *P. aeruginosa* in the MF membrane and MF-50mA-120s was investigated during a cross-flow experiment in order to evaluate their performance in a filtration process. The results revealed 0.66 ± 0.02 log unit reduction of viable sessile cells, and as a consequence, 22% of bacterial viability for MF-50mA-120s (log MF-membrane = 6.63, log MF-50mA-120s = 5.97). These results show an outstanding maintenance of the effectiveness of the MF-50mA-120s in comparison to the MF membrane, even with a continuous flowrate through the membrane.

Liu et al. (2013) [53] observed the deposition of *E. coli* in polysulfone membranes impregnated with silver nanoparticles synthesized by chemical route. The authors concluded that the AgNps did not affect the kinetics of bacterial deposition. However, the bacterial detachment ratio during rising is large in the presence of silver nanoparticles because the bacteria become inactivated after the contact with these nanoparticles, enhancing the detachment rate.

Thereby, in this study, the adhesion of *P. aeruginosa* in MF-AgNps was verified and quantified as the total sessile cells. Nevertheless, the viable sessile cells decreased in comparison with the MF membrane, indicating that AgNps inactivated the microorganisms, which would facilitate the detachment.

Even though the exact mechanism of antibacterial activity of AgNps is not fully understood and further studies are needed to explain this gap, several researchers agree on a synergistic action between contact killing of nanoparticles and the releasing of silver ions from the membranes [15,54,55].

In this context, the antibacterial effect of MF-AgNps membranes produced in this work is affected by its lower concentration of silver leached; however, for applications in water treatment, for example, it is important to minimize this leaching due to the concern with human health and environment protection. Furthermore, the development of membranes with biocidal advantages associated with a long lifespan is needed.

Therefore, the MF-AgNps membrane demonstrated potential for these applications because of its antibacterial properties and its expected longer lifespan due to its low silver leaching.

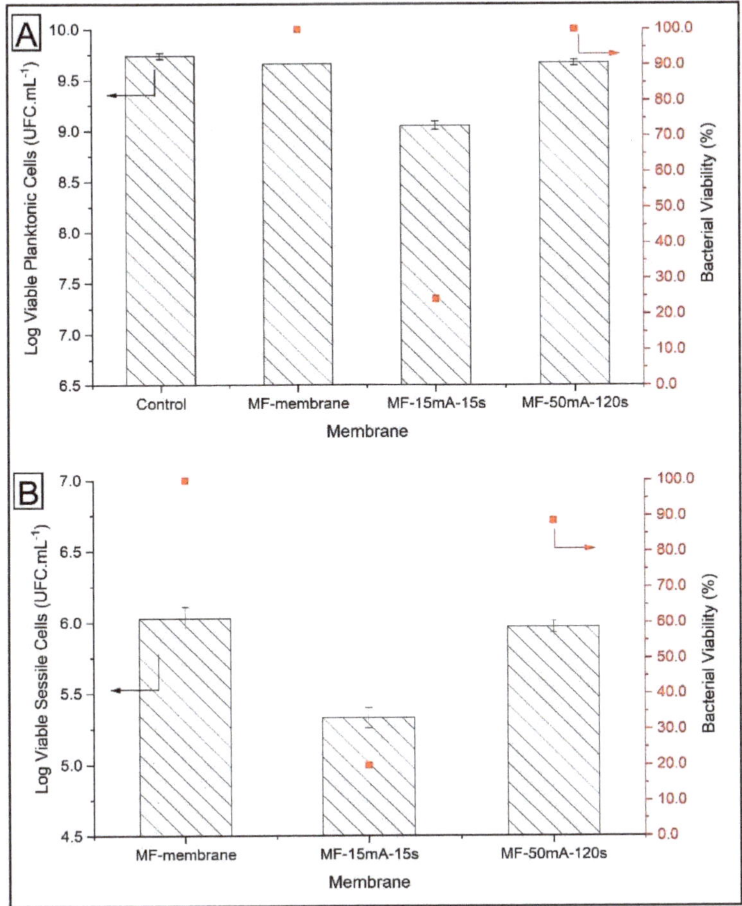

Figure 9. *P. aeruginosa* cells viabilities after their exposure to MF-membrane and MF-AgNps: (**A**) viable planktonic cells and (**B**) viable sessile cells. Number of bacteria initially inoculated on each sample: log CFU mL^{-1} = 7.0 at 30 °C, 200 rpm for 24 h of incubation.

4. Conclusions

Impregnation of silver nanoparticles on the microfiltration membranes by sputtering technique was confirmed by different and complementary analyses. From FESEM images, average diameters of 88 and 50 nm were estimated for MF-15mA-15s and MF-50mA-120s, respectively, suggesting that the increase in sputtering time and sputtering current reduces the diameter of silver particles. The diffractogram of the MF-50mA-120s membrane exhibited a sharp peak at 2 θ = 38°, which is attributed to the crystallinity of silver, proving the presence of silver on the surface of this membrane and corroborating the FESEM and EDS observations.

One of the main advantages of the sputtering technique is the good adhesion between AgNps and membranes, which allows the release of silver ions lower than the maximum limit of silver in drinking water by World Health Organization and the U.S. Environmental Protection Agency. After 10 months of immersion, there was still approximately 93.0 and 87.9% of silver initially impregnated onto MF-15mA-15s and MF-50mA-120s membranes, respectively. Furthermore, after 24 h of filtration test in full-recycle set-up with high flowrate, MF-50mA-120s showed that 98.8% of silver remained on the membrane. These results indicate the efficiency of sputtering technique to entrapped silver nanoparticles on the membranes.

The disc diffusion method demonstrated the antibacterial activity of silver nanoparticles impregnated onto membranes by an inhibition zone of approximately 0.5 and 0.8 mm for MF-15mA-15s and MF-50mA-120s, respectively. The microbial inhibition occurs with the diffusion of silver nanoparticles from the membrane into the agar layer. Thereby, these small inhibition zones can be explained by the lower release of silver nanoparticles impregnated on the microfiltration membranes by sputtering method.

The biofouling resistance test for MF-15mA-15s showed 0.6 and 0.7 log unit reductions in CFU mL^{-1} when compared with the MF membrane (without AgNps) for viable planktonic cells and viable sessile cells, respectively. After exposure to this membrane for 24 h, the planktonic cells and sessile cells viabilities were both under 25%, confirming the biocidal property of the membrane with silver nanoparticles.

Although there was adhesion of *P. aeruginosa* in microfiltration membranes impregnated with AgNps, the viable sessile cells decreased, indicating a great potential of silver nanoparticles to reduce biofouling and its consequences.

The highlight of this work is the occurrence of bacterial inhibition in membranes impregnated with AgNps with reduced silver leaching, which allows a longer lifespan for these membranes. Less silver loss was observed in this work, even though it was analyzed in long-term immersions and in filtration experiments with higher flowrate when compared with other studies in the literature.

Supplementary Materials: The following are available online at http://www.mdpi.com/2073-4360/12/8/1686/s1, Figure S1: FTIR spectra of MF-membrane, MF-15mA-15s and MF-50mA-120s.

Author Contributions: Conceptualization, C.P.B. and F.V.F.; Methodology, A.M.F.L.; Validation, A.M.F.L.; Formal analysis, A.M.F.L.; Investigation, A.M.F.L.; Resources, C.P.B. and F.V.F.; Data curation, A.M.F.L.; Writing—original draft preparation, A.M.F.L.; writing—review and editing, C.P.B. and F.V.F.; Visualization, C.P.B. and F.V.F.; Supervision, C.P.B. and F.V.F.; Project administration, F.V.F.; Funding acquisition, C.P.B. and F.V.F. All authors have read and agreed to the published version of the manuscript.

Funding: This study was funded by Conselho Nacional de Desenvolvimento Científico e Tecnológico—Brasil (CNPq).

Conflicts of Interest: The authors declare no conflicts of interest. The funders had no role in the design of the study; in the collection, analyses, or interpretation of data; in the writing of the manuscript; or in the decision to publish the results.

References

1. BCC Research. The Global Market for Membrane Microfiltration. Available online: https://www.bccresearch. com/market-research/membrane-and-separation-technology/membrane-microfiltration.html (accessed on 20 April 2020).
2. Zhang, M.; Zhang, K.; De Gusseme, B.; Verstraete, W. Biogenic silver nanoparticles (bio-Ag 0) decrease biofouling of bio-Ag 0/PES nanocomposite membranes. *Water Res.* **2012**, *46*, 2077–2087. [PubMed]
3. Kochkodan, V.; Hilal, N. A comprehensive review on surface modified polymer membranes for biofouling mitigation. *Desalination* **2015**, *356*, 187–207. [CrossRef]
4. Jiang, S.; Li, Y.; Ladewig, B.P. A review of reverse osmosis membrane fouling and control strategies. *Sci. Total Environ.* **2017**, *595*, 567–583. [CrossRef] [PubMed]
5. Zodrow, K.; Brunet, L.; Mahendra, S.; Li, D.; Zhang, A.; Li, Q.; Alvarez, P.J.J. Polysulfone ultrafiltration membranes impregnated with silver nanoparticles show improved biofouling resistance and virus removal. *Water Res.* **2009**, *43*, 715–723. [CrossRef] [PubMed]
6. Basri, H.; Ismail, A.F.; Aziz, M.; Nagai, K.; Matsuura, T.; Abdullah, M.S.; Ng, B.C. Silver-filled polyethersulfone membranes for antibacterial applications—Effect of PVP and TAP addition on silver dispersion. *Desalination* **2010**, *261*, 264–271. [CrossRef]
7. Dong, C.; Wang, Z.; Wu, J.; Wang, Y.; Wang, J.; Wang, S. A green strategy to immobilize silver nanoparticles onto reverse osmosis membrane for enhanced anti-biofouling property. *Desalination* **2017**, *401*, 32–41. [CrossRef]

8. Feng, Q.L.; Wu, J.; Chen, G.Q.; Cui, F.Z.; Kim, T.N.; Kim, J.O. A mechanistic study of the antibacterial effect of silver ions on Escherichia coli and Staphylococcus aureus. *J. Biomed. Mater. Res.* **2000**, *52*, 662–668. [CrossRef]

9. Matsumura, Y.; Yoshikata, K.; Kunisaki, S.; Tsuchido, T. Mode of bactericidal action of silver zeolite and its comparison with that of silver nitrate. *Appl. Environ. Microbiol.* **2003**, *69*, 4278–4281. [CrossRef]

10. Sondi, I.; Salopek-Sondi, B. Silver nanoparticles as antimicrobial agent: A case study on E. coli as a model for Gram-negative bacteria. *J. Colloid Interface Sci.* **2004**, *275*, 177–182. [CrossRef]

11. Morones, J.R.; Elechiguerra, J.L.; Camacho, A.; Holt, K.; Kouri, J.B.; Ramírez, J.T.; Yacaman, M.J. The bactericidal effect of silver nanoparticles. *Nanotechnology* **2005**, *16*, 2346–2353. [CrossRef]

12. Su, H.L.; Chou, C.C.; Hung, D.J.; Lin, S.H.; Pao, I.C.; Lin, J.H.; Huang, F.L.; Dong, R.X.; Lin, J.J. The disruption of bacterial membrane integrity through ROS generation induced by nanohybrids of silver and clay. *Biomaterials* **2009**, *30*, 5979–5987. [CrossRef] [PubMed]

13. Dastjerdi, R.; Montazer, M. A review on the application of inorganic nano-structured materials in the modification of textiles: Focus on anti-microbial properties. *Colloids Surf. Biointerfaces* **2010**, *79*, 5–18. [CrossRef] [PubMed]

14. Ferdous, Z.; Nemmar, A. Health impact of silver nanoparticles: A review of the biodistribution and toxicity following various routes of exposure. *Int. J. Mol. Sci.* **2020**, *21*, 2375. [CrossRef] [PubMed]

15. Li, Q.; Mahendra, S.; Lyon, D.Y.; Brunet, L.; Liga, M.V.; Li, D.; Alvarez, P.J.J. Antimicrobial nanomaterials for water disinfection and microbial control: Potential applications and implications. *Water Res.* **2008**, *42*, 4591–4602. [CrossRef] [PubMed]

16. Nowack, B.; Krug, H.F.; Height, M. 120 years of nanosilver history: Implications for policy makers. *Environ. Sci. Technol.* **2011**, *45*, 1177–1183. [CrossRef]

17. WHO. *Guidelines for Drinking-Water Quality*, 4th ed.; World Heathy Organization: Geneve, Switzerland, 2011; ISBN 978-92-4-154815-1.

18. Siddiqi, K.S.; Husen, A.; Rao, R.A.K. A review on biosynthesis of silver nanoparticles and their biocidal properties. *J. Nanobiotechnol.* **2018**, *16*, 14. [CrossRef]

19. Deshmukh, S.P.; Patil, S.M.; Mullani, S.B.; Delekar, S.D. Silver nanoparticles as an effective disinfectant: A review. *Mater. Sci. Eng.* **2019**, *97*, 954–965. [CrossRef]

20. Park, S.H.; Kim, S.H.; Park, S.J.; Ryoo, S.; Woo, K.; Lee, J.S.; Kim, T.S.; Park, H.D.; Park, H.; Park, Y.I.; et al. Direct incorporation of silver nanoparticles onto thin-film composite membranes via arc plasma deposition for enhanced antibacterial and permeation performance. *J. Memb. Sci.* **2016**, *513*, 226–235. [CrossRef]

21. Cao, G.; Wang, Y. *Nanostructures and Nanomaterials: Synthesis, Properties and Applications*, 2nd ed.; World Scientific: London, UK, 2011; ISBN 978-981-4322-50-8.

22. Wasa, K.; Kanno, I.; Kotera, H. *Handbook of Sputter Deposition Technology: Fundamentals and Applications for Functional Thin Films, Nano-Materials and MEMS*, 2nd ed.; Elsevier: Oxford, UK, 2012; ISBN 978-143-7734-836.

23. Liu, S.; Fang, F.; Wu, J.; Zhang, K. The anti-biofouling properties of thin-film composite nanofiltration membranes grafted with biogenic silver nanoparticles. *Desalination* **2015**, *375*, 121–128. [CrossRef]

24. Dong, X.; Shannon, H.D.; Amirsoleimani, A.; Brion, G.M.; Escobar, I.C. Thiol-affinity immobilization of casein-coated silver nanoparticles on polymeric membranes for biofouling control. *Polymers* **2019**, *11*, 2057. [CrossRef]

25. Alpatova, A.; Kim, E.S.; Sun, X.; Hwang, G.; Liu, Y.; Gamal El-Din, M. Fabrication of porous polymeric nanocomposite membranes with enhanced anti-fouling properties: Effect of casting composition. *J. Memb. Sci.* **2013**, *444*, 449–460. [CrossRef]

26. Li, X.; Sotto, A.; Li, J.; Van der Bruggen, B. Progress and perspectives for synthesis of sustainable antifouling composite membranes containing in situ generated nanoparticles. *J. Memb. Sci.* **2017**, *524*, 502–528. [CrossRef]

27. Ferreira, A.M.; Roque, É.B.; da Fonseca, F.V.; Borges, C.P. High flux microfiltration membranes with silver nanoparticles for water disinfection. *Desalin. Water Treat.* **2015**, *56*, 3590–3598. [CrossRef]

28. Kuzminova, A.; Beranová, J.; Polonskyi, O.; Shelemin, A.; Kylián, O.; Choukourov, A.; Slavínská, D.; Biederman, H. Antibacterial nanocomposite coatings produced by means of gas aggregation source of silver nanoparticles. *Surf. Coat. Technol.* **2016**, *294*, 225–230. [CrossRef]

29. Haider, M.S.; Shao, G.N.; Imran, S.M.; Park, S.S.; Abbas, N.; Tahir, M.S.; Hussain, M.; Bae, W.; Kim, H.T. Aminated polyethersulfone-silver nanoparticles (AgNPs-APES) composite membranes with controlled silver ion release for antibacterial and water treatment applications. *Mater. Sci. Eng.* **2016**, *62*, 732–745. [CrossRef]

30. Pasmore, M.; Todd, P.; Smith, S.; Baker, D.; Silverstein, J.A.; Coons, D.; Bowman, C.N. Effects of ultrafiltration membrane surface properties on Pseudomonas aeruginosa biofilm initiation for the purpose of reducing biofouling. *J. Memb. Sci.* **2001**, *194*, 15–32. [CrossRef]

31. De Prijck, K.; Nelis, H.; Coenye, T. Efficacy of silver-releasing rubber for the prevention of Pseudomonas aeruginosa biofilm formation in water. *Biofouling* **2007**, *23*, 405–411. [CrossRef]

32. Vaughan, J.; Benson, R.; Vaughan, K. Assessing the effectiveness of antimicrobial wound dressings in vitro. In *Advanced Wound Repair Therapies*; Woodhead Publishing Limited: Cambridge, UK, 2011; pp. 227–246, ISBN 978-184-5697-006.

33. Khanna, P.K.; Singh, N.; Charan, S.; Subbarao, V.V.V.S.; Gokhale, R.; Mulik, U.P. Synthesis and characterization of Ag/PVA nanocomposite by chemical reduction method. *Mater. Chem. Phys.* **2005**, *93*, 117–121. [CrossRef]

34. RRUFF Database. Available online: http://rruff.info/silver/display=default/R070416 (accessed on 4 April 2018).

35. Toroghi, M.; Raisi, A.; Aroujalian, A. Preparation and characterization of polyethersulfone/silver nanocomposite ultrafiltration membrane for antibacterial applications. *Polym. Adv. Technol.* **2014**, *25*, 711–722. [CrossRef]

36. Gracia-Pinilla, M.Á.; Ferrer, D.; Mejía-Rosales, S.; Pérez-Tijerina, E. Size-selected Ag nanoparticles with five-fold symmetry. *Nanoscale Res. Lett.* **2009**, *4*, 896–902. [CrossRef]

37. Asanithi, P.; Chaiyakun, S.; Limsuwan, P. Growth of silver nanoparticles by DC magnetron sputtering. *J. Nanomater.* **2012**, *2012*, 1–8. [CrossRef]

38. Dutka, M.V.; Turkin, A.A.; Vainchtein, D.I.; De Hosson, J.T.M. On the formation of copper nanoparticles in nanocluster aggregation source. *J. Vac. Sci. Technol.* **2015**, *33*, 031509. [CrossRef]

39. Morel, R.; Brenac, A.; Bayle-Guillemaud, P.; Portement, C.; La Rizza, F. Growth and properties of cobalt clusters made by sputtering gas-aggregation. *Eur. Phys. J.* **2003**, *24*, 287–290. [CrossRef]

40. Bray, K.R.; Jiao, C.Q.; Decerbo, J.N. Nucleation and growth of Nb nanoclusters during plasma gas condensation. *J. Appl. Phys.* **2013**, *113*, 234307. [CrossRef]

41. Ayesh, A.I.; Qamhieh, N.; Ghamlouche, H.; Thaker, S.; El-Shaer, M. Fabrication of size-selected Pd nanoclusters using a magnetron plasma sputtering source. *J. Appl. Phys.* **2010**, *107*, 034317. [CrossRef]

42. Hirsch, U.M.; Teuscher, N.; Rühl, M.; Heilmann, A. Plasma-enhanced magnetron sputtering of silver nanoparticles on reverse osmosis membranes for improved antifouling properties. *Surf. Interfaces* **2019**, *16*, 1–7. [CrossRef]

43. Cruz, M.C.; Ruano, G.; Wolf, M.; Hecker, D.; Castro Vidaurre, E.; Schmittgens, R.; Rajal, V.B. Plasma deposition of silver nanoparticles on ultrafiltration membranes: Antibacterial and anti-biofouling properties. *Chem. Eng. Res. Des.* **2015**, *94*, 524–537. [CrossRef]

44. Saxena, N.; Prabhavathy, C.; De, S.; DasGupta, S. Flux enhancement by argon-oxygen plasma treatment of polyethersulfone membranes. *Sep. Purif. Technol.* **2009**, *70*, 160–165. [CrossRef]

45. Sprick, C.; Chede, S.; Oyanedel-Craver, V.; Escobar, I.C. Bio-inspired immobilization of casein-coated silver nanoparticles on cellulose acetate membranes for biofouling control. *J. Environ. Chem. Eng.* **2018**, *6*, 2480–2491. [CrossRef]

46. Belfer, S.; Fainchtain, R.; Purinson, Y.; Kedem, O. Surface characterization by FTIR-ATR spectroscopy of polyethersulfone membranes-unmodified, modified and protein fouled. *J. Memb. Sci.* **2000**, *172*, 113–124. [CrossRef]

47. Vatsha, B.; Ngila, J.C.; Moutloali, R.M. Preparation of antifouling polyvinylpyrrolidone (PVP 40K) modified polyethersulfone (PES) ultrafiltration (UF) membrane for water purification. *Phys. Chem. Earth* **2014**, *67–69*, 125–131. [CrossRef]

48. Miyano, T.; Matsuura, T.; Carlsson, D.J.; Sourirajan, S. Retention of polyvinylpyrrolidone swelling agent in the poly(ether p-phenylenesulfone) ultrafiltration membrane. *J. Appl. Polym. Sci.* **1990**, *41*, 407–417. [CrossRef]

49. Moarefian, A.; Golestani, H.A.; Bahmanpour, H. Removal of amoxicillin from wastewater by self-made polyethersulfone membrane using nanofiltration. *J. Environ. Health Sci. Eng.* **2014**, *12*, 1–10. [CrossRef] [PubMed]

50. Koloti, L.E.; Gule, N.P.; Arotiba, O.A.; Malinga, S.P. Laccase-immobilized dendritic nanofibrous membranes as a novel approach towards the removal of bisphenol A. *Environ. Technol.* **2018**, *39*, 392–404. [CrossRef] [PubMed]

51. Cao, X.; Tang, M.; Liu, F.; Nie, Y.; Zhao, C. Immobilization of silver nanoparticles onto sulfonated polyethersulfone membranes as antibacterial materials. *Colloids Surf. Biointerfaces* **2010**, *81*, 555–562. [CrossRef]

52. Yin, J.; Yang, Y.; Hu, Z.; Deng, B. Attachment of silver nanoparticles (AgNPs) onto thin-film composite (TFC) membranes through covalent bonding to reduce membrane biofouling. *J. Memb. Sci.* **2013**, *441*, 73–82. [CrossRef]
53. Liu, Y.; Rosenfield, E.; Hu, M.; Mi, B. Direct observation of bacterial deposition on and detachment from nanocomposite membranes embedded with silver nanoparticles. *Water Res.* **2013**, *47*, 2949–2958. [CrossRef]
54. Agnihotri, S.; Mukherji, S.; Mukherji, S. Size-controlled silver nanoparticles synthesized over the range 5–100 nm using the same protocol and their antibacterial efficacy. *RSC Adv.* **2014**, *4*, 3974–3983. [CrossRef]
55. Cobos, M.; De-La-Pinta, I.; Guillermo, Q.; Fernández, M.J.; Fernández, M.D. Synthesis, physical, mechanical and antibacterial properties of nanocomposites based on poly(vinyl alcohol)/graphene oxide–silver nanoparticles. *Polymers* **2020**, *12*, 723. [CrossRef]

Article

A Biodegradable Magnetic Nanocomposite as a Superabsorbent for the Simultaneous Removal of Selected Fluoroquinolones from Environmental Water Matrices: Isotherm, Kinetics, Thermodynamic Studies and Cost Analysis

Geaneth Pertunia Mashile [1,2], Kgokgobi Mogolodi Dimpe [1,2] and Philiswa Nosizo Nomngongo [1,2,3,*]

1 Department of Chemical Sciences, University of Johannesburg, Doornfontein Campus, P.O. Box 17011, Doornfontein 2028, South Africa; petmashile2009@hotmail.com (G.P.M.); mdimpe@uj.ac.za (K.M.D.)
2 DSI/NRF SARChI Chair: Nanotechnology for Water, University of Johannesburg, P.O. Box 17011, Doornfontein 2028, South Africa
3 DSI/Mintek Nanotechnology Innovation Centre, University of Johannesburg, P.O. Box 17011, Doornfontein 2028, South Africa
* Correspondence: pnnomngongo@uj.ac.za; Tel.: +27-115596187

Received: 1 March 2020; Accepted: 31 March 2020; Published: 12 May 2020

Abstract: The application of a magnetic mesoporous carbon/β-cyclodextrin–chitosan (MMPC/Cyc-Chit) nanocomposite for the adsorptive removal of danofloxacin (DANO), enrofloxacin (ENRO) and levofloxacin (LEVO) from aqueous and environmental samples is reported in this study. The morphology and surface characteristics of the magnetic nanocomposite were investigated by X-ray diffraction (XRD), Brunauer–Emmett–Teller (BET) adsorption–desorption and Fourier transform infrared spectroscopy (FTIR). The N_2 adsorption–desorption results revealed that the prepared nanocomposite was mesoporous and the BET surface area was 1435 $m^2 \, g^{-1}$. The equilibrium data for adsorption isotherms were analyzed using two and three isotherm parameters. Based on the correlation coefficients (R^2), the Langmuir and Sips isotherm described the data better than others. The maximum monolayer adsorption capacities of MMPC/Cyc-Chit nanocomposite for DANO, ENRO and LEVO were 130, 195 and 165 $mg \, g^{-1}$, respectively. Adsorption thermodynamic studies performed proved that the adsorption process was endothermic and was dominated by chemisorption.

Keywords: fluoroquinolones; ultrasound radiation; mesoporous carbon; desirability function; thermodynamics; wastewater; cost analysis

1. Introduction

The presence of pharmaceuticals in aquatic environments has become a subject of interest for environmental chemists [1]. Their wide distribution owes itself to the growing need for treatments and cures for human and animals diseases [2]. They are introduced into the aquatic environments through effluents of urban wastewater treatment plants (WWTPs) [3]. This is a result of their extensive use and their ineffective removal processes by wastewater transport and treatment [4]. Among various pharmaceuticals, antibiotics residues have proved to be the most commonly detected in the aquatic environment for both surface and ground waters [5]. Although they may occur in fairly low concentrations in environmental waters, their different modes of action and particular chemical and physical characteristics may pose a risk to the aquatic system [6]. Thus, there is a need to monitor and evaluate their persistent presence, which even at a low level can further increase antibiotic resistance [7]. The focus of this work is mainly on fluoroquinolones which are an important emerging group of synthetic antibacterials [8]. They have been used extensively for both human and veterinary medicine due to their effectiveness against both gram-positive and negative bacteria for the treatment of bacterial infections [2]. Moreover, different antibiotics have different half-lives; therefore, others may be more persistent in the environment which may result in increased levels of contamination to the environment [9].

Studies have shown that they are introduced to environmental bodies by either direct or indirect pathways [4,10,11]. Furthermore, they have been found to occur in surface waters at concentrations ranging from ng L^{-1} to μg L^{-1} [10,12]; ng L^{-1} to mg L^{-1} in groundwater [13]; and mg L^{-1} in soil [14]. Since they are continuously introduced into the environment they have been identified as pseudo-persistent organic pollutants [11]. The greatest challenge is the removal of antibiotics from wastewater before discharge into the environment due to the high costs associated with it [9]. Techniques such as advanced oxidation processes (AOPs), multi-treatment processes, separation processes and biological processes have been applied in the removal of antibiotics from wastewater [15]. However, they prove to be very expensive and require high maintenance for the complete removal of compounds, including antibiotics, at a larger scale [16].

Adsorption processes are of significant interest in removal applications of organic compounds such as antibiotics due to their simplicity in design [17], flexibility, cost and friendliness towards potential the toxicity of biological base processes [18]. The adsorption is a technique based on the removal of contaminants from a matrix onto an adsorbent surface [19]. The effectiveness of the technique is highly dependent on the adsorbate properties, adsorbent type and composition of matrix analyzed [20]. To date, various adsorptive material has been used, such as zeolites [21], graphene oxide (GO) [22,23], activated carbon (AC) [24–28], metal-organic frameworks (MOFs) [29], carbon nanotubes (CNT) [30] and clay [31], amongst others for adsorption removal of pharmaceuticals [32]. However, for antibiotic removal, CNTs, ACs, mesoporous clay material, exchange resins and bentonite are the most widely reported adsorbents [9]. Despite their widespread use, these sorbents also present some limitations, such as inefficient extraction, low antibiotic adsorption properties and costliness (high generation costs) [9]. Mesoporous carbon from carbon-based material, on the other hand, can serve as an artery for adsorbates and also contribute greatly towards adsorption [33,34]. It can boast advantageous features, such as a large surface area; a high adsorption capacity; a large and ordered pore size and structure; and chemical and mechanical stability [28,33–39]. Furthermore, mesoporous carbon can be made from cheaper materials, such as starch and waste biomass [28,33,37–39]. In addition, the incorporation of magnetic nanoparticles to mesoporous carbon facilitates ease during separation, and functionalizing the material enables for reduction of its hydrophobic nature [38,39].

Furthermore, the natural polymers such as chitosan and beta-cyclodextrin have gained prominence in recent years due to their advantageous features [40–46]. They possess similar features, such as biocompatibility [44,47] and biodegradability [46]. Their non-toxicity has proven that they are less harmful to humans and the environment, and thus they are often selected as solid phase materials for adsorptions of various pollutants, including pharmaceutical ones [43,48–50]. Moreover, they are formed from environmentally friendly sources; chitosan is formed from naturally existing resources, such as the exoskeletons of anthropoids, like shellfish, crabs and prawns [51], whereas beta-cyclodextrins can be derived from enzymatic degradation of starch [46]. Great attention has been focused on the immobilization of cyclodextrins on chitosan; their combination improves the adsorption capacity of chitosan [42,44].

Recently, coupling of adsorption processes and ultrasound irradiation have gained considerable attention due to their numerous advantages [26,52–55]. These include faster chemical reactions and mass transfer as a result of acoustic cavitation with the establishment of new adsorption sites on the adsorbent surface [26,52–55]. The influences of ultrasonic irradiation on the adsorptive removals of numerous pollutants from aqueous solutions have been reported in the literature [26,52–58].

Therefore, the objective of the present study was to synthesise magnetic mesoporous carbon/β-cyclodextrin–chitosan (MMPC/Cyc-Chit) nanocomposite as a sorbent for the elimination of fluoroquinolones. Factors that play a role in the adsorptive removal of the fluoroquinolones by MMPC/Cyc-Chit nanocomposite were examined; namely, sonication power level, sample pH and initial concentration of DANO, LEVO and ENRO. The overall process was to utilize cheap and readily available material for nanocomposite synthesis and ultrasonic radiation for superior removal efficiency. The incorporation of biodegradable polymers such as chitosan and β-cyclodextrin to magnetic mesoporous carbon resulted in a nanocomposite with super-adsorbent activities considering high surface area and adsorption capacities. The application of MMPC/Cyc-Chit nanocomposite for removal of fluoroquinolones has been reported for the first time.

2. Materials and Methods

2.1. Materials and Reagents

Chemicals reagents used for this study were of analytical grade, and Ultra-pure water (Direct-Q® 3UV-R purifier system Millipore, Merck, Darmstadt, Germany) was used throughout the duration of the experiments. Danofloxacin (99.7%) (DANO), enrofloxacin (99.0%) (ENRO), levofloxacin (99%) (LEVO), HPLC grade ethanol, methanol and acetonitrile were used, along with acetic acid, sodium hydroxide, ammonium hydroxide, ferrous chloride, ferric chloride, starch, chitosan, β-cyclodextrin and ortho-phosphoric acid purchased from Sigma-Aldrich (St. Loius, MO, USA). A synthetic sample mixture of the fluoroquinolones (FQs) stock solution was prepared by dissolving appropriate amounts of DANO, ENRO and LEVO in small amounts of methanol. The mixture was then diluted with ultra-pure water to a final volume of 100 mL. The solution were stored in to refrigerator at 4–8 °C.

2.2. Instrumentation

The synthesized adsorbent material was analyzed utilizing different techniques of characterization in order to determine its structural suitability for adsorption of the fluoroquinolones (DANO, ENRO LEVO). X-ray diffraction (XRD) patterns were recorded using a PANalytical X'Pert X-ray diffractometer (PANalytical BV, Almelo, Netherlands) utilizing Cu Kα radiation ($\lambda = 0.15406$ nm) in the 2θ range 4–90 at room temperature. The Fourier transform infrared (FT-IR) Perking–Elmer spectrum 100 spectrometer (Perkin-Elmer, Shelton, CT, USA) using the potassium bromide (KBr) pellet technique in a region of 4000–400 cm^{-1} was used to report the infrared spectrum for the prepared material. Surface characteristics such as porosity and area of the as-prepared material were analyzed by using the Brunauer–Emmett–Teller (BET) 77 K using an ASAP2020 porosity and surface area analyzer (Micrometrics Instrument Corp., Norcross, GA, USA).

The samples were degassed was at 100 °C for 3 h using N$_2$ gas before analysis. Adjustments for pH where necessary were performed using an OHAUS starter 2100 pH meter (Pine Brook, NJ, USA). The surface charge/point of zero charge was evaluated for the as-prepared material using a Nano-ZS Zetasizer (Malvern Instruments, Malvern, UK). The pH was adjusted within the range of 2.0–11.0 by the addition of 0.1 mol L^{-1} acetic acid and ammonium solution to each solution with 37 mg of adsorbent material. A Scientech Ultrasonic cleaner (Labotec, Midrand, South Africa) with a volume of 5.7 L (internal dimensions: 300 × 153 × 150 mm) was used to facilitate the adsorption process. The ultrasonic system was equipped with a variable frequency and power setting. In this study, the frequency was fixed at 50 Hz and the emission power of 150 W. The system has 5 power levels (1 (weakest) to 5 (strongest)), this power setting is used to reduce or increase the size of the cavitation bubble implosion force. Therefore, the sonication power levels were varied. The analysis of the antibiotics was performed using an Agilent HPLC 1200 Infinity series, equipped with a photodiode array detector (Agilent Technologies, Waldbronn, Germany). Chromatograms were recorded at 290 nm. An Agilent Zorbax Eclipse Plus C18 column (3.5 μm × 150 mm × 4.6 mm) (Agilent, Newport, CA, USA) was operated at an oven temperature of 25 °C. The mobile phase (water with 10 mmol L^{-1} of phosphoric acid; the pH adjusted to 3.29 with triethylamine): acetonitrile (85.7:14.3, *v/v*) at a flow rate of 1.5 mL min^{-1}. All chromatographic experiments were carried out 25 ± 3 °C while the injection volume was 10 L for all samples.

2.3. Preparation of the Nanocomposite

2.3.1. Synthesis of Mesoporous Carbon (MPC)

Modified version of the hard templating method adapted from literature was used in the synthesis of mesoporous carbon [59]. Briefly, in a 100 mL beaker containing deionized water and equipped with magnetic stirrer for easy dissolving starch was used. The mixture was then heated over an oil bath at 120 °C to form a homogenous solution with continuous stirring at 200 rpm. Silica solution was added dropwise at approximately 1 drop per second using a burette with continuous stirring until the starch had completely dissolved. Thereafter, the solution was transferred onto a glass petri dish and left to cool at an ambient room temperature. A gel-like material was formed and dried at 60 °C in an oven for 1 h and further carbonized with the gentle flow of nitrogen gas at 500 °C for 3 h. Once carbonized the material was stirred for 24 h at 70 °C in a sodium hydroxide (30 wt %) solution to remove silica. The formed product was washed with a mixture of ethanol and water (1:1) and filtered under vacuum. The filtered product was then oven dried at 60 °C for 2 h.

2.3.2. Preparation of Magnetic Mesoporous Carbon Coated with Chitosan and β-CD

Ferrous and ferric chloride solutions were dissolved in ultrapure water at a Fe^{2+}/Fe^{3+} ratio of 1:2 and stirred for 5 min. Then 3 g of β-CD and 4 g MPC were added into the iron solution with vigorous stirring along with the addition of diluted sodium hydroxide solution (1.0 mol L^{-1}) while heating at 80 °C for 1 h. That solution was then filtered by vacuum filter and washed with methanol plus water. The filtrate was then dried in an oven at 60 °C for 24 h. Chitosan flakes were modified based on a method described by [42]. Briefly, 3 g of chitosan flakes was dissolved in 50 mL of 3% acetic acid. Prepared magnetic material was then added to the solution of chitosan and this mixture was transferred to a round bottom flask. These were ultra-sonicated to facilitate dispersion were the pH of the prepared mixture was adjusted to 8.0–9.0 by means of diluted sodium hydroxide solution. Thereafter, it was filtered and washed with mixture of ethanol (50:50) plus water until the pH reached about 7, and oven dried at 40 °C.

2.4. Batch Adsorption Studies

Batch adsorption method was employed for adsorption studies. This was achieved by adding a specific mass of adsorbent (10–30 mg) to 25 mL synthetic sample solutions containing a mixture of FQ antibiotics (that is DANO, ENRO and LEVO) at a concentrations of 10 mg L^{-1}. The pH of the synthetic sample solutions (5–9) were adjusted using 0.1 mol L^{-1} HCl and 0.1 mol L^{-1} NaOH. The adsorption process was carried out using an ultrasonic bath. The frequency of the ultrasonic bath was fixed at high 50 Hz while the sonication power level was varied between 2 (60 W or 40% of total power and 5 (150 W or 100% of total power). Once the adsorption processes was completed, the adsorbent and sample solution were separated using an external magnet. The supernatant was filtered by using 0.22 μm syringe filters and the residual FQ antibiotic concentration in the solution was determined HPLC-PDA.

A response surface methodology constructed by a central composite design (CCD) was used for the optimization of the most influential parameters for the removal of FQ antibiotics. These factors include sample pH, mass of adsorbent (MA) and sonication power level (SP). The removal efficiency (%RE) was used as an analytical response. The optimization process was carried out using Statistica version 13. When the optimal conditions were achieved, the adsorption isotherm and kinetics for the removal of FQ antibiotics were examined.

Under optimum conditions, Langmuir, Freundlich, Hill and Langmuir-Freundlich isotherm models (Table 1) were used to study the interaction between the prepared MMPC/Cyc-Chit nanocomposite and

FQ antibiotic mixture. To achieve this, model solutions containing different concentrations selected FQs antibiotics mixture (5–80 mg L^{-1}) were used.

Table 1. Adsorption isotherms and kinetics models equations.

Isotherm Models	Isotherm Expression	Definition of Terms
Langmuir	$$\frac{C_e}{q_e} = \frac{1}{q_{max}K_L} + \frac{C_e}{q_{max}}$$ $$R_L = \frac{1}{1+K_LC_0}$$	q_{max}: theoretical monolayer adsorption capacity (mg g^{-1}) C_e: equilibrium concentration (mg L^{-1}), q_e: the amount of adsorbate adsorbed per unit weight of adsorbent (mg g^{-1}) C_0: initial concentration (mg g^{-1}) K_L: Langmuir equilibrium constant (L mg^{-1}) R_L: separation factor
Freundlich	$$\ln q_e = \ln K_F + \ln C_e$$	K_F: Freundlich constant (L g^{-1}) n: is the Freundlich exponent (g L^{-1})
Hill	$$q_e = q_H \frac{C_e^{n_H}}{K_D + C_e^{n_H}}$$	n_H and K_D: Hill isotherm constants q_H: maximum equilibrium adsorption capacity (mg/g).
Langmuir–Freundlich	$$\frac{1}{q_e} = \frac{1}{Q_{max}K_s}\left(\frac{1}{C_e}\right)^{\frac{1}{n}} + \frac{1}{Q_{max}}$$	K_S: Sips equilibrium constant (1/mg) Q_{max}: maximum adsorption capacity (mg g^{-1}) n: surface heterogeneity
Pseudo-first order	$$\ln(q_e - q_t) = \ln q_e - k_i t$$	K_1: rate constant (min^{-1}) q–q_e: amount of absorbate at equilibrium (mg g^{-1})
Pseudo-second order	$$\frac{1}{q_t} = \frac{1}{k_2\,q_e^2} + \frac{1}{q_e}t$$	K_2: Equilibrium rate constant (g mg^{-1} min^{-1}) q–q_e: amount of adsorbent at equilibrium (mg g^{-1})
Intra-particle diffusion	$$Q_t = k_i t^{\frac{1}{2}} + C$$	Q_t: amount of solute on surface of sorbent at time t (mg g^{-1}) K_i: intraparticle diffusion constant (mg g^{-1} min$^{1/2}$)

The kinetic studies performed using an initial concentration of 50 mg L^{-1} were used to explain the rate and mechanism of the adsorption process. The kinetics models, such as pseudo-first-order, pseudo-second-order and intraparticle diffusion, were employed to analyze the equilibrium kinetic data. The thermodynamic studies were carried out using a concentration of 50 mg L^{-1} at different temperatures: 25, 35 and 40 °C.

2.5. Regeneration and Reusability (Recyclability) of the Nanocomposite

To investigate the regeneration capability of the MMPC/Cyc-Chit nanocomposite, 36 mg of adsorbent was placed into 25 mL of 10 mg L^{-1} FQ antibiotic solution. The mixture was sonicated for 30 min, and after the adsorption process had been completed, the separation of adsorbent by an external magnet was done. The adsorbent was then sonicated with a mixture of 10 mL of acidified water and acetonitrile mixture (55:45 ratio) for 10 min to remove the adsorbed FQs. The water was obtained by adjusted ultrapure deionized to pH 3 using *ortho*-phosphoric acid. It should be noted that 10 min desorption time was enough to remove all the analytes adsorbed. An external magnet was applied to facilitate the decantation of the desorption solvent. Desorption solution containing the FQs was analyzed using HPLC-PDA. After decantation, the adsorbent was washed with the desorption solvent; filtered; and finally, washed two times with ultrapure water and dried at 60 °C for 2 h. The above procedure was repeated 10 times.

2.6. Application in Real Water Samples

Wastewater (influent and effluent) samples were collected from a wastewater treatment plant (WWTP) in Pretoria, South Africa. River water and tap water were collected from the Apies River (Pretoria, South Africa) and the University of Johannesburg laboratory (Johannesburg, South Africa). The sample collection was performed during October 2019. The wastewater and river water samples were kept in 1 L glass amber bottles and transported to the laboratory to be stored at 4 °C before adsorption studies. The physicochemical characteristics, such as pH; conductivity; total dissolved solids (TDS); and dissolved organic carbon of wastewater, laboratory tap water and river water, are presented in summarized in Table A1. In addition, the concentrations of major elements such as calcium, magnesium, sodium and iron are presented in Table A1.

3. Results and Discussion

3.1. Characterization

3.1.1. X-ray Diffraction Spectroscopy

Figure 1 shows the XRD patterns of the (a) mesoporous carbon, (b) β-cyclodextrin, (c) chitosan and (d) MMPC/Cyc-Chit nanocomposite. The XRD patterns for chitosan, β-cyclodextrin and mesoporous carbon are comparable with those reported in the literature [50,59,60]. The XRD pattern for MMPC/Cyc-Chit nanocomposite shows diffraction peaks at 2θ = 31.3°, 35.7°, 42.8°, 54.1°, 56.8° and 63.2°. These diffraction peaks correspond to the magnetite planes indexed to (220), (311), (400), (422), (511) and (440). These results confirmed the importation of iron oxide nanoparticles (Fe_3O_4) in the nanocomposites. Moreover, they were in agreement with other results in literature [42,45].

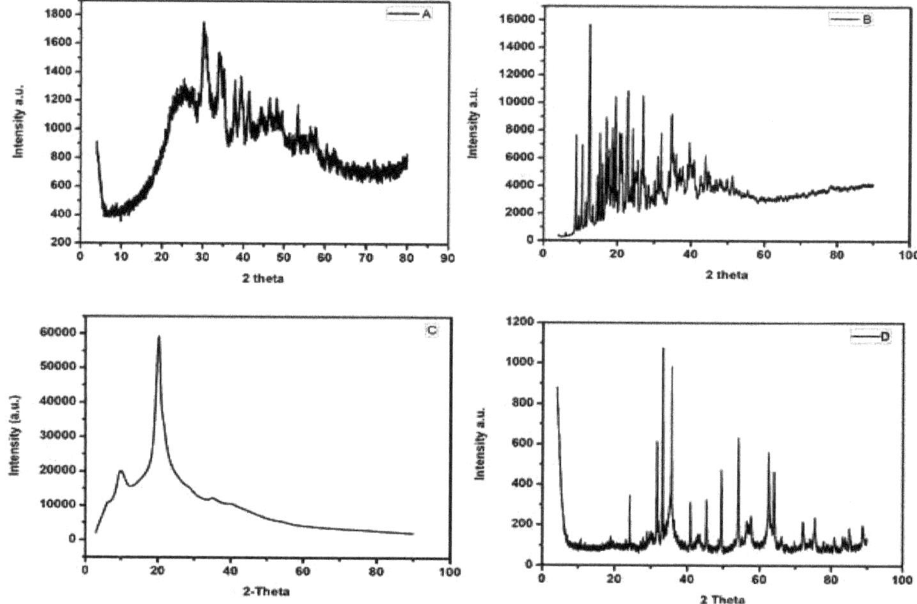

Figure 1. XRD of (**A**) mesoporous carbon, (**B**) beta-cyclodextrin, (**C**) chitosan and (**D**) MMPC/Cyc-Chit nanocomposite composite.

3.1.2. Fourier Transform Infrared Spectroscopy

FTIR spectra for mesoporous carbon (MPC), β-cyclodextrin (β-CD), chitosan (Chi) and MMPC/Cyc-Chit nanocomposite are presented in Figure 2. The FTIR spectrum of MPC (Figure 2) reveals the peaks at 2924–2889 cm^{-1} and 1384 cm^{-1} which were ascribed to the stretching and bending of CH_3 and CH_2 stretching [37], whereas the broad peak at 3439 cm^{-1} was attributed to the O–H stretching. The band at 1615 cm^{-1} was assigned to the C=O vibration of carbonyl groups [39,61]. In addition, the CH_3 stretching and unsaturated sites were observed at 2361 cm^{-1} [37]. The major bands for β-cyclodextrin and chitosan (Figure 2) were allocated as follows: 1024 cm^{-1} for (R-1, 4-bond skeleton vibration of β-CD); 1649–1656 cm^{-1} for C–N and C=O (NHCO (amide I)) stretching vibrations; and 3280–3353 cm^{-1} (O–H and N–H stretching vibrations) [42,45]. In addition, the peaks at 1586 and 1153 cm^{-1} were assigned to the N–H stretching vibration (primary amine) and antisymmetric glycosidic linkages [42]. The MMPC/Cyc-Chit nanocomposite shows two characteristic absorbance bands centered at 1652 and 1597 cm^{-1}, which correspond the C=O stretching vibration of NHCO (amide I) and N–H bending of NH_2, respectively [42].

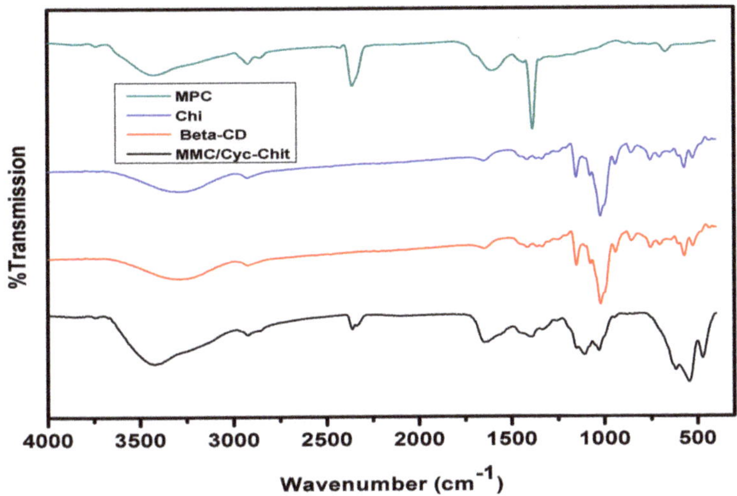

Figure 2. FT-IR spectra of MPC, Chi, beta-CD and MMPC/Cyc-Chit composite.

3.1.3. Nitrogen Adsorption–Desorption

The important textural properties that influence the quality and application of an adsorbent, especially for adsorptive removal of pollutants in matrices that are complex (such as wastewater and polluted river waters), are the porosity and specific surface area [24,28,36,62,63]. It has been reported that the two properties are significant because they are strongly related to the maximum adsorption capacity of the adsorbent [24,27,28,36,62–64]. Textural properties of the nanocomposite are presented in Table 2. The results confirmed that incorporating chitosan and β-cyclodextrin into magnetic mesoporous carbon resulted in a superabsorbent with high specific surface area (1264 $m^2\ g^{-1}$). The micropore and mesopore surface areas of the nanocomposite in comparison with mesoporous carbon were used to analyse the textual properties of the prepared material. As seen in Table 2, the percentage of the surface comprised of mesopores was 60%, suggesting that the nanocomposite is predominantly a mesoporous material [28,34–37,39]. These characteristics validate the applicability of the nanocomposite for adsorption processes. According to the results in Table 2, it was anticipated that during the adsorption process, the investigated FQ antibiotics would percolate through pores of the adsorbents. These findings were in agreement with SEM results, and they both confirm that the prepared adsorbent possesses outstanding characteristics which endorse it for wastewater treatment using adsorption technology.

Table 2. Characteristics of adsorbent material; BET surface area; pore volume parameters of MPC and MMPC/Cyc-Chit.

Surface Properties	Mesoporous Carbon	Nanocomposite
S_{BET} ($m^2\ g^{-1}$)	1181	1264
Total pore volume ($cm^3\ g^{-1}$)	2.54	4.65
Average pore size (nm)	7.93	8.61
t-Plot $S_{mesopore}$ ($m^2\ g^{-1}$)	728	755
t-Plot $S_{micropore}$ ($m^2\ g^{-1}$)	453	509

3.1.4. Point of Zero Charge

The pH of the FQs solution might have an effect on their adsorption on the surface of the MMPC/Cyc-Chit nanocomposite. Moreover, the pH of the sample solution was used to assess the distribution percentage of the investigated FQ species during their adsorption process. For example, subject to the pH of the sample solution, the surface of the nanocomposite could be protonated or deprotonated, thereby changing the surface charge of an adsorbent. Therefore, it is important to investigate the pH at which negative and positive charges are equal, also known as pH at point of zero charge (pH_{pzc}). This point will as assist in the determination of the possible adsorption mechanism. Therefore, the influence of pH onto the zeta potential of MMPC/Cyc-Chit nanocomposite was evaluated and results are shown in Figure 3. The surface MMPC/Cyc-Chit nanocomposite was positively charged at pH values lower than 8 and the pH_{pzc} value was estimated as 8.0. This implied that MMPC/Cyc-Chit nanocomposite has a negative charge above pH = 8.0.

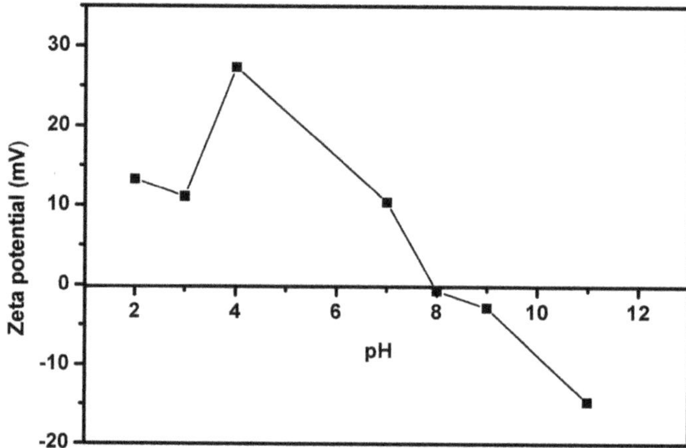

Figure 3. Determination of pHpzc of nanocomposite.

3.2. Optimization

The batch adsorption process was optimized using the RSM-ased CCD, and the design matrix together with respective responses obtained at the equivalent experimental conditions are indicated in Table 3. Statistica software was used to generate second-order polynomial model which used to explain the adsorption process of FQ antibiotics onto the MMPC/Cyc-Chit nanocomposite. The removal efficiency was used as the dependent variable or analytical response. The R^2 values were used to assess the performance of the RSM model, and they were found to be 0.9985, 0.99876 and 0.9975 for DANO, ENRO and LEVO, respectively. These findings revealed the best agreement between the actual and predicted responses. Moreover, these results proposed that about 99% of the total variation in removal efficiency was attributable to the experimental factors.

The validity and appropriateness of the RSM model, as well as the estimation of the most significant independent variables and their interactions, were examined by analysis of variance (ANOVA). The ANOVA results are reproduced in the form of Pareto charts (Figure 4). The importance of an independent variable was evaluated by the magnitude of the bar length. If the length of the bar passes the red line (0.05 confidence level line), this phenomenon suggests that the corresponding independent factor is significant at a 95% confidence level. As seen in Figure 4, the mass of the adsorbent and sample pH were significant at the 95% confidence level for every sample investigated. This implied that they had more influence on the analytical response.

Table 3. The design matrix and the results of the two-level fractional factorial design.

Run	pH	MA	SP	DANO	ENRO	LEVO
1	5	10	2	33.7	33.7	36.6
2	5	10	4	36.7	39.5	37.5
3	5	30	2	49.1	49.1	51.4
4	5	30	4	45.3	45.3	46.0
5	9	10	2	70.8	65.0	66.8
6	9	10	4	71.0	71.0	71.8
7	9	30	2	71.9	74.7	76.0
8	9	30	4	75.1	75.1	72.7
9	4.1	20	3	45.4	45.4	47.3
10	9.9	20	3	59.1	59.1	66.9
11	7	5.3	3	29.5	29.5	27.3
12	7	35	3	96.2	96.2	97.7
13	7	20	1.5	85.4	85.4	81.5
14	7	20	4.5	96.2	99.1	97.6
15 (C)	7	20	3	97.9	97.9	94.8
16 (C)	7	20	3	97.8	97.8	95.2
17 (C)	7	20	3	96.9	99.8	96.2
18 (C)	7	20	3	98.3	98.3	94.9
19 (C)	7	20	3	96.1	96.1	95.0

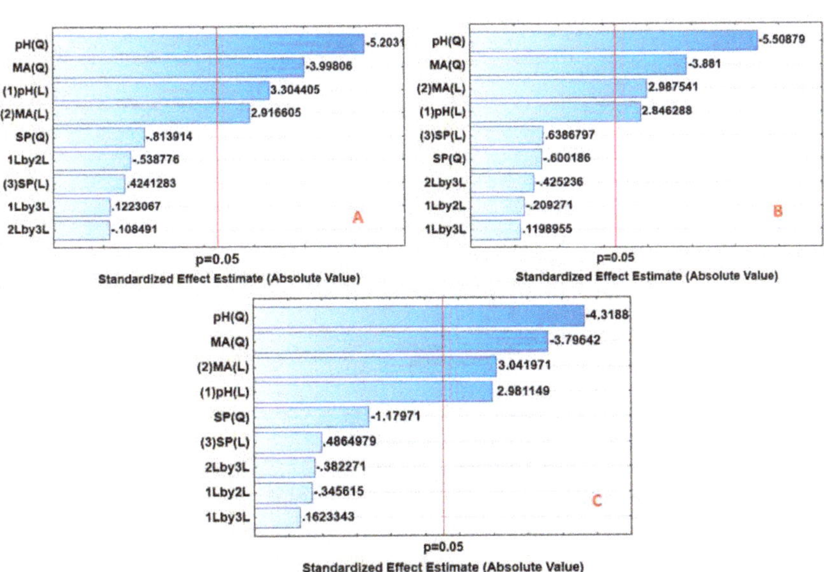

Figure 4. Pareto chart of standardized effects for adsorption of (**A**) DANO, (**B**) ENRO and (**C**) LEVO). MA = mass of adsorbent; SP = sonication power; 1Lby2L shows the interaction between pH and MA; 2Lby3L shows the interaction between MA and SP; 1Lby3L shows the interaction between pH and SP.

3.2.1. Response Surface Methodology

Three-dimensional (3D) response surface plots were constructed to investigate the effect of each variable on the removal efficiency, and their interactions (Figures 5, A1 and A2). The effects of sample pH, mass of adsorbent (MA) and sonication power level (SP) were concurrently examined for the adsorptive removal of FQs from synthetic samples. Figure 5A shows the 3D plot of sample pH versus mass of adsorbent. As seen in Figure 5, both mass of adsorbent and the sample pH played a critical role in removal of FQ antibiotics from aqueous solutions. This might be because sample pH affects the ionization of analytes and the charge on surface of adsorbent. Based on Figure 5A,B, the removal efficiency increased with increasing sample pH, and the maximum removal was achieved between pH 6 and 8. Below and above these values, a decrease in analytical response was observed. This is because DANO, ENDRO and LEVO can exist in three forms in aqueous systems, that is, cationic (pH > pK_{a2}), zwitterionic ($pKa1 \leq pH \leq pK_{a2}$) and anionic (pH < pK_{a1}), and these forms are pH-dependent [65–68]. Consequently, the adsorption mechanism is also dependent on the adsorbent surface charge. For instance, the FQ antibiotics can be adsorbed by a negatively or positively charged adsorbent using cation exchange through protonation of amine group or electrostatic interaction due to the deprotonation of carboxylic groups [39,61,65–67,69–72].

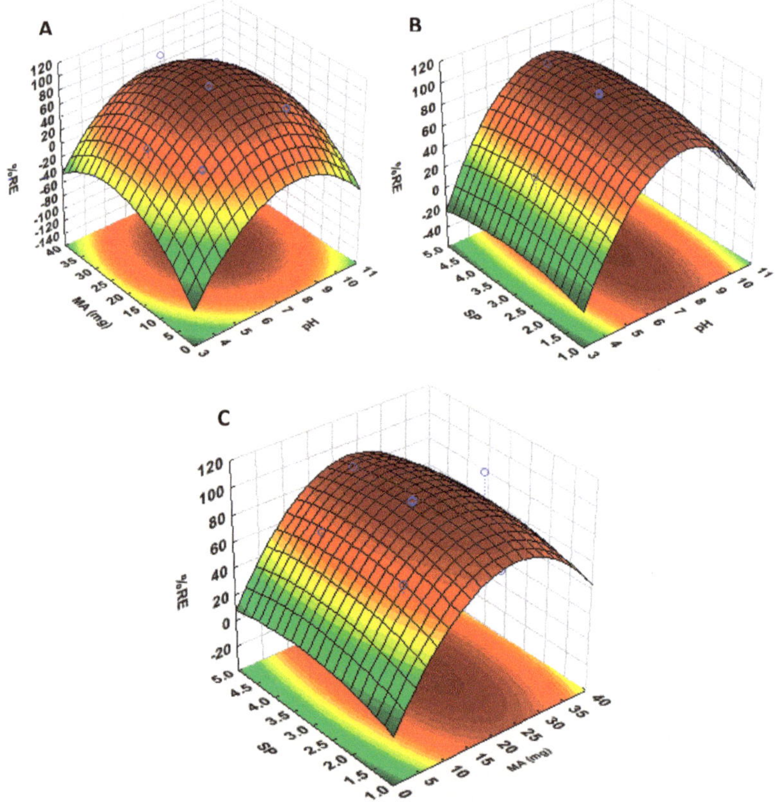

Figure 5. The 3D surface response plots describing the interactions of the parameters investigated. (**A**) interaction between sample pH and mass of adsorbent (MA); (**B**) interactions between sonication power (SP) and sample pH and (**C**) interactions between SP and MA.

At lower pH values, the FQs are predominately in cationic forms due to a high concentration of hydronium ions [61,68,73,74]. This results in lower removal efficiencies due to the competition between the adsorbate and small molecules of hydronium ions which can fill the available active sites. Additionally, the pH_{pzc} of the material was found to be 8, indicating that the charges on surface of the nanocomposite are positive charges. Therefore, lower removal efficiencies can also be attributed to electrostatic repulsion between positively charged nanocomposite and cationic forms of FQs. As the sample pH increases, the electrostatic interaction between the adsorbate/analytes and the surface of the adsorbent occurs, resulting in higher removal efficiencies. However, at pH values > pH_{pzc} value of 8, a decline in the removal efficiency was observed. These could be attributed to electrostatic repulsion between negatively charged FQs and negatively charged nanocomposite. Several researchers in the literature have observed similar findings with respect to the adsorption behavior of FQs at low and high pH values [61,65–71,73,74]. The results for the effect of sonication power level are shown in Figure 5B,C; it was not significant at the 95% confidence level. However, the 3D response surface plots reveal that as the sonication power levels increases, the removal efficiency also increases. As seen in Figure 5B,C, %RE values above 80% were obtained when the sonication power was 3 (90 W or 60% of the total power) and above. The increased removal efficiency can be attributed to the increase in adsorbate–adsorbent interactions due to turbulence produced by implosion of the cavitation bubbles.

3.2.2. Desirability Function

The desirability profile was used to estimate the optimum experimental conditions obtained using RSM optimization approach (Figures 6, A3 and A4). The optimal conditions for the removal of fluoroquinolones were sample pH: 7.0, mass of adsorbent: 36 mg and sonication power level 3. The sonication or contact time, initial concentration and sonication frequency were fixed at 30 min, 10 mg L^{-1} and 50 Hz. Under the abovementioned conditions, the predicted removal efficiencies of the model for the adsorption of DANO, ENRO and LEVO were 97.2%, 98.3 and 95.3%, respectively. To certify the acceptability of the RSM model and to confirm the agreement between the predicted and experimental removal efficiency, six replicates were carried out at the abovementioned conditions. The obtained experimental results showed removal efficiencies of 98.7 ± 1.3%, 99.1 ± 0.9% and 96.8 ± 1.2% for DANO, ENRO and LEVO, respectively. These results showed that the RSM model could be considered an accurate and valid procedure for the optimization of the adsorption process.

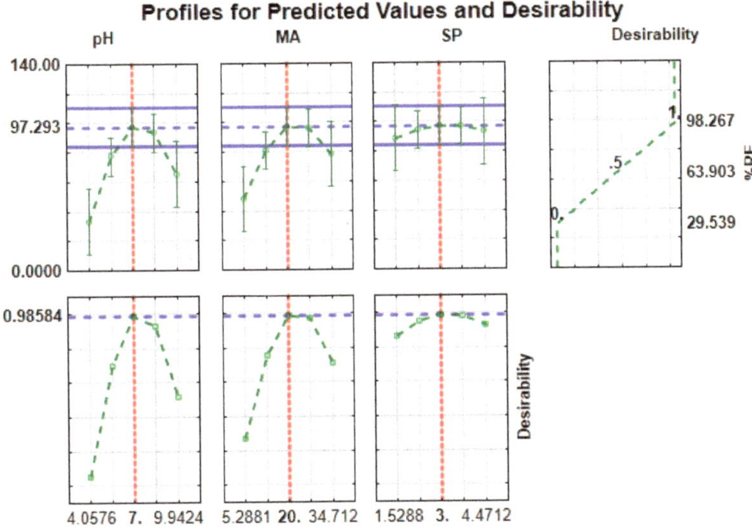

Figure 6. Profiles for predicated values and desirability function for removal of fluoroquinolones.

3.3. Adsorption Kinetics

The adsorption kinetics data (Figure 7) were used to study the adsorption process of FQ antibiotics onto the surface of the nanocomposite. The data were analysed using three commonly used kinetic models; namely, pseudo-first-order, pseudo-second-order and intraparticle diffusion. The equations of these kinetic models are widely reported (See Table 1), and they were adapted from the literature [66,67,69].

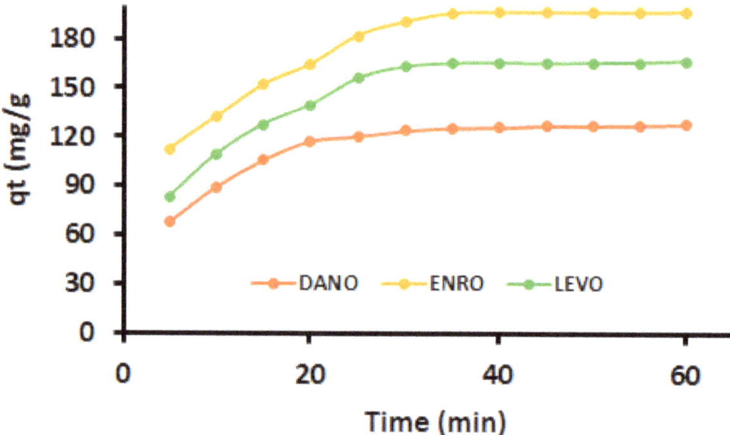

Figure 7. Effects of contact time on DANO, ENRO and LEVO by MMPC/Cyc-Chit kinetic modelling.

The estimated parameters are presented in Table 4. As it is indicated, the R^2 values achieved for pseudo-second-order were constantly higher compared to those of pseudo-first-order. In addition, the adsorption capacities obtained using the pseudo-second-order kinetic model were in agreement with experimental values. These outcomes suggested that the rate-determining step might be dominated by chemical interactions of FQ antibiotics with the homogenous surface of the adsorbent. The chemisorption mechanism might be driven by electrostatic attraction between the adsorbent and FQs. The dissociated forms FQ antibiotics have carboxylate and nitrogen functional groups that can bind on the positive or negative adsorbent surface.

Table 4. Parameters for the various kinetic models fitted onto data obtained for adsorbate solutions and results.

Kinetic Models	Parameters	DANO	ENRO	LEVO
	q_{expt}	130	195	165
Pseudo-first order	q_e	93.2	161	128
	k_1	0.0636	0.11	0.099
	R^2	0.9400	0.9201	0.9372
Pseudo-second order	q_e	130	196	167
	k_2	0.0018	0.0011	0.0011
	R^2	0.9986	0.9971	0.9967
Intraparticle diffusion	k_{id1}	17.7	24.6	24.8
	C_1	32.8	56.3	29.7
	R^2_1	0.9843	0.9866	0.9754
	k_{id2}	1.31	0.348	0.613
	C_2	118	194	162

To further understand the adsorption mechanism and the rate-controlling step, the adsorption data were fitted to the intraparticle diffusion model [74]. The plots of q_t versus $t^{1/2}$ for the investigated adsorbates showed multi-linearity (Figure A5). These plots indicated that there were two adsorption steps that took take place. According to the literature, the steeper first-step is due to diffusion of FQ antibiotics through the solution to the mesoporous nanososorbent. The second stage is attributed to transfer of the DANO, ENRO and LEVO charged molecules into intraparticle active sites or pores of the nanocomposite. Furthermore, it was noticed that the linear part of the first step did not pass through the origin. This signified that intraparticle diffusion was not the only rate-determining step [74,75]. Therefore, it can be concluded that adsorption processes were driven by both surface adsorption and intra-particle diffusion. The intraparticle diffusion rate constants for the first and second stages (k_{id1}, k_{id2}), correlation coefficients and intercept, C are indicated in Table 4. The R^2 values suggested that the adsorption of FQs on MMPC/Cyc-Chit nanocomposite may be dominated by intra-particle diffusion.

3.4. Adsorption Isotherms

To study the relationship between the concentration of FQs retained by the surface of the adsorbent and that of residual FQs in the bulk solution, the equilibrium studies were performed. The adsorption data were determined using Langmuir, Freundlich, Hill and Langmuir–Freundlich (Sips) isotherm models, and the model expressions are summarized in Table 1. The adsorption isotherms of FQs using the nanocomposite were carried out at 25 °C, and the pH of the solution, mass of adsorbent and contact time were set at 7, 30.0 mg and 30 min, respectively. Figure 8 demonstrates the adsorption isotherms of FQs onto nanocomposite from aqueous solutions. The isotherm models were used to derive various parameters related to the adsorption process.

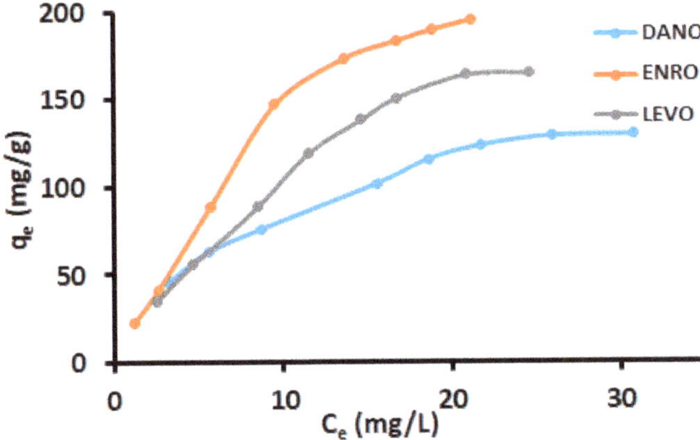

Figure 8. Sorption isotherms—modeling with Langmuir, Freundlich, Sips and Hill (MA: 34 mg; sonication time: 25 min; pH 7; temperature: 25 ± 3 °C).

Table 5 shows the summary of parameters derived from Langmuir, Freundlich, Hill and Langmuir–Freundlich (Sips) isotherm model plots. Comparing the correlation coefficients (R^2) values for the Langmuir and Freundlich isotherm models, it was detected that DANO, ENRO and LEVO were better suited to the Langmuir model. These findings demonstrated that the adsorption process took place as a monolayer of FQs on the surface of the adsorbent. The maximum DANO, ENRO and LEVO adsorption capacities for the adsorbent were 130, 196 and 194 mg g^{-1}, respectively. As seen in Table 5, the Freundlich isotherm model was also used to some extent; however, it was not as good as the Langmuir isotherm model.

The separation factor (R_L) values for each adsorbate (Table 5) were used to examine wherever the adsorption process was favourable. The values were calculated from the Langmuir isotherm and they suggested that the selected FQ antibiotics were easily adsorbed onto the nanocomposite because R_L values were less than 1. In addition, the observation was also made that R_L values decrease with an increase in the initial concentration, stipulating that the adsorption of FQs was more favourable at high concentrations [76]. The equilibrium adsorption data were also modelled using three-parameter isotherms expressions (Hill and Langmuir–Freundlich isotherm models) and parameter values are illustrated in Table 5. As seen, both models confirmed that the adsorption process assumes the homogeneous monolayer on the heterogeneous surface of the nanocomposite. In addition, the Hill model exponent n_H values for DANO, ENRO and LEVO were greater than 1, indicating that the binding interaction between FQ antibiotics and nanocomposite was in the form of positive cooperativity [72].

Table 5. Isothermal parameters of DANO, ENRO and LEVO on MMPC/Cyc-Chit.

	Two Parameter Models				Three Parameter Models		
	Langmuir isotherm				Hill isotherm		
Parameters	DANO	ENDRO	LEVO	Parameters	DANO	ENRO	LEVO
q_{max} (mg g^{-1})	130	196	165	q_H	129	195	164
K_L (L g^{-1})	0.14	0.24	0.086	n_H	2.7	2.3	2.1
R_L	0.13–0.42	0.087–0.22	0.11–0.54	K_d	115	27.6	84.9
R^2	0.9911	0.9950	0.9934	R^2	0.9946	0.9859	0.9858
	Freundlich isotherm				Langmuir-Freundlich (Sips) isotherm		
K_F	27.2	20.7	18.2	$Qmax$	135	208	170
n	2.1	1.3	1.4	n	1.1	1.1	1.3
R^2	0.9884	0.9810	0.9806	K_s	0.097	0.065	0.050
				R^2	0.9902	0.9884	0.9878

3.5. Adsorption Thermodynamics

The effect of temperature in the removal of DANO, ENRO and LEVO using the nanocomposite was investigated. The thermodynamic parameters, such as Gibs energy ($\Delta G°$) enthalpy ($\Delta H°$) and entropy ($\Delta S°$) are presented in Table 6. The values of $\Delta G°$ were calculated using Equation (4), whereas the $\Delta H°$ and $\Delta S°$ values were estimated from the slopes and intercepts of the plots that were obtained using Equation (4). As seen, the $\Delta G°$ values were negative at all investigated temperatures. This phenomenon suggested that the adsorption was spontaneous [65,66,77]. Furthermore, the positive values of $\Delta H°$ demonstrated that the adsorption interaction between the antibiotics and the nanocomposite was characterised by endothermic nature [24,65,66,77]. The values of $\Delta H°$ were higher than 20.9 kJ/mol, confirming that the adsorption processes of FQ antibiotics were dominated by a chemisorption mechanism [78]. Moreover, the positive values of $\Delta S°$ suggested that there is an increase in randomness at the boundary of solid/liquid phases, which might reveal the possible structural variations of the analyte and adsorbent [65,66,77].

Table 6. Thermodynamic parameters for DANO, ENRO and LEVO sorption on MMPC/Cyc-Chit.

Analytes	T (K)	ΔG (kJ mol^{-1})	ΔH (kJ mol^{-1})	ΔS (J mol K^{-1})
DANO	298	−13.57	59.4	55.3
	308	−13.69		
	313	−13.74		
ENRO	298	−15.52	61.8	96.4
	308	−15.71		
	313	−15.81		
LEVO	298	−14.70	71.1	104
	308	−14.87		
	313	−14.94		

3.6. Comparison of Sorption Capacities for Various Adsorbents

To compare the performance of the nanocomposite for the adsorption of FQ antibiotics, adsorption capacities of DANO, ENRO and LEVO on various adsorbents is presented in Table 7. As observed in Table 7, the adsorption capacity of the nanocomposite was comparable even better than other adsorbents reported elsewhere [25,61,65,67,69–71,74]. However, the adsorption capacity was lower than those reported by references [66,73].

Table 7. Comparison of sorption capacities for DANO, ENRO and LEVO fluoroquinolones with various composite sorbents at 25 ± 1 °C.

Adsorbent	Adsorbate	Adsorption Capacity (mg/g)	Refs
Alkalized biochar	ENRO	40.91	[71]
Magnetic biochar-based manganese oxide composite	ENRO	7.19	[74]
iron-pillared montmorillonite	LEVO	48.61	[67]
MIL-100(Fe)	LEVO	87.34	[61]
Chitosan derived granular hydrogel with 3D structure	ENRO	388	[73]
Tb/Eu-Loaded Garlic Peels	ENRO	769	[66]
Co-modified MCM-41	LEVO	108	[65]
Fe_3O_4 and $Fe_3O_4@SiO_2$	LEVO	6.85	[69]
NBent-NTiO$_2$-Chit.	LEVO	90.91	[70]
Activated carbon-decorated polyacrylonitrile nanofibers	DANO, ENRO	99, 112	[25]
MMPC/Cyc-Chit	DANO, ENRO, LEVO	130, 196, 165	This work

3.7. Regeneration and Reusability Studies

Regeneration and reusability for spent adsorbent are two of most crucial factors from the cost-effective point of view. This study investigated the possibility of regenerating and reusing the spent nanocomposites-loaded with FQ antibiotics. The regeneration and reusability process was performed according to the procedure described in Section 2.5. As seen in Figure 9, the regenerated nanocomposite retained 90–100% of its adsorption capacity toward the removal of DANO, ENRO and LEVO, after five cycles of the desorption–adsorption. Furthermore, the adsorption capacities of the spent adsorbent for removal of DANO, ENRO and LEVO remained at 88, 122 and 116 mg g^{-1}, respectively, after the eighth cycle. Furthermore, the spent adsorbent after the eighth cycle was used for the removal of FQ antibiotics. It was observed that even though the adsorption capacities decreased, the removal efficiency remained above 95%. These results demonstrated that the nanocomposite can be reused several times without affecting its removal efficiency. Additionally, it was then concluded that the prepared nanocomposite had relatively high chemical and thermal stability.

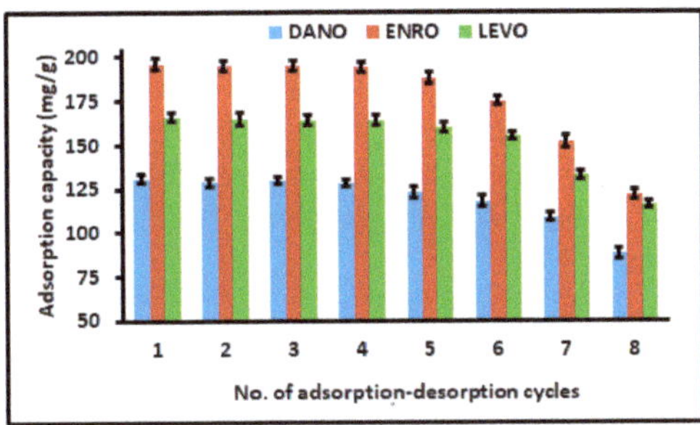

Figure 9. Regeneration of MMC/Cyc-Chit nanocomposite for eight successive adsorption–desorption cycles. Experimental conditions: C_0 = 50 mg L^{-1}; extraction time = 180 min; pH = 3.0; mass adsorbent = 36 mg; desorption solvent: acidified water and acetonitrile mixture (55:45 ratio); desorption time = 10 min.

3.8. Application to Real Samples

The applicability of the synthesized nanocomposite was assessed for the adsorptive removal DANO, ENRO and LEVO from real water samples; i.e., tap water, river water, influent and effluent wastewater. The river water, influent and wastewater samples were filtered using a 0.22 µm syringe filter. The target analytes were detected in influent and effluent wastewater and their concentrations ranged from 58 to 1230 µg L^{-1}, whereas only traces of ENRO were detected in river water samples (Table 8). As seen, the overall removal efficiencies of DANO, DANO and LEVO in spiked water samples ranged from 90–99% and the concentration of the target analyte reduced significantly. These outcomes demonstrate the good performance of the adsorbent for water and wastewater treatment.

Table 8. Adsorptive removal of fluoroquinolones in wastewater, river water and tap water samples.

Samples	Added (mg L^{-1})	DANO		ENRO		LEVO	
		Found (mg L^{-1})	%RE	Found (mg L^{-1})	%RE	Found (mg L^{-1})	%RE
Tap water	0	ND	-	ND	-	ND	-
	2.0	2.13 ± 0.11	99.5	1.96 ± 0.08	99.7	2.03±0.09	99.5
	5.0	5.03 ± 0.09	98.9	4.98 ± 0.05	99.0	5.11±0.02	98.9
River water	0	ND	-	0.098 ± 0.009	-	ND	-
	2.0	1.97 ± 0.12	97.1	2.11 ± 0.08	98.3	2.04 ± 0.02	97.7
	5.0	5.11 ± 0.09	96.8	4.95 ± 0.11	97.9	5.08 ± 0.04	95.7
Influent	0	0.148 ± 0.012	-	1.23 ± 0.07	-	0.573 ± 0.007	-
	2.0	2.15 ± 0.09	97.2	3.32 ± 0.04	96.3	2.58 ± 0.06	95.4
	5.0	5.21 ± 0.07	94.7	6.29 ± 0.06	90.7	5.60 ± 0.06	93.6
Effluent	0	0.058 ± 0.011	-	0.443 ± 0.010	-	0.078 ± 0.011	-
	2.0	2.11 ± 0.05	98.7	2.40 ± 0.06	99.0	2.13 ± 0.07	98.3
	5.0	5.07 ± 0.07	95.7	5.51 ± 0.04	98.7	5.06 ± 0.09	97.9

3.9. Cost Analysis for the Preparation of Adsorbent

The cost of the adsorption process is predominantly dependent on the cost of adsorbent used for the removal of organic and inorganic pollutants from wastewater [79]. Therefore, relatively low-cost materials with properties that are comparable to commercially available adsorbents are required. The cost estimation breakdown for the preparation of the mesoporous carbon and nanocomposite is presented in Table 9. In comparison with the other commercially available nanomaterials, such as multi-walled carbon nanotubes (R2354/g, Sigma-Aldrich), graphene oxide (R2163/g, Sigma-Aldrich) and mesoporous carbon (R2623/5 g, Sigma-Aldrich), the cost of mesoporous carbon and nanocomposite is much cheaper. A kilogram of the prepared nanocomposite will cost about R23262.50 ($1324.85). The regeneration and reusability studies of the nanocomposite further reduce the cost of the adsorbent, since one batch can be reused at least five times. This confirms that the production of the MMPC/Cyc-Chit nanocomposite is economical and sustainable. Furthermore, regeneration and reusability are value-added properties of MMPC/Cyc-Chit nanocomposite as a promising adsorbent in the treatment of wastewater contaminated with emerging contaminants. The incorporation of magnetic nanoparticles led to the easy and fast separation (using external magnet) of adsorbent from aqueous solutions. The spent adsorbent can be first treated by the Fenton process (advanced oxidation processes, AOPs) before degrading the adsorbed pollutants. Furthermore, chitosan and β-cyclodextrin are types of fully biodegradable natural materials. This means that once the organic pollutants have been degraded by Fenton process, the adsorbent can be buried in the soil to allow biodegradation process.

Table 9. Cost estimation breakdown for the production of magnetic mesoporous carbon/β-cyclodextrin–chitosan (MMPC/Cyc-Chit) nanocomposite.

Processes	Cost Breakdown	Temperature/Time/Mass/Volume	Unit cost (R)	Power Rating (kWh)	Price (R)
Preparation of MPC	Starch	15 g	1917 (2 kg)		14.37
	Sodium hydroxide	3 g	844 (1 Kg)		2.53
	Silica	20 g	1779 (500 g)		71.16
	Heating	120 °C (power = 205 W), 30 min	71.65	0.06	4.30
	Carbonization	500 °C (power = 520W), 3 h	71.65	1.56	111.77
	Cleaning	70 °C, (power = 120 W), 24 h	71.65	2.9	207.79
	Drying	60 °C (power = 240 W), 1 + 2 h	71.65	0.72	51.56
Subtotal for 10–13 g MPC					463.48
Preparation of the nanocomposite	Ferrous chloride	1g	925 (250 g)		3.70
	Ferric chloride	2 g	936 (1 kg)		1.87
	MPC	3 g	463.48 (10–13 g)		106.96
	Chitosan	3 g	1939 (100 g)		58.17
	Beta-cyclodextrin	3 g	2769.00 (100 g)		83.07
	Acetic acid	1.5 mL	1006 (2.5 L)		0.60
Net amount of 12 g nanocomposite					253.77
Overhead cost (10% of net cost)					25.38
Total cost					279.15

4. Conclusions

A magnetic mesoporous carbon/β-cyclodextrin–chitosan (MMPC/Cyc-Chit) hybrid nanocomposite adsorbent was synthesized by the facile hydrothermal method. The prepared MMPC/Cyc-Chit adsorbent was characterized using BET, XRD, TEM and FTIR. The adsorption capabilities of the synthesized nanocomposite were studied in a multicomponent system employing the ultrasound-aided removal process. The effects of independent variables (sample pH, mass of adsorbent and sonication power level) were investigated and optimized using RSM based on the CCD. The use of the ultrasound system led to rapid achievement of equilibrium and improved the adsorption process due to intensified mass transfer as well as the enhanced affinity between adsorbate and adsorbent due to acoustic cavitation effects. The adsorption isotherm equilibrium data followed the Langmuir model, suggesting that the surface of the adsorbent is coated as monolayer coverage by DANO, ENRO and LEVO molecules. Furthermore, the three-parameter models confirmed that the adsorption process assumes the homogeneous monolayer on the heterogeneous surface of the MMPC/Cyc-Chit nanocomposite. The kinetic data were best described by the pseudo-second-order model proposing that the adsorptive removal process was dominated by chemisorption. The thermodynamic parameters which include $\Delta G°$, $\Delta H°$, and $\Delta S°$ indicated the adsorption process was feasible, spontaneous and endothermic in nature. In addition, the magnitude of $\Delta H°$ suggested that the removal of FQ antibiotics was via chemisorption and these findings agreed with the kinetic data. The synthesized MMPC/Cyc-Chit nanocomposite showed relatively high chemical and thermal stability and reusability over five adsorption–desorption cycles. The adsorption process was also applied in the removal of fluoroquinolones from real wastewater, tap water and river water samples. The results obtained demonstrated that MMPC/Cyc-Chit nanocomposite can be applied in water and wastewater treatment process.

Author Contributions: Formulated the research idea, G.P.M., K.M.D. and P.N.N.; designed the experiments, G.P.M. and P.N.N.; performed the actual experiments and data collection, G.P.M.; carried out the analysis of data, G.P.M. and P.N.N.; wrote the first draft of the manuscript, G.P.M.; reviewed and edited the final version of the manuscript, P.N.N.; collected real water samples, G.P.M. and P.N.N.; supervision K.M.D. and P.N.N. All authors have read and agreed to the published version of the manuscript.

Funding: This study was supported by the University of Johannesburg, South Africa (Department of Chemical Sciences) and the National Research Foundation (grants 91230 and 99270, South Africa).

Acknowledgments: The authors wish to thank the University of Johannesburg, department of Chemical Sciences for providing laboratory space.

Conflicts of Interest: The authors declare no conflict of interest.

Appendix A

Table A1. Physicochemical properties of water samples.

Parameters	Tap Water	River Water	Influent	Effluent
pH	7.7	7.6	6.64	7.11
Conductivity (μS/cm)	150	338	854	513
Total dissolved solid (TDS, mg/L)	50.6	208	432	235
Dissolved organic carbon (DOC, mg/L)	1.94	15.5	20.5	12.7
Calcium (mg/L)	2.46	6.65	19,652	18,509
Mg (mg/L)	1.33	4.30	15,231	12,861
Fe (mg/L)	0.98	3.20	634	339
Na (mg/L)	2.26	6.78	23,435	16,784
K (mg/L)	1.08	1.54	9277	6745

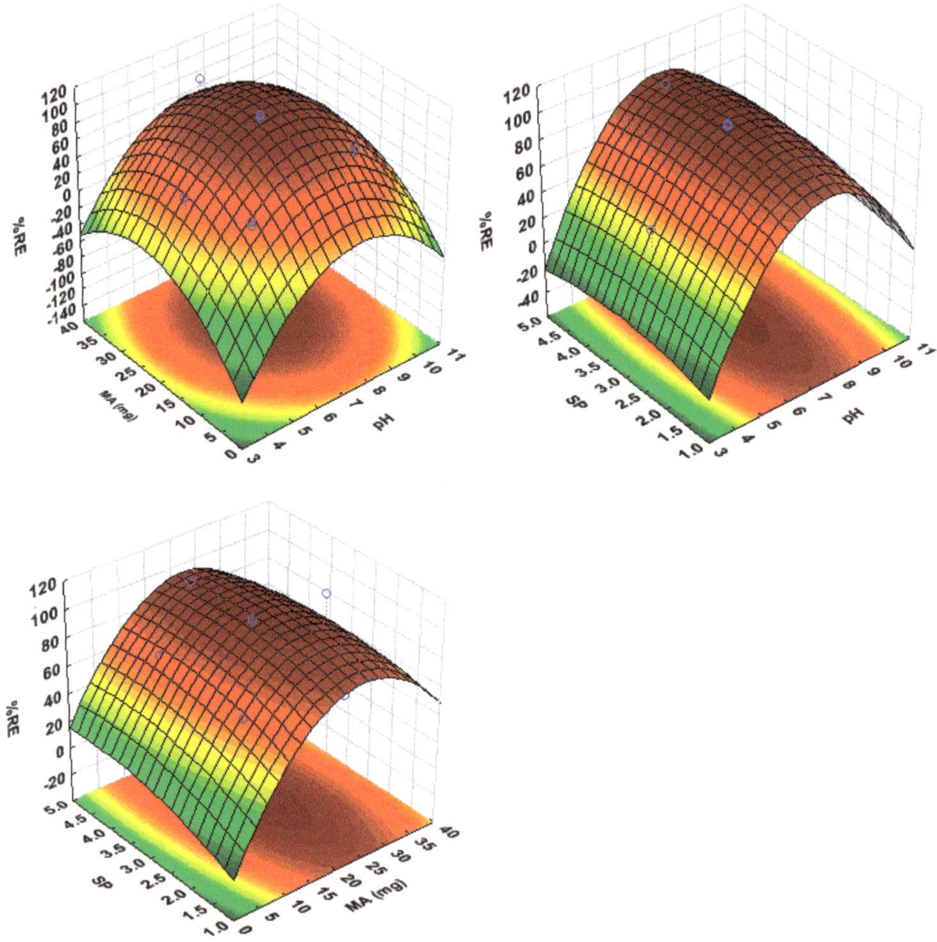

Figure A1. The 3D surface response plots describing the interactions of investigated parameters (ENRO).

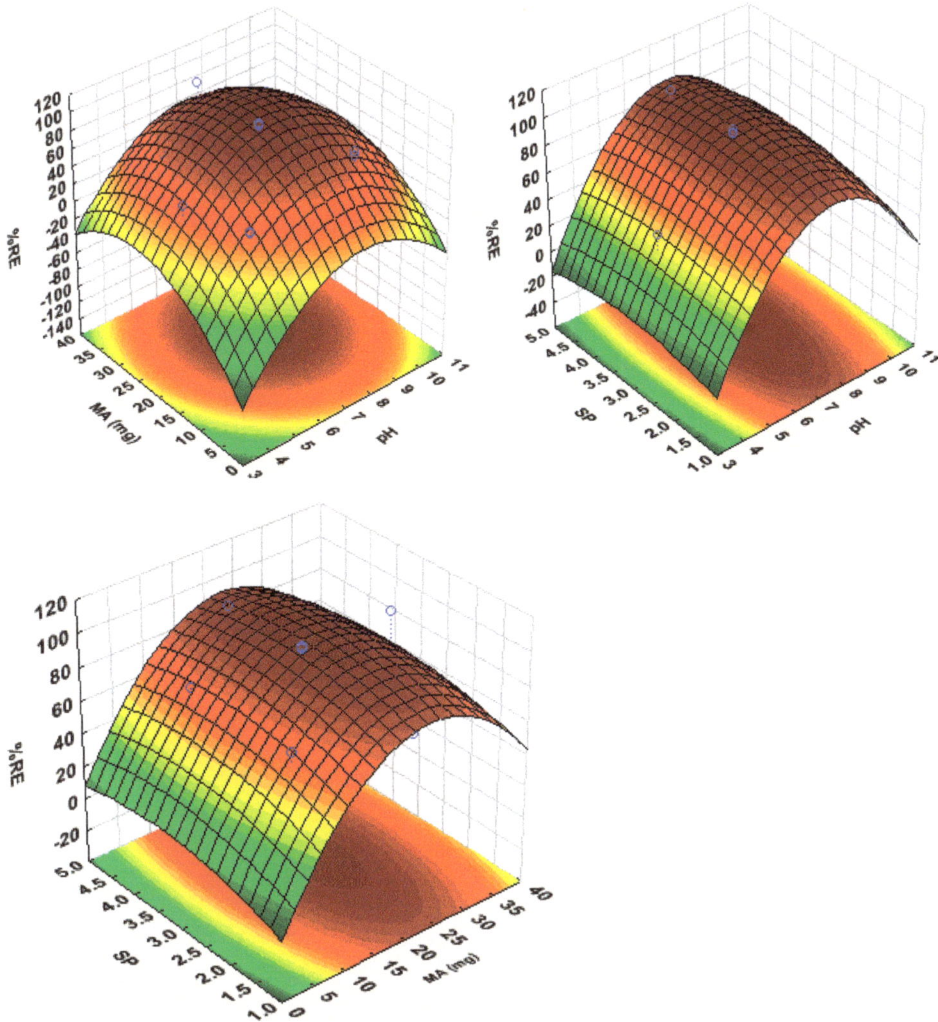

Figure A2. The 3D surface response plots describing the interactions of investigated parameters (LEVO).

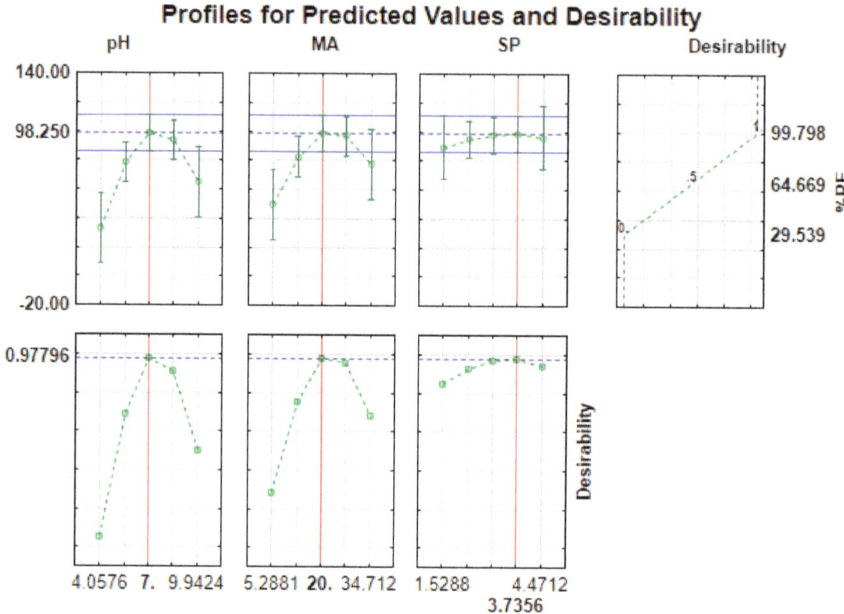

Figure A3. Profiles for predicated values and desirability function for removal of ENRO.

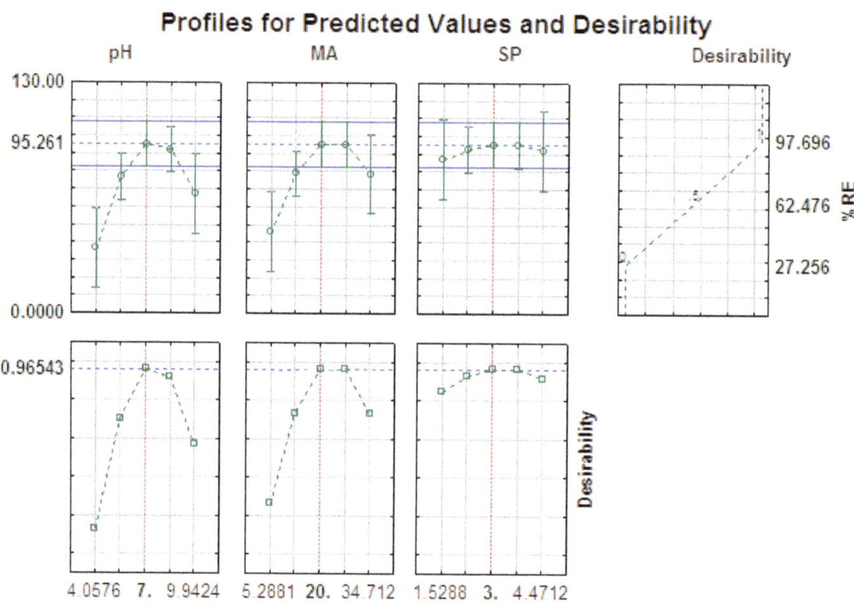

Figure A4. Profiles for predicated values and desirability function for removal of LEVO.

Intraparticle Diffusion Model

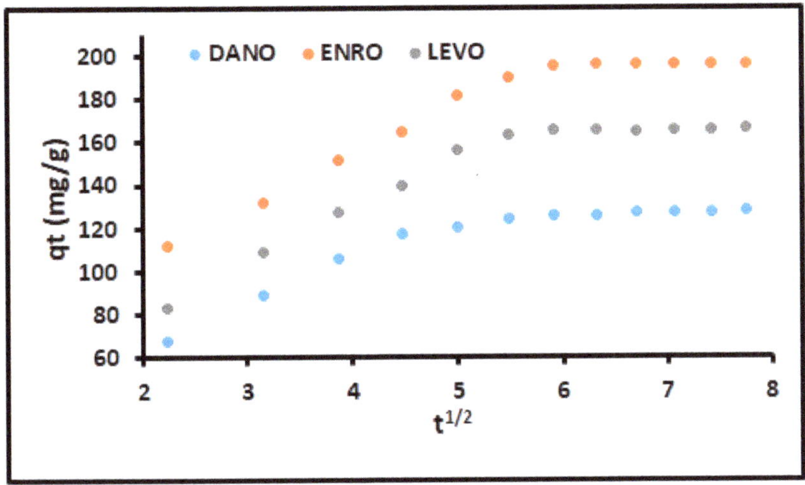

Figure A5. Intraparticle diffusion model.

References

1. Esponda, S.M.; Padrón, M.E.T.; Ferrera, Z.S.; Rodríguez, J.J.S. Solid-phase microextraction with micellar desorption and HPLC-fluorescence detection for the analysis of fluoroquinolones residues in water samples. *Anal. Bioanal. Chem.* **2009**, *394*, 927–935. [CrossRef] [PubMed]
2. de Oliveira, H.L.; da Silva Anacleto, S.; da Silva, A.T.M.; Pereira, A.C.; de Souza Borges, W.; Figueiredo, E.C.; Borges, K.B. Molecularly imprinted pipette-tip solid phase extraction for selective determination of fluoroquinolones in human urine using HPLC-DAD. *J. Chromatogr. B* **2016**, *1033*, 27–39. [CrossRef] [PubMed]
3. Baghdadi, M.; Ghaffari, E.; Aminzadeh, B. Removal of carbamazepine from municipal wastewater effluent using optimally synthesized magnetic activated carbon: Adsorption and sedimentation kinetic studies. *J. Environ. Chem. Eng.* **2016**, *4*, 3309–3321. [CrossRef]
4. Wang, J.; Wang, S. Removal of pharmaceuticals and personal care products (PPCPs) from wastewater: A review. *J. Environ. Manag.* **2016**, *182*, 620–640. [CrossRef]
5. Denadai, M.; Cass, Q.B. Simultaneous determination of fluoroquinolones in environmental water by liquid chromatography–tandem mass spectrometry with direct injection: A green approach. *J. Chromatogr. A* **2015**, *1418*, 177–184. [CrossRef]
6. Fent, K.; Weston, A.A.; Caminada, D. Ecotoxicology of human pharmaceuticals. *Aquat. Toxicol.* **2006**, *76*, 122–159. [CrossRef]
7. Michael, I.; Rizzo, L.; McArdell, C.; Manaia, C.; Merlin, C.; Schwartz, T.; Dagot, C.; Fatta-Kassinos, D. Urban wastewater treatment plants as hotspots for the release of antibiotics in the environment: A review. *Water Res.* **2013**, *47*, 957–995. [CrossRef]
8. Espinosa-Mansilla, A.; Girón, A.J.; De La Peña, A.M. Simultaneous determination of the residues of fourteen quinolones and fluoroquinolones in fish samples using liquid chromatography with photometric and fluorescence detection. *Czech J. Food Sci.* **2012**, *30*, 74–82.
9. Ahmed, M.B.; Zhou, J.L.; Ngo, H.H.; Guo, W. Adsorptive removal of antibiotics from water and wastewater: Progress and challenges. *Sci. Total Environ.* **2015**, *532*, 112–126. [CrossRef]
10. Roberts, J.; Kumar, A.; Du, J.; Hepplewhite, C.; Ellis, D.J.; Christy, A.G.; Beavis, S.G. Pharmaceuticals and personal care products (PPCPs) in Australia's largest inland sewage treatment plant, and its contribution to a major Australian river during high and low flow. *Sci. Total Environ.* **2016**, *541*, 1625–1637. [CrossRef]

11. Shi, Y.; Gao, L.; Li, W.; Liu, J.; Cai, Y. Investigation of fluoroquinolones, sulfonamides and macrolides in long-term wastewater irrigation soil in Tianjin, China. *Bull. Environ. Contam. Toxicol.* **2012**, *89*, 857–861. [CrossRef] [PubMed]

12. Archer, E.; Petrie, B.; Kasprzyk-Hordern, B.; Wolfaardt, G.M. The fate of pharmaceuticals and personal care products (PPCPs), endocrine disrupting contaminants (EDCs), metabolites and illicit drugs in a WWTW and environmental waters. *Chemosphere* **2017**, *174*, 437–446. [CrossRef] [PubMed]

13. Cabeza, Y.; Candela, L.; Ronen, D.; Teijon, G. Monitoring the occurrence of emerging contaminants in treated wastewater and groundwater between 2008 and 2010. The Baix Llobregat (Barcelona, Spain). *J. Hazard. Mater.* **2012**, *239*, 32–39. [CrossRef] [PubMed]

14. Karnjanapiboonwong, A.; Suski, J.G.; Shah, A.A.; Cai, Q.; Morse, A.N.; Anderson, T.A. Occurrence of PPCPs at a wastewater treatment plant and in soil and groundwater at a land application site. *Water Air Soil Pollut.* **2011**, *216*, 257–273. [CrossRef]

15. Xu, Y.; Liu, T.; Zhang, Y.; Ge, F.; Steel, R.M.; Sun, L. Advances in technologies for pharmaceuticals and personal care products removal. *J. Mater. Chem. A* **2017**, *5*, 12001–12014. [CrossRef]

16. Mehrjouei, M.; Müller, S.; Möller, D. Energy consumption of three different advanced oxidation methods for water treatment: A cost-effectiveness study. *J. Clean. Prod.* **2014**, *65*, 178–183. [CrossRef]

17. Jiang, J.-Q.; Ashekuzzaman, S. Development of novel inorganic adsorbent for water treatment. *Curr. Opin. Chem. Eng.* **2012**, *1*, 191–199. [CrossRef]

18. Ahmaruzzaman, M. Adsorption of phenolic compounds on low-cost adsorbents: A review. *Adv. Colloid Interface Sci.* **2008**, *143*, 48–67. [CrossRef]

19. Ramos, S.; Homem, V.; Alves, A.; Santos, L. Advances in analytical methods and occurrence of organic UV-filters in the environment—A review. *Sci. Total Environ.* **2015**, *526*, 278–311. [CrossRef]

20. Aksu, Z. Application of biosorption for the removal of organic pollutants: A review. *Process Biochem.* **2005**, *40*, 997–1026. [CrossRef]

21. Cavazos-Rocha, N.; Carmona-Alvarado, I.; Vera-Cabrera, L.; Waksman-de-Torres, N.; Salazar-Cavazos, M.D.L.L. HPLC method for the simultaneous analysis of fluoroquinolones and oxazolidinones in plasma. *J. Chromatogr. Sci.* **2014**, *52*, 1281–1287. [CrossRef] [PubMed]

22. Khan, A.; Wang, J.; Li, J.; Wang, X.; Chen, Z.; Alsaedi, A.; Hayat, T.; Chen, Y.; Wang, X. The role of graphene oxide and graphene oxide-based nanomaterials in the removal of pharmaceuticals from aqueous media: A review. *Environ. Sci. Pollut. Res.* **2017**, *24*, 7938–7958. [CrossRef] [PubMed]

23. Ji, L.; Chen, W.; Xu, Z.; Zheng, S.; Zhu, D. Graphene nanosheets and graphite oxide as promising adsorbents for removal of organic contaminants from aqueous solution. *J. Environ. Qual.* **2013**, *42*, 191–198. [CrossRef] [PubMed]

24. de Oliveira Carvalho, C.; Rodrigues, D.L.C.; Lima, É.C.; Umpierres, C.S.; Chaguezac, D.F.C.; Machado, F.M. Kinetic, equilibrium, and thermodynamic studies on the adsorption of ciprofloxacin by activated carbon produced from Jerivá (Syagrus romanzoffiana). *Environ. Sci. Pollut. Res.* **2019**, *26*, 4690–4702. [CrossRef] [PubMed]

25. Dimpe, K.M.; Nomngongo, P.N. Application of activated carbon-decorated polyacrylonitrile nanofibers as an adsorbent in dispersive solid-phase extraction of fluoroquinolones from wastewater. *J. Pharm. Anal.* **2019**, *9*, 117–126. [CrossRef] [PubMed]

26. Dil, E.A.; Ghaedi, M.; Asfaram, A.; Bazrafshan, A.A. Ultrasound wave assisted adsorption of congo red using gold-magnetic nanocomposite loaded on activated carbon: Optimization of process parameters. *Ultrason. Sonochem.* **2018**, *46*, 99–105. [CrossRef] [PubMed]

27. Pezoti, O.; Cazetta, A.L.; Bedin, K.C.; Souza, L.S.; Martins, A.C.; Silva, T.L.; Júnior, O.O.S.; Visentainer, J.V.; Almeida, V.C. NaOH-activated carbon of high surface area produced from guava seeds as a high-efficiency adsorbent for amoxicillin removal: Kinetic, isotherm and thermodynamic studies. *Chem. Eng. J.* **2016**, *288*, 778–788. [CrossRef]

28. Jain, A.; Balasubramanian, R.; Srinivasan, M. Production of high surface area mesoporous activated carbons from waste biomass using hydrogen peroxide-mediated hydrothermal treatment for adsorption applications. *Chem. Eng. J.* **2015**, *273*, 622–629. [CrossRef]

29. Khan, N.A.; Hasan, Z.; Jhung, S.H. Adsorptive removal of hazardous materials using metal-organic frameworks (MOFs): A review. *J. Hazard. Mater.* **2013**, *244*, 444–456. [CrossRef]

30. Jung, C.; Son, A.; Her, N.; Zoh, K.-D.; Cho, J.; Yoon, Y. Removal of endocrine disrupting compounds, pharmaceuticals, and personal care products in water using carbon nanotubes: A review. *J. Ind. Eng. Chem.* **2015**, *27*, 1–11. [CrossRef]

31. Sun, K.; Shi, Y.; Chen, H.; Wang, X.; Li, Z. Extending surfactant-modified 2: 1 clay minerals for the uptake and removal of diclofenac from water. *J. Hazard. Mater.* **2017**, *323*, 567–574. [CrossRef] [PubMed]

32. Yu, Y.; Liu, Y.; Wu, L. Sorption and degradation of pharmaceuticals and personal care products (PPCPs) in soils. *Environ. Sci. Pollut. Res.* **2013**, *20*, 4261–4267. [CrossRef] [PubMed]

33. Hu, Z.; Srinivasan, M.P.; Ni, Y. Preparation of mesoporous high-surface-area activated carbon. *Adv. Mater.* **2000**, *12*, 62–65. [CrossRef]

34. Gao, X.; Du, D.; Li, S.; Yan, X.; Xing, W.; Bai, P.; Xue, Q.; Yan, Z. Outstanding capacitive performance of ordered mesoporous carbon modified by anthraquinone. *Electrochim. Acta* **2018**, *259*, 110–121. [CrossRef]

35. Phan, T.N.; Gong, M.K.; Thangavel, R.; Lee, Y.S.; Ko, C.H. Enhanced electrochemical performance for EDLC using ordered mesoporous carbons (CMK-3 and CMK-8): Role of mesopores and mesopore structures. *J. Alloys Compd.* **2019**, *780*, 90–97. [CrossRef]

36. Saygılı, H.; Güzel, F. High surface area mesoporous activated carbon from tomato processing solid waste by zinc chloride activation: Process optimization, characterization and dyes adsorption. *J. Clean. Prod.* **2016**, *113*, 995–1004. [CrossRef]

37. Baikousi, M.; Georgiou, Y.; Daikopoulos, C.; Bourlinos, A.B.; Filip, J.; Zbořil, R.; Deligiannakis, Y.; Karakassides, M.A. Synthesis and characterization of robust zero valent iron/mesoporous carbon composites and their applications in arsenic removal. *Carbon* **2015**, *93*, 636–647. [CrossRef]

38. Goel, C.; Bhunia, H.; Bajpai, P.K. Mesoporous carbon adsorbents from melamine–formaldehyde resin using nanocasting technique for CO2 adsorption. *J. Environ. Sci.* **2015**, *32*, 238–248. [CrossRef]

39. Shi, S.; Fan, Y.; Huang, Y. Facile low temperature hydrothermal synthesis of magnetic mesoporous carbon nanocomposite for adsorption removal of ciprofloxacin antibiotics. *Ind. Eng. Chem. Res.* **2013**, *52*, 2604–2612. [CrossRef]

40. Chiu, C.-W.; Wu, M.-T.; Lee, J.C.-M.; Cheng, T.-Y. Isothermal adsorption properties for the adsorption and removal of reactive blue 221 dye from aqueous solutions by cross-linked β-chitosan glycan as acid-resistant adsorbent. *Polymers* **2018**, *10*, 1328. [CrossRef]

41. Tzereme, A.; Christodoulou, E.; Kyzas, G.Z.; Kostoglou, M.; Bikiaris, D.N.; Lambropoulou, D.A. Chitosan Grafted Adsorbents for Diclofenac Pharmaceutical Compound Removal from Single-Component Aqueous Solutions and Mixtures. *Polymers* **2019**, *11*, 497. [CrossRef]

42. Fan, L.; Luo, C.; Sun, M.; Qiu, H.; Li, X. Synthesis of magnetic β-cyclodextrin–chitosan/graphene oxide as nanoadsorbent and its application in dye adsorption and removal. *Colloids Surf. B Biointerfaces* **2013**, *103*, 601–607. [CrossRef]

43. Kyzas, G.Z.; Kostoglou, M.; Lazaridis, N.K.; Lambropoulou, D.A.; Bikiaris, D.N. Environmental friendly technology for the removal of pharmaceutical contaminants from wastewaters using modified chitosan adsorbents. *Chem. Eng. J.* **2013**, *222*, 248–258. [CrossRef]

44. Fan, L.; Li, M.; Lv, Z.; Sun, M.; Luo, C.; Lu, F.; Qiu, H. Fabrication of magnetic chitosan nanoparticles grafted with β-cyclodextrin as effective adsorbents toward hydroquinol. *Colloids Surf. B Biointerfaces* **2012**, *95*, 42–49. [CrossRef] [PubMed]

45. Wang, C.; Li, B.; Niu, W.; Hong, S.; Saif, B.; Wang, S.; Dong, C.; Shuang, S. β-Cyclodextrin modified graphene oxide–magnetic nanocomposite for targeted delivery and pH-sensitive release of stereoisomeric anti-cancer drugs. *RSC Adv.* **2015**, *5*, 89299–89308. [CrossRef]

46. Lu, D.; Yang, L.; Zhou, T.; Lei, Z. Synthesis, characterization and properties of biodegradable polylactic acid-β-cyclodextrin cross-linked copolymer microgels. *Eur. Polym. J.* **2008**, *44*, 2140–2145. [CrossRef]

47. Shukla, S.K.; Mishra, A.K.; Arotiba, O.A.; Mamba, B.B. Chitosan-based nanomaterials: A state-of-the-art review. *Int. J. Biol. Macromol.* **2013**, *59*, 46–58. [CrossRef]

48. Kyzas, G.Z.; Kostoglou, M.; Lazaridis, N.K.; Bikiaris, D.N. N-(2-Carboxybenzyl) grafted chitosan as adsorptive agent for simultaneous removal of positively and negatively charged toxic metal ions. *J. Hazard. Mater.* **2013**, *244*, 29–38. [CrossRef]

49. Lessa, E.F.; Nunes, M.L.; Fajardo, A.R. Chitosan/waste coffee-grounds composite: An efficient and eco-friendly adsorbent for removal of pharmaceutical contaminants from water. *Carbohydr. Polym.* **2018**, *189*, 257–266. [CrossRef]

50. Danahoğlu, S.T.; Bayazit, Ş.S.; Kuyumcu, Ö.K.; Salam, M.A. Efficient removal of antibiotics by a novel magnetic adsorbent: Magnetic activated carbon/chitosan (MACC) nanocomposite. *J. Mol. Liq.* **2017**, *240*, 589–596. [CrossRef]

51. Hirano, S.; Seino, H.; Akiyama, Y.; Nonaka, I. Chitosan: A biocompatible material for oral and intravenous administrations. In *Progress in Biomedical Polymers*; Springer: Berlin/Heidelberg, Germany, 1990; pp. 283–290.

52. Bhanvase, B.; Veer, A.; Shirsath, S.; Sonawane, S. Ultrasound assisted preparation, characterization and adsorption study of ternary chitosan-ZnO-TiO2 nanocomposite: Advantage over conventional method. *Ultrason. Sonochem.* **2019**, *52*, 120–130. [CrossRef] [PubMed]

53. Oveisi, M.; Asli, M.A.; Mahmoodi, N.M. MIL-Ti metal-organic frameworks (MOFs) nanomaterials as superior adsorbents: Synthesis and ultrasound-aided dye adsorption from multicomponent wastewater systems. *J. Hazard. Mater.* **2018**, *347*, 123–140. [CrossRef] [PubMed]

54. Wu, Y.; Han, Y.; Tao, Y.; Fan, S.; Chu, D.-T.; Ye, X.; Ye, M.; Xie, G. Ultrasound assisted adsorption and desorption of blueberry anthocyanins using macroporous resins. *Ultrason. Sonochem.* **2018**, *48*, 311–320. [CrossRef] [PubMed]

55. Ali, A.; Bilal, M.; Khan, R.; Farooq, R.; Siddique, M. Ultrasound-assisted adsorption of phenol from aqueous solution by using spent black tea leaves. *Environ. Sci. Pollut. Res.* **2018**, *25*, 22920–22930. [CrossRef]

56. Hamza, W.; Dammak, N.; Hadjltaief, H.B.; Eloussaief, M.; Benzina, M. Sono-assisted adsorption of cristal violet dye onto tunisian smectite clay: Characterization, kinetics and adsorption isotherms. *Ecotoxicol. Environ. Saf.* **2018**, *163*, 365–371. [CrossRef]

57. Dehghan, A.; Mohammadi, A.A.; Yousefi, M.; Najafpoor, A.A.; Shams, M.; Rezania, S. Enhanced Kinetic Removal of Ciprofloxacin onto Metal-Organic Frameworks by Sonication, Process Optimization and Metal Leaching Study. *Nanomaterials* **2019**, *9*, 1422. [CrossRef]

58. Hassani, A.; Khataee, A.; Karaca, S.; Karaca, C.; Gholami, P. Sonocatalytic degradation of ciprofloxacin using synthesized TiO2 nanoparticles on montmorillonite. *Ultrason. Sonochem.* **2017**, *35*, 251–262. [CrossRef]

59. Mpupa, A.; Mashile, G.P.; Nomngongo, P.N. Ultrasound-assisted dispersive solid phase nanoextraction of selected personal care products in wastewater followed by their determination using high performance liquid chromatography-diode array detector. *J. Hazard. Mater.* **2019**, *370*, 33–41. [CrossRef]

60. Cortés, M.E.; Sinisterra, R.D.; Avila-Campos, M.J.; Tortamano, N.; Rocha, R.G. The chlorhexidine: Beta;-cyclodextrin inclusion compound: Preparation, characterization and microbiological evaluation. *J. Incl. Phenom. Macrocycl. Chem.* **2001**, *40*, 297–302. [CrossRef]

61. Chaturvedi, G.; Kaur, A.; Umar, A.; Khan, M.A.; Algarni, H.; Kansal, S.K. Removal of fluoroquinolone drug, levofloxacin, from aqueous phase over iron based MOFs, MIL-100 (Fe). *J. Solid State Chem.* **2019**, 121029. [CrossRef]

62. Singh, G.; Kim, I.Y.; Lakhi, K.S.; Srivastava, P.; Naidu, R.; Vinu, A. Single step synthesis of activated bio-carbons with a high surface area and their excellent CO$_2$ adsorption capacity. *Carbon* **2017**, *116*, 448–455. [CrossRef]

63. Cunha, G.d.C.; Silva, I.A.A.; Alves, J.R.; Oliveira, R.V.M.; Menezes, T.H.S.; Romão, L.P. Magnetic hybrids synthesized from agroindustrial byproducts for highly efficient removal of total chromium from tannery effluent and catalytic reduction of 4-nitrophenol. *Cellulose* **2018**, *25*, 7409–7422. [CrossRef]

64. Cunha, M.R.; Lima, E.C.; Cimirro, N.F.; Thue, P.S.; Dias, S.L.; Gelesky, M.A.; Dotto, G.L.; dos Reis, G.S.; Pavan, F.A. Conversion of Eragrostis plana Nees leaves to activated carbon by microwave-assisted pyrolysis for the removal of organic emerging contaminants from aqueous solutions. *Environ. Sci. Pollut. Res.* **2018**, *25*, 23315–23327. [CrossRef] [PubMed]

65. Jin, T.; Yuan, W.; Xue, Y.; Wei, H.; Zhang, C.; Li, K. Co-modified MCM-41 as an effective adsorbent for levofloxacin removal from aqueous solution: Optimization of process parameters, isotherm, and thermodynamic studies. *Environ. Sci. Pollut. Res.* **2017**, *24*, 5238–5248. [CrossRef] [PubMed]

66. Zhao, Y.; Li, W.; Liu, Z.; Liu, J.; Zhu, L.; Liu, X.; Huang, K. Renewable Tb/Eu-Loaded Garlic Peels for Enhanced Adsorption of Enrofloxacin: Kinetics, Isotherms, Thermodynamics, and Mechanism. *ACS Sustain. Chem. Eng.* **2018**, *6*, 15264–15272. [CrossRef]

67. Liu, Y.n.; Dong, C.; Wei, H.; Yuan, W.; Li, K. Adsorption of levofloxacin onto an iron-pillared montmorillonite (clay mineral): Kinetics, equilibrium and mechanism. *Appl. Clay Sci.* **2015**, *118*, 301–307. [CrossRef]

68. Yadav, S.; Goel, N.; Kumar, V.; Tikoo, K.; Singhal, S. Removal of fluoroquinolone from aqueous solution using graphene oxide: Experimental and computational elucidation. *Environ. Sci. Pollut. Res.* **2018**, *25*, 2942–2957. [CrossRef]

69. Al-Jabari, M.H.; Sulaiman, S.; Ali, S.; Barakat, R.; Mubarak, A.; Khan, S.A. Adsorption study of levofloxacin on reusable magnetic nanoparticles: Kinetics and antibacterial activity. *J. Mol. Liq.* **2019**, *291*, 111249. [CrossRef]

70. Mahmoud, M.E.; El-Ghanam, A.M.; Mohamed, R.H.A.; Saad, S.R. Enhanced adsorption of Levofloxacin and Ceftriaxone antibiotics from water by assembled composite of nanotitanium oxide/chitosan/nano-bentonite. *Mater. Sci. Eng. C* **2019**, 110199. [CrossRef]

71. Wang, W.; Ma, X.; Sun, J.; Chen, J.; Zhang, J.; Wang, Y.; Wang, J.; Zhang, H. Adsorption of enrofloxacin on acid/alkali-modified corn stalk biochar. *Spectrosc. Lett.* **2019**, *52*, 367–375. [CrossRef]

72. Ashiq, A.; Sarkar, B.; Adassooriya, N.; Walpita, J.; Rajapaksha, A.U.; Ok, Y.S.; Vithanage, M. Sorption process of municipal solid waste biochar-montmorillonite composite for ciprofloxacin removal in aqueous media. *Chemosphere* **2019**, *236*, 124384. [CrossRef]

73. Wang, N.; Xiao, W.; Niu, B.; Duan, W.; Zhou, L.; Zheng, Y. Highly efficient adsorption of fluoroquinolone antibiotics using chitosan derived granular hydrogel with 3D structure. *J. Mol. Liq.* **2019**, *281*, 307–314. [CrossRef]

74. Li, R.; Wang, Z.; Zhao, X.; Li, X.; Xie, X. Magnetic biochar-based manganese oxide composite for enhanced fluoroquinolone antibiotic removal from water. *Environ. Sci. Pollut. Res.* **2018**, *25*, 31136–31148. [CrossRef] [PubMed]

75. Zhang, C.-L.; Qiao, G.-L.; Zhao, F.; Wang, Y. Thermodynamic and kinetic parameters of ciprofloxacin adsorption onto modified coal fly ash from aqueous solution. *J. Mol. Liq.* **2011**, *163*, 53–56. [CrossRef]

76. Xiong, W.; Zeng, Z.; Li, X.; Zeng, G.; Xiao, R.; Yang, Z.; Zhou, Y.; Zhang, C.; Cheng, M.; Hu, L. Multi-walled carbon nanotube/amino-functionalized MIL-53 (Fe) composites: Remarkable adsorptive removal of antibiotics from aqueous solutions. *Chemosphere* **2018**, *210*, 1061–1069. [CrossRef]

77. Miraboutalebi, S.M.; Nikouzad, S.K.; Peydayesh, M.; Allahgholi, N.; Vafajoo, L.; McKay, G. Methylene blue adsorption via maize silk powder: Kinetic, equilibrium, thermodynamic studies and residual error analysis. *Process Saf. Environ. Prot.* **2017**, *106*, 191–202. [CrossRef]

78. Mahmoud, M.E.; Fekry, N.A. Fabrication of engineered silica-functionalized-polyanilines nanocomposites for water decontamination of cadmium and lead. *J. Polym. Environ.* **2018**, *26*, 3858–3876. [CrossRef]

79. Mohanta, D.; Ahmaruzzaman, M. Bio-inspired adsorption of arsenite and fluoride from aqueous solutions using activated carbon@ SnO$_2$ nanocomposites: Isotherms, kinetics, thermodynamics, cost estimation and regeneration studies. *J. Environ. Chem. Eng.* **2019**, *6*, 356–366. [CrossRef]

 *polymers*

Article

Dopamine Grafted Iron-Loaded Waste Silk for Fenton-Like Removal of Toxic Water Pollutants

Md Shipan Mia [†], Biaobiao Yan [†], Xiaowei Zhu, Tieling Xing * and Guoqiang Chen

National Engineering Laboratory for Modern Silk, Soochow University, Suzhou 215123, China; shipan0143@gmail.com (M.S.M.); yanbiao136@163.com (B.Y.); 18771093295@163.com (X.Z.); chenguojiang@suda.edu.cn (G.C.)
* Correspondence: xingtieling@suda.edu.cn; Tel.: +86-512-6706-1175
† These authors contributed equally to this work.

Received: 24 October 2019; Accepted: 4 December 2019; Published: 9 December 2019

Abstract: Dispersion of iron was achieved on waste silk fibers (wSF) after grafting of polydopamine (PDA). The catalytic activity of the resulting material (wSF-DA/Fe) was investigated in Fenton-like removal of toxic aromatic dyes (Methylene Blue, Cationic Violet X-5BLN, and Reactive Orange GRN) water. The dye removal yield reached 98%, 99%, and 98% in 10–40 min for Methylene Blue, Cationic Violet X-5BLN, and Reactive Orange GRN, respectively. The catalytic activity was explained in terms of the effects of temperature, dyes, and electrolytes. In addition, the kinetic study showed that the removal of dyes followed pseudo-1st order adsorption kinetics. These findings allow envisaging the preparation of fiber-based catalysts for potential uses in environmental and green chemistry.

Keywords: waste silk; dopamine; iron particles; wastewater treatment

1. Introduction

The control of the water pollution is becoming one of the major challenges worldwide. As for extensive industrialization and subsequently the massive discharge of numerous types of organic toxicants, such as dyes, phenols, and nitroaromatic compounds [1–3] releasing in water supplies is becoming a serious issue. However, all of the pollutants are receiving tremendous attention from the water researchers, among them the aromatic dyes received one of the highest concerns due to their high toxic nature, chemical stability, and their resistance to conventional treatment methods [1]. Many approaches have been introduced by the several researchers where hazardous aromatic dyes were removed by various processes like biodegradation, chemical oxidation, adsorption, and so on [4–8]. Various heterogeneous catalytic systems with metal oxides or hydroxides like CuO, ZnO, TiO$_2$, and FeOOH as catalysts, [5,9–11] as well as advanced oxidation systems have also been introduced all over the decades [12–14].

Fenton and Fenton-like treatments showed superior pollutant degradation/reduction efficiency compared with microbial oxidative metabolism, and physical adsorption also has some drawbacks [15–18]. Several new treatment configurations were introduced to optimize the Fenton process including a photo-assisting process to reduce iron supply [19,20], use of solid iron, or avoid modifying the natural pH value of wastewater [21], and so on. However, one of the drawbacks of this system still remained the separation of iron sludge in the solution after the treatment and recycling before discharging the treated wastewater. This is a time consuming and costly procedure [22]. Immobilization of iron particles (Fe) may become an appropriate and permanent remediation technique to solve this problem. However, it is important to immobilize Fe onto a carrier that can be easily separated from the contaminated water solution as well as provide no/less harm on subsequent management. The Fe immobilized inpolyacrylic acid(PAA) [23,24], starch [25], and polyglycol [26] reported in different literature are difficult to recycle once they are used to treat contaminants. To avoid secondary water contamination, immobilization of zero-valent iron (ZVI) onto solid supports, for example, polymeric membranes [27] and activated carbon [28], could be an ideal option. Morshed et al. [29] immobilized and stabilized ZVI nanoparticles on fibrous polyester. Very few literatures have been found on the use of textile fibers or fabrics to immobilize iron particles aiming heterogeneous catalytic application.

Silk fibers are one of the most discussed fibers derived from natural sources due to mechanical and thermomechanical properties. However, due to their exceptional properties and complex production procedures, raw silk fiber spun out from silkworm cocoons is a luxurious material in many applications. Hence, the material constructed by silk would not be cost effective compared with other plant based natural fibers. On the contrary, during industrial process, lots of waste silk fibers or scrap silk fibers are produced, which are considered as waste and available at extremely low cost. Therefore, the increasing challenge of recycling them is another concerning issue. So utilizing waste silk into a sustainable remediation application can be an ultimate alternative.

Due to strong adhesion property, dopamine, a component of marine mussel gaining tremendous attention for surface modification of different materials [17–19], is being extensively used to prepare super-hydrophobic, antimicrobial, UV-blocking, conductive, as well as dye adsorption materials. Our previous study reported use of dopamine to prepare silk fabric with hydrophobicity, flame retardancy, and UV shielding properties [30]. However, no study reported the effectiveness of dopamine grafted waste silk fibers to prepare heterogeneous catalyst for Fenton-like wastewater treatment application.

In this work, dopamine was grafted on the surface of silk fibers by rapid oxidative polymerization followed by loading/immobilization of Fe particles on PDA grafted waste silk fibers. Hence, a fiber-based catalyst was prepared for heterogeneous Fenton-like removal of toxic aromatic dyes. The surface morphology and chemical composition of resultant catalyst were characterized by scanning electron microscopy (SEM), energy dispersive spectroscopy (EDS), and Fourier transform infrared (FT-IR) spectroscopy. The effectiveness of catalytic property was assessed under the influence of H_2O_2 concentration, dye concentration, temperature, and electrolytes.

2. Experimental

2.1. Materials

Analytical grade dopamine hydrochloride (98.5%), sodium perborate($NaBO_3{\cdot}4H_2O$), ferric chloride hexahydrate (FeCl3·6H$_2$O), hydrogen peroxide (H_2O_2, 33%), and ferrous sulfate heptahydrate ($FeSO_4{\cdot}7H_2O$) were purchased from Shanghai Lingfeng Chemical Reagent Co., Ltd. (Shanghai, China) and used as received. Waste silk (average fiber diameter 10 μm) were obtained from Nantong Nafuer Clothing Co., Ltd (Nantong, China). All the dyes (Methylene Blue, Cationic Violet X-5BLN and Reactive Orange GRN) were purchased from Tianjin Tianshun Chemical Dyestuff Co., Ltd (Tianjin, China).

2.2. Methods

2.2.1. Preparation of Dopamine Grafted Waste Silk

Scheme 1 illustrates the pathways used for grafting of dopamine on waste silk. Typically, 1.60 g of waste silk was wetted by deionized water as a pre-treatment. Then, 0.6 g of dopamine hydrochloride and 0.16 g of FeCl$_3$·6H$_2$O were dispersed in 300 mL deionized water. Wet waste silk fibers were added in dopamine solution and placed in a shaking water bath for 20 min followed by addition of 0.55 g of $NaBO_3{\cdot}4H_2O$, the final solution was stirred at 50 °C for 50 min. The resultant dopamine grafted waste silk fibers (wSF-PDA) were rinsed and dried overnight in ambient condition.

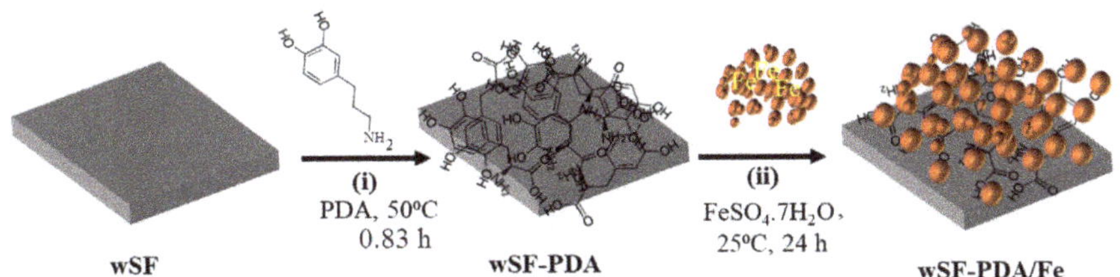

Scheme 1. Schematic illustration of the preparation process of (**i**) grafting of polydopamine (PDA) on waste silk fibers (wSF) and (**ii**) loading of iron on dopamine grafted waste silk fibers (wSF-PDA).

2.2.2. Preparation of wSF-PDA/Fe

The loading of iron on dopamine grafted waste silk fibers were carried out using chemical adsorption and incorporation method as illustrated in Scheme 1. Here, 15.54 g of $FeSO_4 \cdot 7H_2O$ were dissolved in 1400 mL deionized water in a ceramic tray and stirred until uniform dispersion. Then, the prepared (as Section 2.2.1 described) dopamine-grafted waste silk fibers were immersed in the iron solution and kept at 25 °C for 24 h. The resultant wSF-PDA/Fe undergone successive rinse and stored in a desiccator before analysis and use.

2.3. Material Characterizations

Fourier transform IR spectroscopy of pristine and iron-loaded wSF was carried out using a Nicolet-5700 Fourier transform infrared spectrometer (MA, USA). The sample to be tested was cut into powder and sampled by potassium bromide. The scanning range was 600–4000 cm^{-1} and the number of scans was 120 times. The samples were drilled before IR analysis, and background spectra were recorded on air. Surface morphologies of all wSF (before and after dopamine grafting and iron loading) were analyzed using desktop scanning electron microscope (Hitachi Ltd., Tokyo, Japan) at an accelerating voltage of 15 kV. Prior to SEM, samples were sprayed with a conducting resin followed by sputter coating with carbon films having a deposition depth of about 10 nm. Energy dispersive spectroscopy (EDS) was carried out using BRUKNER axes EDS analyzer mounted with SEM.

2.4. Catalytic Activity of wSF-DA/Fe

To study the catalytic performance of wSF-PDA/Fe, the heterogeneous Fenton-like removal of dyes (methylene blue, cationic violet X-5BLN, and reactive orange GRN) in presence of hydrogen peroxide (H_2O_2) was investigated. Typically, 70 mL of dye solution (10–81 mg/L) were treated by using 0.1 mg wSF-PDA/Fe 8 μL (0.05–5 mmol/L) of H_2O_2 at a specific temperature (25, 50, or 75 °C). A small amount of solution was taken through a Lab Sphere UV-1000F transmission analyzer (Lab sphere, Inc., North Sutton, VA, USA) to determine the specific absorbance at the characteristic peak of the dye, and the corresponding dye residual rate is calculated as per the Equation (1) [31]:

$$\text{Removal rate\%} = (C_0 - C)/C_0 \times 100 \tag{1}$$

where C is the concentration of the dye during removal at different time interval and C_0 is the initial concentration of the dye. C/C_0 was calculated at the maximum absorption wavelength of the dye in visible region. For example, methylene blue at 665 nm, cationic violet X-5BLN at 590 nm, and reactive orange GRN at 480 nm. Control experiments using wSF and wSF-PDA with/or without hydrogen peroxide were conducted to investigate the adsorption and catalytic property of the catalyst.

2.4.1. Effect of H_2O_2 Concentration on Removal Performance

The effects of different H_2O_2 concentrations on dye degradation were investigated. The reaction conditions were as follows: the reaction temperature was 50 °C, the pH of the reaction was about 7, the concentration of the dye solution was 20 mg/L, and the wSF-PDA/Fe was 0.1 g. The concentration of hydrogen peroxide studied was 0.05, 0.1, 0.5, 1, 3, and 5 mmol/L.

2.4.2. Effect of Pollutant Concentration on Removal Performance

The effects of dye concentration on removal efficiency using wSF-PDA/Fe catalyst were examined. All the experiments were carried out at 50 °C, pH = 7 in a shaking water bath where the initial concentration of H_2O_2 reagent was 1 mmol/L, and the dopamine grafted iron-loaded waste silk (wSF-PDA/Fe) was 0.1 g. The dye concentrations were studied at 10, 20, 40, 60, and 80 mg/L, respectively.

2.4.3. Effect of Reaction Temperature on Pollutant Removal Performance

The pollutant removal performances of wSF-PDA/Fe as function of different reaction temperatures were studied. Standard dye (20 mg/L) and H_2O_2 (1 mmol/L) concentration was used for 0.1 mg of wSF-PDA/Fe catalyst. Three-reaction temperatures (25, 50, and 75 °C) were studied.

2.4.4. Effect of Different Electrolytes (NaCl, Na_2SO_4) of Pollutant Removal Performance

In industrial production, dyeing wastewater contains not only dyes, but also many other inorganic and organic additives. Therefore, the effects of different electrolytes (NaCl, Na_2SO_4) on the removal efficiency of wSF-PDA/Fe were also investigated. All the experiments were carried out at 50 °C, pH = 7 in a shaking water bath for the 20 mg/L dye using 1 mmol/L H_2O_2, and 0.1 g wSF-PDA/Fe catalyst.

3. Results and Discussion

The results were presented and discussed in two separate parts, where the first part focused on analysis of waste silk fibers before and after PDA incorporation and iron loading, and the second part conferred the catalytic behavior of prepared Fe loaded waste silk fibers towards removal of various dyes.

Part 1: Analysis of Waste Silk Fibers Before and After Dopamine Incorporation and Iron Loading

Morphological analysis using SEM, as well as elemental and functional group analysis by means of EDS X-ray and infrared spectroscopy was used to characterize waste silk samples before and after PDA grafting and iron loading immobilization.

3.1. Morphological Analysis

The changes in surface morphology of the waste silk after different treatment were investigated (see Figure 1). The untreated waste silk (Figure 1a) exhibits a network of a randomly overlapping fibers with a smooth surface. However, a uniform layer of cluster can be observed in PDA grafted waste silk (see Figure 1b), indicating that the PDA is successfully grafted onto the fibers surface through rapid oxidative polymerization. A large amount of particles are found successfully loaded on the surface of wSF-PDA and formed wSF-PDA/Fe. In the case of Fe loaded waste silk fibers (see Figure 1c), several patches appear as regularly shaped clusters [30,32] in the form of layers as demonstrated in Scheme 2. The presence of functional groups on treated waste silk has been further established by EDS X-ray and infrared analysis (see proceeding sections).

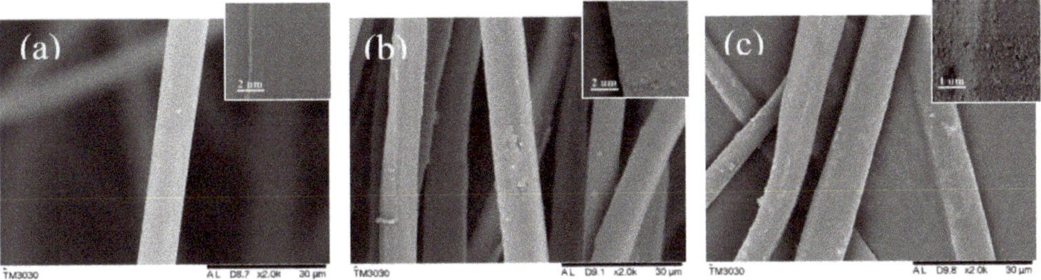

Figure 1. Morphology of waste silk after different treatments: (**a**) wSF; (**b**) wSF-PDA; (**c**) wSF-PDA/Fe.

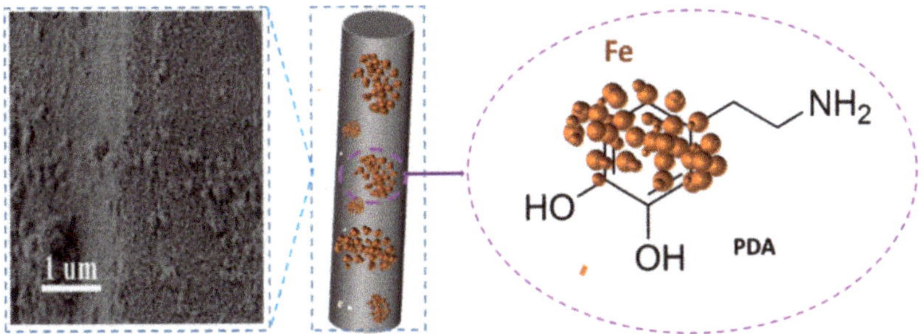

Scheme 2. Schematic illustration of iron loading on waste silk.

3.2. Fourier Transform Infrared (FT-IR) Analysis

In order to further investigate the changes in the structure of the waste silk surface after various surface modifications, the infrared spectrum analysis was carried out as shown in Figure 2. The untreated and treated waste silk demonstrated rough bands in the region 1660–1630 cm^{-1}, which were attributed to the stretching vibration of amide I. Other characteristics peaks found at 1545–1525 cm^{-1} are the amide II and 1265–1235 cm^{-1} is the amide III [33,34]. The FT-IR spectra of sample wSF-PDA and wSF-PDA/Fe showed a characteristic bending at 570 cm^{-1} [35,36], indicating the formation of Fe–O due to the presence of PDA [30] and subsequent loading of iron particles.

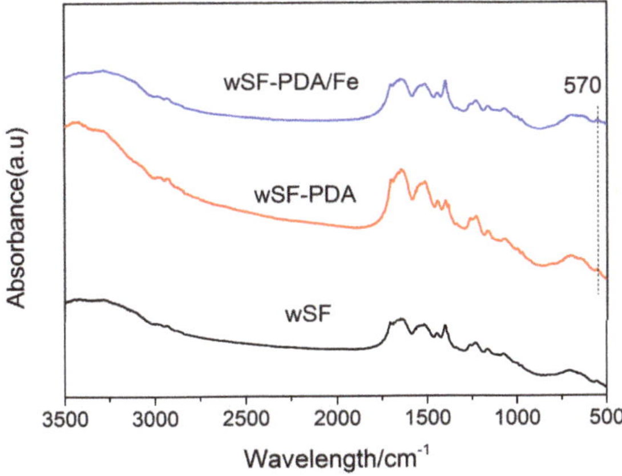

Figure 2. Infrared spectrum of untreated and functionalized waste silk.

3.3. Loading Analysis

EDS X-ray analysis further confirmed the changes in the chemical composition of the fabric surface before and after treatment (see Figures 3 and 4). The surface elemental composition of untreated and functionalized silk fibers is presented in Table 1. It can be concluded that the surface element ratio of N/C of the treated fabric is significantly smaller than the untreated sample [30], indicating successful grafting of PDA. After loading of iron, the peak intensity of Fe element (4.23%) can be noticed in wSF-PDA/Fe. A small quantity of iron element was noticed in wSF-PDA derived from the presence of some Fe-oxides in PDA.

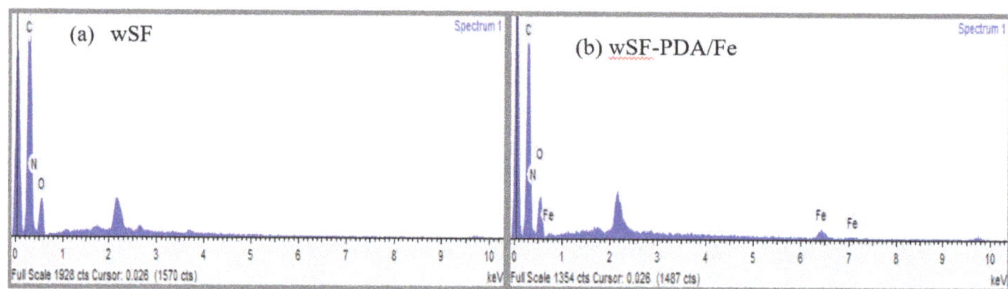

Figure 3. Energy dispersive spectra of (**a**) wSF and (**b**) wSF-PDA/Fe.

Table 1. Surface elemental composition and loading analysis of samples (energy dispersive spectroscopy (EDS) analysis).

Samples	Atomic Percentage of Surface Elements			
	C (%)	N (%)	O (%)	Fe (%)
wSF	71.07	10.99	17.94	0
wSF-PDA	76.14	5.56	17.13	1.17
wSF-PDA/Fe	74.47	0.12	21.18	4.23

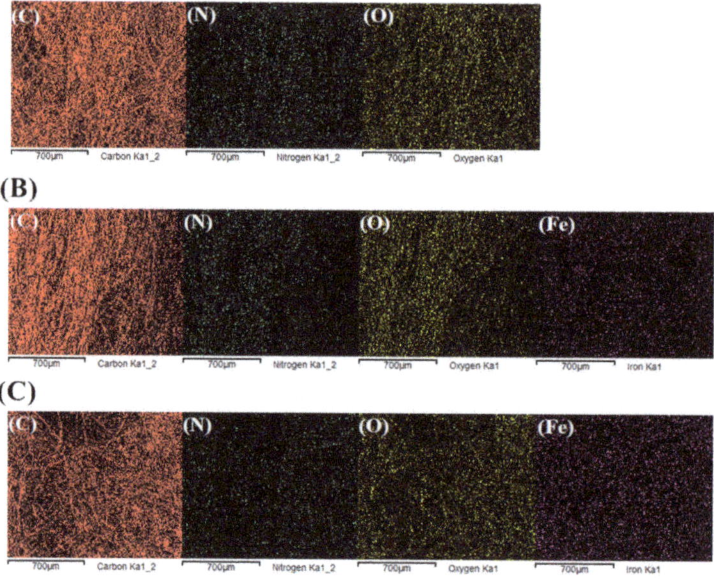

Figure 4. Energy Dispersive X-ray Spectroscopy (**A**) wS, (**B**)wSF-PDA, and (**C**) wSF-PDA/Fe.

Part 2: Analysis of Catalytic Fenton-Like Removal Toxic Water Pollutants

Three different dyes (methylene blue, cationic violet X-5BLN, and reactive orange GRN) were adopted to analyze the catalytic Fenton-like property of prepared fibrous catalyst (wSF-PDA/Fe). Dye removal as a function of time and influence of several factors, such as concentration of pollutants, reactants, temperature, and presence of electrolytes, has been investigated by means of UV-Vis colorimetric analysis.

3.4. Degradation of Dyes as a Function of Time

In order to investigate catalytic behavior of prepared iron loaded waste silk fibrous catalyst, Fenton-like degradation of methylene blue, cationic violet X-5BLN, and reactive orange GRN dye as a function of time was studied and analyzed by UV-Vis spectrophotometer (see Figure 5). Fenton reaction involves the combined use of ferrous ions and hydrogen peroxide (H_2O_2) to produce advanced oxidation potential active oxygen species capable of degrading organic contaminants. The dyes selected in this study are common industrial dyes containing an aromatic ring, which are resistant to traditional chemical and biological removal methods. As shown in Figure 5, the characteristic absorption peak intensity of methylene blue at 285 and 665 nm [35] decreased with the function of the reaction time. Once the reaction reached 10 min, the color almost completely disappeared, hence most degradation was achieved by 40 min of catalytic oxidation. Similar phenomena were found in the other two (cationic violet X-5BLN and reactive orange GRN) dyes, where characteristic absorption peak intensity of cationic violet X-5BLN at 275 and 590 nm and reactive orange GRN at 260 and 480 nm disappeared in 20 and 40 min, respectively. Iron loaded wSF fibrous catalyst showed good color removal efficiency, but no visible color removal was observed in control experiments performed using wSF-PDA or wSF-PDA/Fe without one or more reactants (Fe or H_2O_2) necessary to initiate Fenton reaction. Nonetheless, trivial reduction in color concentration was observed in various sample (wSF-PDA, wSF-PDA + H_2O_2, and wSF-PDA/Fe), which can be due to the adsorption characteristics of heterogeneous catalysts consistent with the literature [29,37,38].

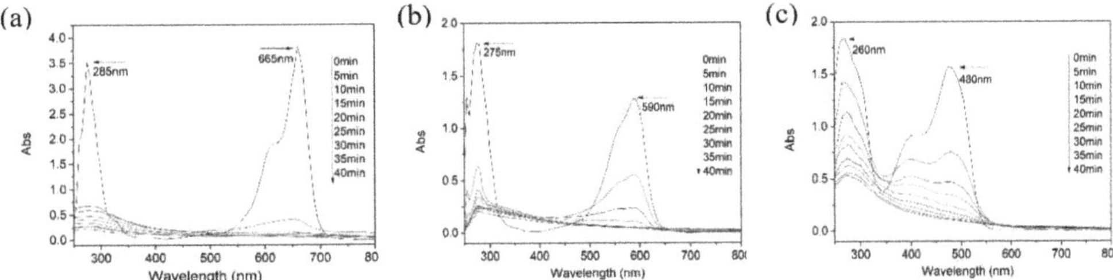

Figure 5. UV-visible absorption spectrum of (**a**) methylene blue, (**b**) cationic violet X-5BLN, and (**c**) reactive orange GRN removal as a function of time. (wSF-PDA/Fe 0.1 g, dye concentration 20 mg/L, H_2O_2 concentration 1.0 mmol/L, T = 50 °C, pH = 7).

3.4.1. Kinetics of Dye Degradation

The [instant/initial] absorbance ratio of the methylene blue, cationic violet X-5BLN, and reactive orange GRN dye at λ = 665, 590, and 480 nm (A_t/A_0) respectively, which accounts for the corresponding concentration ratio (C/C_0) and allows plotting of $\ln(C/C_0)$ as a function of time. Model validation of the pseudo-first-order kinetics for color removal with the catalyst is obtained by the linear evolution in time of $\ln(C/C_0)$, as supported by R^2 values 0.98, 0.99, and 0.98, respectively (see Table 2). Plots summarized in Table 2 show that dye degradation exhibited good linear relationships of $\ln(C/C_0)$ versus reaction time up to a certain time where maximum number of dyes was degraded following pseudo-first-order kinetics. These results are consistent with those found in previous reports [10,24,33,39–42].

Table 2. Pseudo First-order kinetics for the Fenton-like removal of methylene blue, cationic violet X-5BLN, and reactive orange GRN dye.

Samples	Methylene Blue			Cationic Violet X-5BLN			Reactive Orange GRN		
	[a]*Time* (min)	[b]k (min^{-1})	[c]R^2	[a]*Time* (min)	[b]k (min^{-1})	[c]R^2	[a]*Time* (min)	[b]k (min^{-1})	[c]R^2
wSF-PDA/Fe	10	0.576	0.982	25	0.157	0.997	40	0.065	0.984

[a] Reaction time required for color removal; [b] k: rate constant for the 1st order kinetics and is expressed in min^{-1}; [c] R^2: correlation coefficient of the linear regression.

3.4.2. Postulated Mechanism of Dye Removal

Based on the results explained in the above sections, a plausible mechanism for removal of all three dyes has been postulated. The underlying mechanisms of removal of methylene blue, cationic violet X-5BLN, and reactive orange GRN dyes in the system of iron loaded PDA grafted waste silk fibrous catalyst were considered for Fenton-like reaction in presence of hydrogen peroxide. The removal of all dyes was attributed to the synergistic effect caused by free radicals and other reactive species formed though a heterogeneous Fenton reaction [29]. The produced free radicals oxidize the dyes into colorless nontoxic substances as illustrated in Scheme 3.

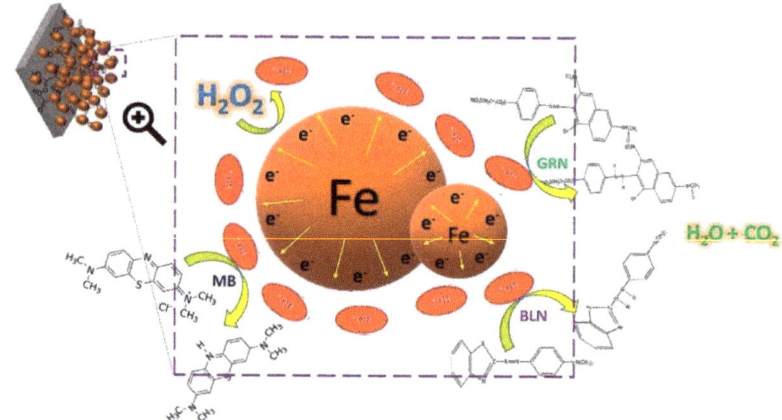

Scheme 3. Schematic postulated main mechanism of degradation of methylene blue, cationic violet X-5BLN, and reactive orange GRN dyes.

The postulated main reaction mechanism involves three steps as follows:

(i)   The process of producing reactive species

$$Fe^{2+} + H_2O_2 \rightarrow Fe^{3+} + HO^- + \bullet OH \tag{1}$$

(ii)   The process of color removal of dyes

$$Dyes + \bullet OH \rightarrow Reaction\ intermediates \tag{2}$$

(iii)   The process of degradation

$$Reaction\ intermediates + \bullet OH \rightarrow CO_2 + H_2O \tag{3}$$

3.5. Factors Influencing the Removal of Dyes

3.5.1. Effect of Different Samples

It can be seen from Figure 6 that different samples have a great influence on the degradation dye, and wSF-PDA/Fe-H_2O_2 is the most effective for degrading dyes. The addition of wSF-PDA mainly brought about the adsorption of dyes by silk fibers and dopamine dominantly, and the dye removal rate could not meet the requirements. The addition of wSF-PDA and H_2O_2 resulted in a lower degradation rate due to the weaker oxidative decomposition of H_2O_2 itself and the weak adsorption of wSF-PDA. The wSF-PDA/Fe also played a dominant role in adsorption, and the dye removal rate was not significant. The addition wSF-PDA/Fe and H_2O_2 generated a strong reaction system, which effectively increased the release rate of hydroxyl radicals and degraded most of the dyes in a short time.

In addition, it is interesting that wSF-PDA and wSF-PDA-H_2O_2 have better degradation for methylene blue and cationic violet X-5BLN. This may be due to the negative charge on the surface of the wSF-PDA sample, which can better absorb cationic dyes and achieve effective degradation.

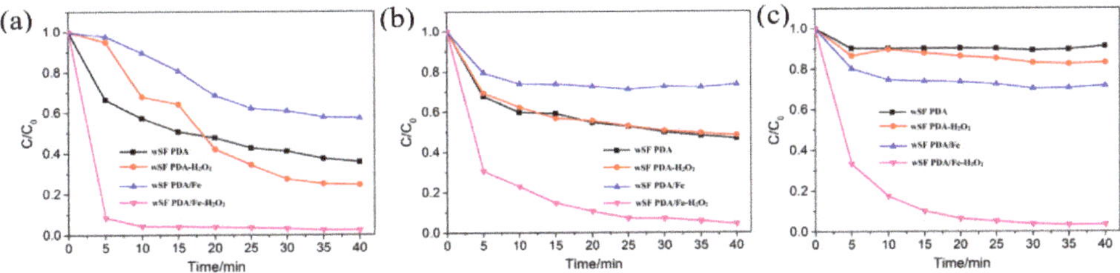

Figure 6. The C/C_0 of different samples in the three dye solutions (**a**) methylene blue, (**b**) cationic violet X-5BLN, and (**c**) active orange GRN. (wSF-PDA/Fe 0.1 g, dye concentration 20 mg/L, H_2O_2 concentration 1.0 mmol/L, T = 50 °C, pH = 7).

3.5.2. Effect of H_2O_2 Concentration

The effect of H_2O_2 concentration (0.05, 0.1, 0.5, 1, 3, and 5 mmol/L) on the removal of dyes was studied. The evaluation of color removal in terms C/C_0 as shown in Figure 7 shows that, H_2O_2 concentration below 0.5 mmol/L showed poor or insufficient dye removal. However, 0.5 to 5 mmol/L of H_2O_2 concentration showed most removal of dyes in similar experimental condition. When the concentration of H_2O_2 was 0.05 and 0.1 mmol/L, the dye degradation rates after reaction for 40 min were 48.1% and 57.5%, respectively. At this point, it can be clearly seen that the low concentration of H_2O_2 was too diluted to react with the mineralized iron to form sufficient hydroxyl radicals to fully degrade methylene blue. When the concentration of H_2O_2 was gradually increased to 0.5, 1, 3, and 5 mmol/L, the final degradation rate reached nearly 97% to 98%, but the efficiency of the reaction was also quite different. The H_2O_2 concentration of 1 mmol/L was the appropriate, and the degradation rate was the faster and saturated within 15–20 min, while the reaction rate of high concentration H_2O_2 was decreased. This phenomenon indicated that high concentration of H_2O_2 cannot increase the degradation rate as linear regression rather showed no improvement after saturation. The similar phenomena can be noticed for other two dyes (cationic violet X-5BLN and reactive orange GRN) as well (see Figure 7b,c).

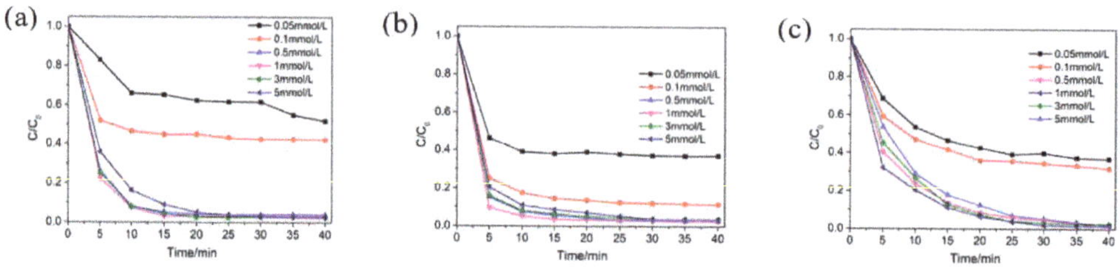

Figure 7. Evolution in time of C/C_0 of (**a**) methylene blue, (**b**) cationic violet X-5BLN, and (**c**) reactive orange GRN removal as a function of H_2O_2 concentration. (wSF-PDA/Fe 0.1 g, dye concentration 20 mg/L, T = 50 °C, pH = 7).

3.5.3. Effect of Dye Concentration

The concentration of dyes is a precarious parameter of color removal rate, which influences the effectiveness of the removal process. The effect of dye concentration on the removal of dyes were studied in terms of different concentration (10, 20, 40, 60, and 80 mg/L) of dye solutions. Results presented in Figure 8 show a certain influence on the removal rate of all three dyes due to the variation in concentration of the dye solution. The results show (see Figure 8) the most removal of dyes at the initial concentrations from 10–20 mg/L, which moderates upon increase in dye concentration for a specific reaction time. However, for the maximum concentration (80 mg/L) of dye solution, the removal percentage remains above 80%. The highest 90% of removal rate on the maximum concentration are recorded for methylene blue. This may be due to the fact that in the case of high dye concentration, PDA grafted iron loaded waste silk catalyst forms a stable reaction system with H_2O_2, at which time the hydroxyl radical is not easily deactivated, and the probability of contact with the dye increases, eventually resulting in a high removal.

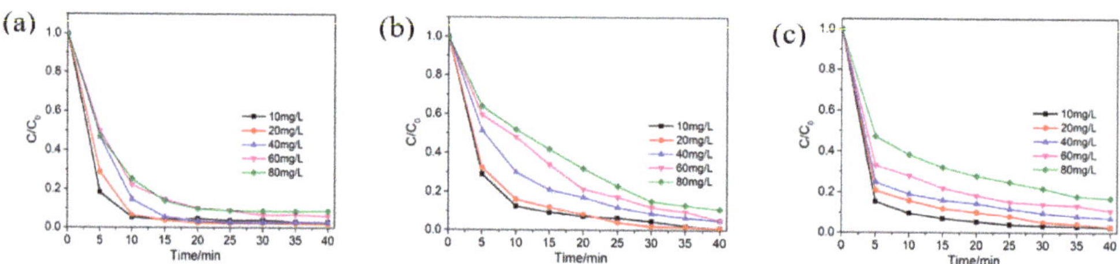

Figure 8. Evolution in time of C/C_0 of (**a**) methylene blue, (**b**) cationic violet X-5BLN, and (**c**) reactive orange GRN removal as a function of dye concentration. (wSF-PDA/Fe 0.1 g, H_2O_2 concentration 1.0 mmol/L, T = 50 °C, pH – 7).

3.5.4. Effect of Reaction Temperature

Temperature in a catalytic reaction might increase or decrease the reaction rate, thus the effect of reaction temperature (25, 50, and 75 °C) on the removal of methylene blue, cationic violet X-5BLN, and reactive orange GRN dyes were studied and the results are shown in Figure 9. It can be seen from Figure 9a–c that the reaction temperature has a significant influence on the removal of dyes. In the range of 25–75 °C, the dye removal rate increases with increasing temperature. A slight increase in removal rate can be found in methylene blue removal (see Figure 9a). Whereas, a significant increase has been noticed in cationic violet X-5BLN and reactive orange GRN dyes. Figure 9c shows that, when the reaction temperature is 25 °C, the final dye removal rate of active orange GRN is 77.2% after 40 min reaction; when the reaction temperature is 50 and 75 °C, the final dye removal rate of active orange GRN is about 98.5% after 40 min reaction. However, it is clear that the reaction rate at 75 °C is significantly faster than 50 °C. This phenomenon indicates that the reaction rate under high temperature conditions accelerates the decomposition of H_2O_2 into hydroxyl radicals, which increases the velocity of dye degradation reaction. Similar phenomena are found in removal of cationic violet X-5BLN (see Figure 9b).

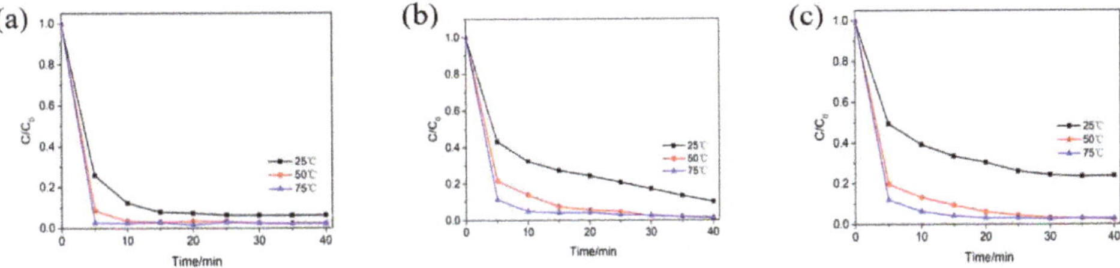

Figure 9. Evolution in time of C/C_0 of (**a**) methylene blue, (**b**) cationic violet X-5BLN, and (**c**) reactive orange GRN removal as a function of reaction temperature. (wSF-PDA/Fe was 0.1 g, dye concentration 20 mg/L, H_2O_2 concentrations 1.0 mmol/L, pH = 7).

3.5.5. Effect of Different Electrolytes

The traditional dye house uses significant amounts of electrolytes during textile processing, so that considerable amounts of electrolytes are present in wastewater. Consequently, the effect of different electrolytes (NaCl, Na_2SO_4) on effectiveness of removal of dyes were studied. Stimulating phenomena are noticed in the results as shown in Figure 10. It can be seen that the electrolyte does influence the removal of degradation of methylene blue in a limited way. For cationic violet X-5BLN and reactive orange GRN, the effect of electrolyte is significant. This may be due to the fact that methylene blue belongs to the class of easily degradable dyes, and the structure of cationic violet X-5BLN and reactive orange GRN are more complicated. Therefore, the electrolyte has a greater influence on the latter two dyes. For different electrolytes, Cl^- and SO_4^{2-} can capture and destroy •OH [43,44], affecting the degradation process of dyes. In general, wSF-PDA/Fe still plays an important role in the degradation of dyes.

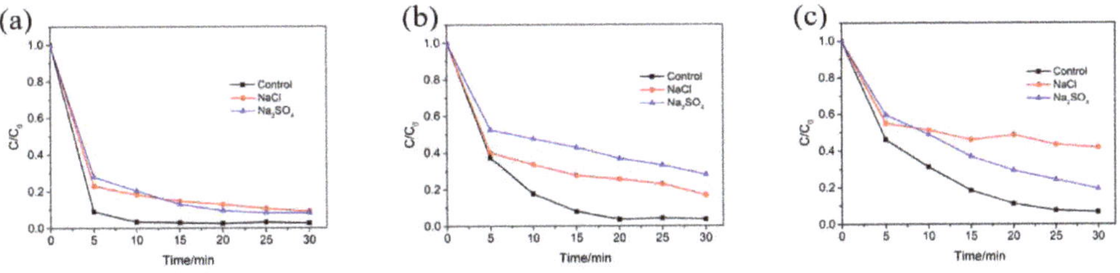

Figure 10. Evolution in time of C/C_0 of (**a**) methylene blue, (**b**) cationic violet X-5BLN, and (**c**) reactive orange GRN removal as a function of electrolyte. (wSF-PDA/Fe 0.1 g, dye concentration 20 mg/L, H_2O_2 concentration 1.0 mmol/L, T = 50 °C, NaCl, Na_2SO_4 concentration 40 g/L, pH = 7).

4. Conclusions

In this report, the loading of iron particle on dopamine grafted waste silk fibers were achieved successfully. The resultant material was investigated by means of physicochemical and catalytic property. SEM and EDS analysis confirmed the loading of PDA and iron particles on the surface of waste silk fibers, which revealed the potentiality of the prepared material as a catalyst in possible Fenton-like removal of aromatic dyes. The results of investigation of H_2O_2 concentration, dye concentration, temperature, and electrolytes on dye removal indicated the prepared catalyst showed significant 98%–99% removal dyes in 10–40 min depending on the concentration of H_2O_2 used. Postulate mechanism showed the most degradation of pollutants into nontoxic substance. Thus, this study can provide rational foundation towards further advancement in fiber based catalyst for Fenton-like removal of toxic pollutants in water.

Author Contributions: T.X. and G.C. conceived and designed the experiments; M.S.M., B.Y., and X.Z. performed the experiments and analyzed the data; M.S.M., B.Y., and T.X. wrote the paper. All authors discussed the results and improved the final text of the paper.

Funding: This work was supported by the National Natural Science Foundation of China (51973144, 51741301); the Major Program of Natural Science Research of Jiangsu Higher Education Institutions of China (18KJA540002); the Priority Academic Program Development of Jiangsu Higher Education Institutions (PAPD).

Conflicts of Interest: Authors declare no conflict of interest.

References

1. Singh, C.; Goyal, A.; Singhal, S. Nickel-doped cobalt ferrite nanoparticles: Efficient catalysts for the reduction of nitroaromatic compounds and photo-oxidative degradation of toxic dyes. *Nanoscale* **2014**, *6*, 7959–7970. [CrossRef] [PubMed]

2. Fernández, C.; Larrechi, M.S.; Callao, M.P. An analytical overview of processes for removing organic dyes from wastewater effluents. *Trends Anal. Chem.* **2010**, *29*, 1202–1211. [CrossRef]

3. Kovacic, P.; Somanathan, R. Nitroaromatic compounds: Environmental toxicity, carcinogenicity, mutagenicity, therapy and mechanism. *J. Appl. Toxicol.* **2014**, *34*, 810–824. [CrossRef] [PubMed]

4. Robinson, T.; McMullan, G.; Marchant, R.; Nigam, P. Remediation of dyes in textile effluent: A critical review on current treatment technologies with a proposed alternative. *Bioresour. Technol.* **2001**, *77*, 247–255. [CrossRef]

5. Wang, Y.; Mo, Z.; Zhang, P.; Zhang, C.; Han, L.; Guo, R.; Gou, H.; Wei, X.; Hu, R. Synthesis of flower-like TiO_2 microsphere/graphene composite for removal of organic dye from water. *Mater. Des.* **2016**, *99*, 378–388. [CrossRef]

6. Azizi, A.; Abouseoud, M.; Amrane, A. Phenol Removal by a Sequential Combined Fenton-Enzymatic Process. *Nat. Environ. Pollut. Technol.* **2017**, *16*, 321.

7. Liang, S.X.; Jia, Z.; Zhang, W.C.; Wang, W.M.; Zhang, L.C. Rapid malachite green degradation using Fe73. 5Si13. 5B9Cu1Nb3 metallic glass for activation of persulfate under UV–Vis light. *Mater. Des.* **2017**, *119*, 244–253. [CrossRef]

8. Deb, H.; Xiao, S.; Morshed, M.N.; Al Azad, S. Immobilization of Cationic Titanium Dioxide ($TiO_2{}^+$) on ElectrospunNanofibrous Mat: Synthesis, Characterization, and Potential Environmental Application. *Fiber Polym.* **2018**, *19*, 1715–1725. [CrossRef]

9. Nidheesh, P.V.; Gandhimathi, R.; Ramesh, S.T. Degradation of dyes from aqueous solution by Fenton processes: A review. *Environ. Sci. Pollut. Res.* **2013**, *20*, 2099–2132. [CrossRef]

10. Nabil, B.; Morshed, M.N.; Nemeshwaree, B.; Christine, C.; Julien, V.; Olivier, T.; Abdelkrim, A. Development of new multifunctional filter based nonwovens for organics pollutants reduction and detoxification: High catalytic and antibacterial activities. *Chem. Eng. J.* **2019**, *356*, 702–716. [CrossRef]

11. Vieillard, J.; Bouazizi, N.; Mohammad, N.M.; Thomas, C.; Florie, D.; Radhouane, B.; Thébault, P.; Lesouhaitier, O.; Le Derf, F.; Azzouz, A. CuONanosheets Modified with Amine and Thiol Grafting for High Catalytic and Antibacterial Activities. *Ind. Eng. Chem. Res.* **2019**, *58*, 10179–10189. [CrossRef]

12. Augugliaro, V.; Palmisano, L.; Sclafani, A.; Minero, C.; Pelizzetti, E. Photocatalytic degradation of phenol in aqueous titanium dioxide dispersions. *Toxicol. Environ. Chem.* **1988**, *16*, 89–109. [CrossRef]

13. Martínez, F.; Calleja, G.; Melero, J.A.; Molina, R. Heterogeneous photo-Fenton degradation of phenolic aqueous solutions over iron-containing SBA-15 catalyst. *Appl. Catal. B* **2005**, *60*, 181–190. [CrossRef]

14. Hameed, B.H.; Lee, T.W. Degradation of malachite green in aqueous solution by Fenton process. *J. Hazard. Mater.* **2009**, *164*, 468–472. [CrossRef]

15. Babuponnusami, A.; Muthukumar, K. Degradation of phenol in aqueous solution by fenton, sono-fenton and sono-photo-fentonmethods. *Clean Soil Air Water* **2011**, *39*, 142–147. [CrossRef]

16. Chen, D.; Chen, S.; Jiang, Y.; Xie, S.; Quan, H.; Hua, L.; Luo, X.; Guo, L. Heterogeneous Fenton-like catalysis of Fe-MOF derived magnetic carbon nanocomposites for degradation of 4-nitrophenol. *RSC Adv.* **2017**, *7*, 49024–49030. [CrossRef]

17. Ma, Y.S.; Huang, S.T.; Lin, J.G. Degradation of 4-nitrophenol using the Fenton process. *Water Sci. Technol.* **2000**, *42*, 155–160. [CrossRef]

18. Karimi, A.; Aghbolaghy, M.; Khataee, A.; ShoaBargh, S. Use of enzymatic bio-Fenton as a new approach in decolorization of malachite green. *Sci. World J.* **2012**, *2012*. [CrossRef]

19. Hermosilla, D.; Cortijo, M.; Huang, C.P. The role of iron on the degradation and mineralization of organic compounds using conventional Fenton and photo-Fenton processes. *Chem. Eng. J.* **2009**, *155*, 637–646. [CrossRef]

20. Morshed, M.N.; Shen, X.L.; Zhang, X.Y.; Deb, H.; Azad, S.A. Effective Removal of Colorants from Aqueous Suspension through Cellulose Nano whisker Templated Titanium Dioxide/Cellulose Nanocomposite. In Proceedings of the 10th Textile Bioengineering and Informatics Symposium, Wuhan, China, 16–19 May 2017; pp. 1166–1173.

21. Hermosilla, D.; Merayo, N.; Ordóñez, R.; Blanco, Á. Optimization of conventional Fenton and ultraviolet-assisted oxidation processes for the treatment of reverse osmosis retentate from a paper mill. *Waste Manag.* **2012**, *32*, 1236–1243. [CrossRef]

22. Hermosilla, D.; Cortijo, M.; Huang, C.P. Optimizing the treatment of landfill leachate by conventional Fenton and photo-Fenton processes. *Sci. Total Environ.* **2009**, *407*, 3473–3481. [CrossRef] [PubMed]

23. Yang, G.C.; Tu, H.C.; Hung, C.H. Stability of nanoiron slurries and their transport in the subsurface environment. *Sep. Purif. Technol.* **2007**, *58*, 166–172. [CrossRef]

24. Deb, H.; Morshed, M.N.; Xiao, S.; Al Azad, S.; Cai, Z.; Ahmed, A. Design and development of TiO_2-Fe^0 nanoparticle-immobilized nanofibrous mat for photocatalytic degradation of hazardous water pollutants. *J. Mater. Sci.* **2019**, *30*, 4842–4854. [CrossRef]

25. He, F.; Zhao, D. Preparation and characterization of a new class of starch-stabilized bimetallic nanoparticles for degradation of chlorinated hydrocarbons in water. *Environ. Sci. Technol.* **2005**, *39*, 3314–3320. [CrossRef] [PubMed]

26. Wang, W.; Jin, Z.H.; Li, T.L.; Zhang, H.; Gao, S. Preparation of spherical iron nanoclusters in ethanol—water solution for nitrate removal. *Chemosphere* **2006**, *65*, 1396–1404. [CrossRef] [PubMed]

27. Kim, H.; Hong, H.J.; Lee, Y.J.; Shin, H.J.; Yang, J.W. Degradation of trichloroethylene by zero-valent iron immobilized in cationic exchange membrane. *Desalination* **2008**, *223*, 212–220. [CrossRef]

28. Choi, H.; Al-Abed, S.R.; Agarwal, S.; Dionysiou, D.D. Synthesis of reactive nano-Fe/Pd bimetallic system-impregnated activated carbon for the simultaneous adsorption and dechlorination of PCBs. *Chem. Mater.* **2008**, *20*, 3649–3655. [CrossRef]

29. Morshed, M.N.; Bouazizi, N.; Behary, N.; Guan, J.; Nierstrasz, V. Stabilization of zero valent iron (Fe^0) on plasma/dendrimer functionalized polyester fabrics for Fenton-like removal of hazardous water pollutant. *Chem. Eng. J.* **2019**, *374*, 658–673. [CrossRef]

30. Yan, B.; Zhou, Q.; Zhu, X.; Guo, J.; Mia, M.S.; Yan, X.; Chen, G.; Xing, T. A superhydrophobic bionic coating on silk fabric with flame retardancy and UV shielding ability. *Appl. Surf. Sci.* **2019**, *483*, 929–939. [CrossRef]

31. Bouazizi, N.; Vieillard, J.; Bargougui, R.; Couvrat, N.; Thoumire, O.; Morin, S.; Ladam, G.; Mofaddel, N.; Brun, N.; Azzouz, A.; et al. Entrapment and stabilization of iron nanoparticles within APTES modified graphene oxide sheets for catalytic activity improvement. *J. Alloy. Compd.* **2019**, *771*, 1090–1102. [CrossRef]

32. Zhou, Q.; Wu, W.; Zhou, S.; Xing, T.; Sun, G.; Chen, G. Polydopamine-induced growth of mineralized γ-FeOOH nanorods for construction of silk fabric with excellent superhydrophobicity, flame retardancy and UV resistance. *Chem. Eng. J.* **2019**, *122988*. [CrossRef]

33. Murphy, A.R.; Kaplan, D.L. Biomedical applications of chemically-modified silk fibroin. *J. Mater. Chem.* **2009**, *19*, 6443–6450. [CrossRef] [PubMed]

34. Huang, F.; Wei, Q.; Liu, Y.; Gao, W.; Huang, Y. Surface functionalization of silk fabric by PTFE sputter coating. *J. Mater. Sci.* **2007**, *42*, 8025–8028. [CrossRef]

35. Morshed, M.N.; Bouazizi, N.; Behary, N.; Vieillard, J.; Thoumire, O.; Nierstrasz, V.; Azzouz, A. Iron-loaded amine/thiol functionalized polyester fibers with high catalytic activities: A comparative study. *Dalton Trans.* **2019**, *48*, 8384–8399. [CrossRef]

36. Silva, V.A.J.; Andrade, P.L.; Silva, M.P.C.; Valladares, L.D.L.S.; Aguiar, J.A. Synthesis and characterization of Fe_3O_4 nanoparticles coated with fucanpolysaccharides. *J. Magn. Magn. Mater.* **2013**, *343*, 138–143. [CrossRef]

37. Ivanets, A.; Roshchina, M.; Srivastava, V.; Prozorovich, V.; Dontsova, T.; Nahirniak, S.; Pankov, V.; Hosseini-Bandegharaei, A.; Tran, H.N.; Sillanpää, M. Effect of metal ions adsorption on the efficiency of Methylene Blue degradation onto $MgFe_2O_4$ as Fenton-like catalysts. *Colloids Surf. A* **2019**, *571*, 17–26. [CrossRef]

38. Tekbaş, M.; Yatmaz, H.C.; Bektaş, N. Heterogeneous photo-Fenton oxidation of reactive azo dye solutions using iron exchanged zeolite as a catalyst. *Microporous Mesoporous Mater.* **2008**, *115*, 594–602. [CrossRef]

39. Dantas, T.L.P.; Mendonca, V.P.; Jose, H.J.; Rodrigues, A.E.; Moreira, R.F.P.M. Treatment of textile wastewater by heterogeneous Fenton process using a new composite Fe_2O_3/carbon. *Chem. Eng. J.* **2006**, *118*, 77–82. [CrossRef]

40. Sharma, G.; Kumar, A.; Sharma, S.; Ala'a, H.; Naushad, M.; Ghfar, A.A.; Ahamad, T.; Stadler, F.J. Fabrication and characterization of novel Fe^0@ Guar gum-crosslinked-soya lecithin nanocomposite hydrogel for photocatalytic degradation of methyl violet dye. *Sep. Purif. Technol.* **2019**, *311*, 895–908. [CrossRef]

41. Muruganandham, M.; Swaminathan, M. Decolourisation of Reactive Orange 4 by Fenton and photo-Fenton oxidation technology. *Dyes Pigment.* **2004**, *63*, 315–321. [CrossRef]

42. Gupta, A.K.; Pal, A.; Sahoo, C. Photocatalytic degradation of a mixture of Crystal Violet (Basic Violet 3) and Methyl Red dye in aqueous suspensions using Ag^+ doped TiO_2. *Dyes Pigment.* **2006**, *69*, 224–232. [CrossRef]

43. Khattri, S.D.; Singh, M.K. Removal of malachite green from dye wastewater using neem sawdust by adsorption. *J. Hazard. Mater.* **2009**, *167*, 1089–1094. [CrossRef] [PubMed]

44. Nadtochenko, V.; Kiwi, J. Photochemical reactions in the photo-Fenton system with ferric chloride. 2. Implications of the precursors formed in the dark. *Environ. Sci. Technol.* **1998**, *32*, 3282–3285. [CrossRef]

Article

Polystyrene Magnetic Nanocomposites as Antibiotic Adsorbents

Leili Mohammadi [1], Abbas Rahdar [2],* , Razieh Khaksefidi [3], Aliyeh Ghamkhari [4], Georgios Fytianos [5] and George Z. Kyzas [5],*

[1] PhD of Environmental Health, Infectious Diseases and Tropical Medicine Research Center, Resistant Tuberculosis Institute, Zahedan University of Medical Sciences, Zahedan 98167-43463, Iran; lailimohamadi@gmail.com

[2] Department of Physics, Faculty of Science, University of Zabol, Zabol 538-98615, Iran

[3] Department of Environmental Health, Zahedan University of Medical Sciences, Zahedan 98167-43463, Iran; r.khaksefidi110@gmail.com

[4] Institute of Polymeric Materials, Faculty of Polymer Engineering, Sahand University of Technology, Tabriz 51335-1996, Iran; aliyeh_ghamkhari@yahoo.com

[5] Department of Chemistry, International Hellenic University, Kavala 65404, Greece; gfytianos@gmail.com

* Correspondence: a.rahdar@uoz.ac.ir (A.R.); kyzas@chem.ihu.gr (G.Z.K.); Tel.: +30-2510-462-218 (G.Z.K.)

Received: 16 May 2020; Accepted: 5 June 2020; Published: 9 June 2020

Abstract: There are different ways for antibiotics to enter the aquatic environment, with wastewater treatment plants (WWTP) considered to be one of the main points of entrance. Even treated wastewater effluent can contain antibiotics, since WWTP cannot eliminate the presence of antibiotics. Therefore, adsorption can be a sustainable option, compared to other tertiary treatments. In this direction, a versatile synthesis of poly(styrene-block-acrylic acid) diblock copolymer/Fe_3O_4 magnetic nanocomposite (abbreviated as P(St-*b*-AAc)/Fe_3O_4)) was achieved for environmental applications, and particularly for the removal of antibiotic compounds. For this reason, the synthesis of the P(St-*b*-AAc) diblock copolymer was conducted with a reversible addition fragmentation transfer (RAFT) method. Monodisperse superparamagnetic nanocomposite with carboxylic acid groups of acrylic acid was adsorbed on the surface of Fe_3O_4 nanoparticles. The nanocomposites were characterized with scanning electron microscopy (SEM), X-ray diffraction (XRD) and vibrating sample magnetometer (VSM) analysis. Then, the nanoparticles were applied to remove ciprofloxacin (antibiotic drug compound) from aqueous solutions. The effects of various parameters, such as initial drug concentration, solution pH, adsorbent dosage, and contact time on the process were extensively studied. Operational parameters and their efficacy in the removal of Ciprofloxacin were studied. Kinetic and adsorption isothermal studies were also carried out. The maximum removal efficiency of ciprofloxacin (97.5%) was found at an initial concentration of 5 mg/L, pH 7, adsorbent's dosage 2 mg/L, contact time equal to 37.5 min. The initial concentration of antibiotic and the dose of the adsorbent presented the highest impact on efficiency. The adsorption of ciprofloxacin was better fitted to Langmuir isotherm (R^2 = 0.9995), while the kinetics were better fitted to second-order kinetic equation (R^2 = 0.9973).

Keywords: ciprofloxacin; Polystyrene nanocomposite; modifications; adsorption; characterizations

---

1. Introduction

Aside from the well-known pollutants and contaminants in the aquatic environment, compounds of emerging concern (CECs) may impact aquatic life even in very low concentrations [1]. Wastewater influents and effluents can contain CECs, due to their presence in everyday products, such as detergents, fabric coatings, pharmaceuticals, cosmetics, beverages and food packaging [2]. Pharmaceuticals are being detected in drinking and surface water, and although not very persistent, the continuous re-entering increases their abundance, and renders them pseudo-persistent [3]. Pharmaceuticals, include diverse types of compounds, e.g., antibiotics and show low biodegradability. CECs cannot be removed completely by wastewater treatment plants (WWTPs) [2], since WWTP were not designed to treat CECS. In some cases, even less than 10% of CECs is removed, making WWTP effluents a major factor for introducing CECs into the environment [2]. Recently, great attention is given to adsorption technique [4–19], which is easily applied to the last stage of wastewater treatment plants (WWTPs), with the aim of removing all residues that were not separated and removed from the previous stages.

In particular, special attention is given to find appropriate ways to effectively treat antibiotics from effluents, due to their strong resistance to various decontamination techniques [20]. Available statistics indicate that 100–200 tn of antibiotics are used annually worldwide. As a result, the risk of water resources contamination by these compounds is very high. The residue of those antibiotics in the form of major constituents or metabolites has also been observed in WWTP.

It is noteworthy to mention that the inability of WWTP to remove antibiotics leads to the discharge of those compounds into surface water and underground waters. The inadequate and incorrect use of those compounds, and their continuous entry into the environment, leads to biodistribution and faulty resistance [21,22]. Of the large antibiotic classes, fluoroquinolones are worth mentioning. Antibiotics in this family include Ciprofloxacin (CIP), epinephrine, and norfloxacin. The presence of fluorine atoms in combination with these antibiotics makes these compounds particularly stable, so they are considered to be very dangerous and toxic pollutants in the environment. CIP is detected in sewage and surface water in medical effluents and pharmaceutical plants. The antibiotic can be adsorbed into the sludge and, if applied as fertilizer, it is accumulated in the soil and enters into plants [23]. CIP was observed in surface waters and wastewaters at concentrations below 1 μg/L, while in medical wastewaters in 150 μg/L. Therefore, it is mandatory to find and apply an efficient method for ciprofloxacin removal.

The most important methods used to remove and separate the drug compounds from water and sewage include ozonation, nanofiltration, electron radiation, ion exchange, chemical coagulation and photocatalytic oxidation, all of which have high performance and operation costs [24–29]. Nowadays, nanotechnologies are mainly used in water and wastewater treatment, using materials like iron nanoparticles, zeolites and magnetic nanomaterials [30,31]. Among the various methods, adsorption is a simple, environmental friendly, fast, highly efficient and low-cost solution, making it one of the most favorable methods [32–37].

The removal of pharmaceuticals by adsorption has been the focus of many studies. So far, as adsorptive materials, activated carbon [38–41] or zeolites [42,43] have been widely used in wastewater treatment. The removal of pharmaceuticals by adsorption shows great potential, due to its easy application into existing water treatment processes. On the other hand, issues regarding adsorbent stability and regeneration costs lead to R&D of innovative and effective adsorbents from polymeric materials. Adsorption processes, such as activated carbon-based have high capital cost, and ineffectiveness and non-selectivity against vat and disperse dyes. Furthermore, saturated carbon regeneration is expensive and leads to adsorbent loss. Depending on the demand, cost, and the nature of the pollutant to be adsorbed, the adsorbents are either disposed or regenerated for future use. The regeneration process of adsorbents needs to be cheap and environmental friendly by recovering valuable adsorbates while reducing the need of virgin adsorbents.

In this study, a versatile synthesis of poly(styrene-block-acrylic acid) diblock copolymer/Fe_3O_4 magnetic nanocomposite (abbreviated as P(St-*b*-AAc)/Fe_3O_4)) was achieved for environmental applications with a focus on the removal of ciprofloxacin. The nanocomposites were characterized with SEM, XRD and VSM analysis. The nanoparticles were then applied to remove ciprofloxacin (antibiotic drug compound) from aqueous solutions, evaluating the effect of certain important parameters such as the solution's pH, initial ciprofloxacin concentration, adsorbent dosage and contact time.

2. Materials and Methods

2.1. Materials

To begin, 4-cyano-4-[(phenylcarbothioyl) sulfanyl] pentanoic acid, as a RAFT agent, was synthesized [32]. Acrylic acid (AAc), styrene (St) monomers, 2, 2-azobisisobutyronitrile (AIBN), and dimethylformamide (DMF), $FeCl_2 \cdot 4H_2O$, 99% and $FeCl_3 \cdot 6H_2O$, 98% were purchased from Merck (Darmstadt, Germany).

The antibiotic model compound used in the present study is ciprofloxacin, purchased from Merck (Germany). Its molecular structure is presented in Figure 1. When it comes to ciprofloxacin's dissociation and isoelectric constants, the isoelectric point has a value of pI = 7.14, which is calculated by the average of pKa_1 = 6.09 and pKa_2 = 8.62. This portrays the two ionizable functional groups of ciprofloxacin; the 6-carboxylic group and the N-4 of the piperazine substituent. pKa1 corresponds to the dissociation of a proton from the carboxyl group, and pKa_2 corresponds to the dissociation of a proton from the N-4 in the piperazinyl group [44].

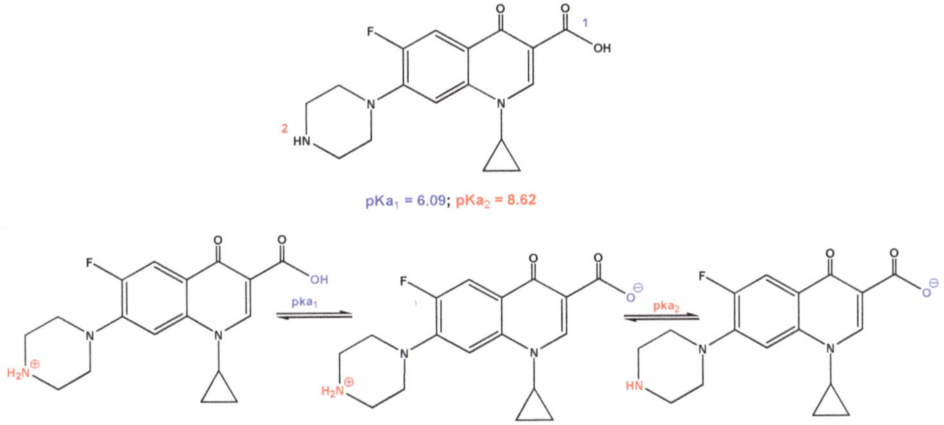

Figure 1. Chemical structure of CIP and its ionizable forms.

2.2. Synthesis of Poly(styrene) Homopolymer

RAFT agents (10 mg, 0.036 mmol), styrene monomer (4 mL, 34.96 mmol) and AIBN (3.0 mg, mmol) were added in a 100-mL flask; the reaction was achieved with three freeze pump-thaw cycles under a nitrogen atmosphere. The solution was put to an oil-bath with a temperature of 75 °C for 24 h. The flask was then quenched by cooling. The polystyrene homopolymer was precipitated in methanol. Finally, drying of the product under vacuum at 25 °C for 24 h took place [22].

2.3. Synthesis of Poly(styrene-block-acrylic acid), Sphere Superparamagnetic Iron Oxide Nanoparticles (SPIONs) and Poly(St-b-AAc)/Fe₃O₄ Supermagnetic Nanocomposite

Macro-RAFT agent (PSt, 200 mg, 19.8 mmol), AAc monomer (1.56 mL, 28.24 mmol), AIBN (3 mg, mmol) and DMF (10 mL) were charged in a two-neck reactor. The reaction was induced using three freeze pump-thaw cycles under a nitrogen atmosphere. The reaction solution was put to an oil-bath with a temperature of 75 °C for 24 h. The reaction mixture was then precipitated in cold diethyl ether (150 mL) and dried under vacuum at 25 °C. The SPIONs were synthesized using a co-precipitation method, as described in literature [22,45]. The poly(St-*b*-AAc)/Fe₃O₄ supermagnetic nanocomposite was synthesized as described in a previous study [46]. The final product is magnetic nanocomposite [22] and its structure is illustrated in Figure 2.

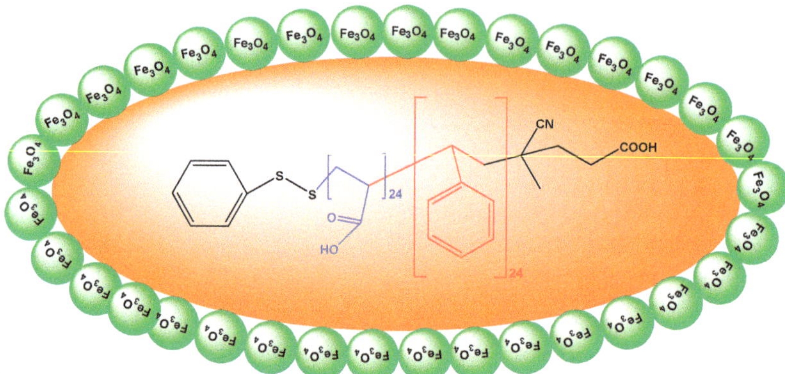

Figure 2. Structure of the prepared poly(St-*b*-AAc)/Fe$_3$O$_4$.

It merits clarification that the objective of using magnetic nanoadsorbents and not common adsorbents was the easier separation of solid adsorbent particles from the solution at the end of the process. Due to magnetic particles, using an external magnetic field, poly(St-*b*-AAc)/Fe$_3$O$_4$ was easily and fast separated from the aqueous solution after adsorption experiments. Also, the preparation of the polysterene nanocomposites was possible (instead of single magnetic particles Fe$_3$O$_4$), because it contains functional groups which increase its adsorption capacity.

2.4. Characterization of Nanoadsorbents

For the XRD patterns, a Bruker XRD diffractometer (Billerica, MA, USA) with CuKα radiation was used. SEM (model Mira 3XMU, TESCAN company, Brno, Czech Republic) was used to study the morphology of nanoparticles.

2.5. Preparation of CIP Solutions

Ciprofloxacin hydrochloride (purity 99.8%) was purchased from Alborz Pharmaceutical Company of Qazvin (Qazvin, Iran), and used to prepare the stock CIP solution (100 mg/L prepared with the fixed pre-weighted amount of CIP and the respective volume of Milli-Q ultra-pure water). The residual concentration of CIP after the adsorption experiments was analyzed by (using a) UV-vis spectrophotometer (model Hach DR5000, Duesseldorf, Germany). The concentration of CIP was measured based on previous studies at a wavelength of $\lambda_{max} = 274$ nm [47].

2.6. Adsorption Experimental Design Method and Data Analysis

In this study, the 7.0.1 Design Expert software was used to determine the number of experiments and the amount of parameters, and to perform the final analysis of the data obtained after the process (Table 1). The measurement of the level of pollutant removal was carried out with the standard design of the statistical model of the CCD (RSM). The main parameters affecting the process are: the initial pH of the medium in the range of 4 to 10, the amount of nanoparticles used in the reaction of 1 to 3 mg/L, the initial concentration of antibiotic ranging from 5 to 25 mg/L, and the reaction time (15 to 60 min).

Table 1. Design parameters together with the values and regions selected.

Parameters	Level of Parameters		
	$-\alpha$	0	$+\alpha$
A: pH	4	7	10
B: Mass (mg)	1	2	3
C: Concentration (mg/L)	5	15	25
D: Reaction time (min)	15	37.5	60

After the determination of optimal conditions and modeling of the process, the rate of CIP removal was investigated. Finally, the process efficiency in CIP removal was determined using the following equation. The removal (R, %) was also calculated based on the following formula:

$$Removal = \left(\frac{C_0 - C_f}{C_0}\right) \times 100\%.\qquad(1)$$

In this relation, R is the efficiency, C_0 (mg/L) is the initial concentration of CIP, and C_f (mg/L) denotes the CIP concentration at the time of t. The amount of adsorbed CIP at equilibrium Q_e (mg/g) was calculated from the following equation. In this relation, C_0 (mg/L)is the initial concentration of CIP, C_e (mg/L) denotes the CIP concentration at the time of t, m (g) is the adsorbent mass, and V (L) is the sample volume:

$$Q_e = \frac{(C_0 - C_e)V}{m}.\qquad(2)$$

3. Results

3.1. Characterizations

The morphologies of the P(St-*b*-AAc)/Fe$_3$O$_4$ nanocomposite are spherical, with $D_{average}$ of 30 nm (Figure 3). It is obvious that the size of spheres is not the same for all particles, due to possible aggregation, but the uniformity regarding the shape is almost the same (spherical).

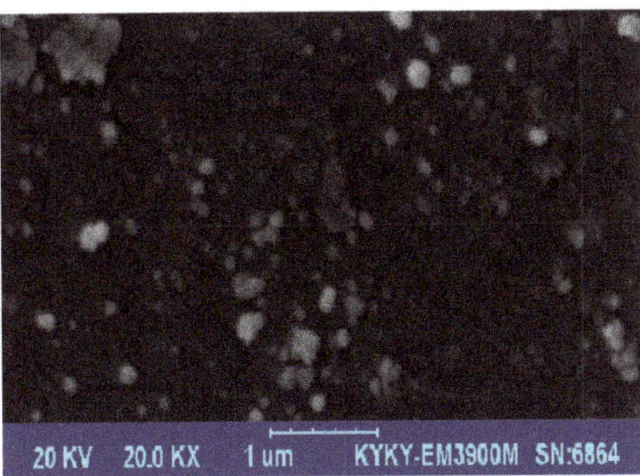

Figure 3. SEM image of P(St-*b*-AAc)/Fe$_3$O$_4$ nanocomposite.

The X-ray diffraction patterns (XRD) resulting from the P(St-*b*-AAc)/Fe$_3$O$_4$ superparamagnetic nanocomposite are indicated in Figure 4. The resulting peaks at 2θ equal to 30.28, 35.48, 43, 53.4, 57.16, and 63.04° correspond to (221), (312), (400), (421), (512), and (440) prisms of P(St-*b*-AAc)/Fe$_3$O$_4$ nanocomposite crystalline structure, respectively (Figure 4) [22].

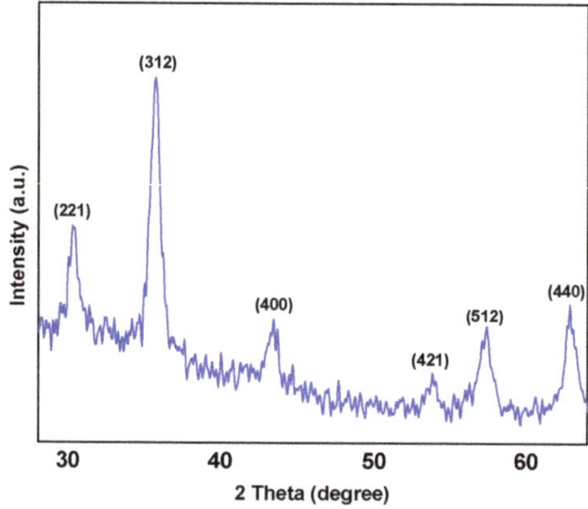

Figure 4. XRD patterns P(St-*b*-AAc)/Fe$_3$O$_4$ magnetic nanocomposite.

The super paramagnetic behavior is demonstrated in Figure 5 with a VSM plot. The saturation magnetization of the P(St-*b*-AAc)/Fe$_3$O$_4$ supermagnetic nanocomposite was around 26 emu/g, which shows that the synthesized magnetic nanocomposite is superparamagnetic.

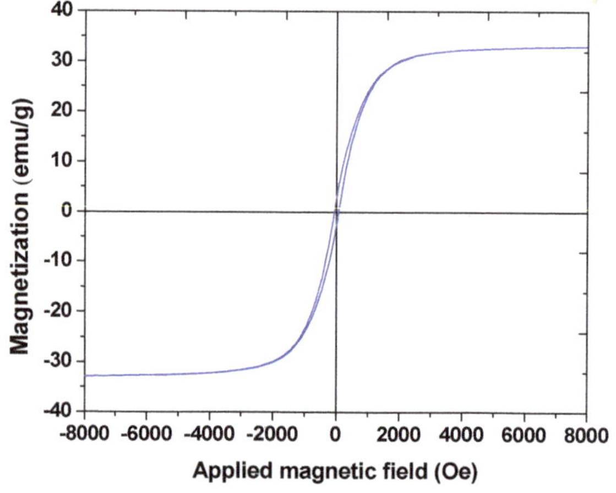

Figure 5. Magnetization curve of P(St-*b*-AAc)/Fe$_3$O$_4$ supermagnetic nanocomposite.

A Fourier-transform infrared spectroscopy of nanocomposites was conducted both prior and after adsorption of CIP, and the spectra are presented in Figure 6. Regarding the FTIR spectrum of CIP, a band around 3400 cm^{-1} represent the vibrational frequency of stretching of the N–H bond of the imino moiety on the piperazine group of CIP. Absorption bands at 1633 cm^{-1} and 1080 cm^{-1} represent a primary amine (N–H) bend of the pyridone moiety and the C–F functional group, respectively. On the other hand, the FTIR spectrum related to the CIP-adsorbed nanoadsorbent is, in turn, related to the addition of the nanocomposite to the CIP solution. The broad peaks at 3463 cm^{-1} are attributed to the stretching vibration of O–H bonds. O–H bonds were weaker and shifted down in the presence of ferrite nanoparticles. Similarly, the slight shift at around 1641 cm^{-1} may be related to the interaction of carboxylic groups of polymer with the amino group of CIP (Figure 6). Also, by comparing the FTIR spectra, the intensity of the peaks after adsorption has increased in comparison to those before adsorption, due to the presence of ferrite nanostructures in the CIP solution.

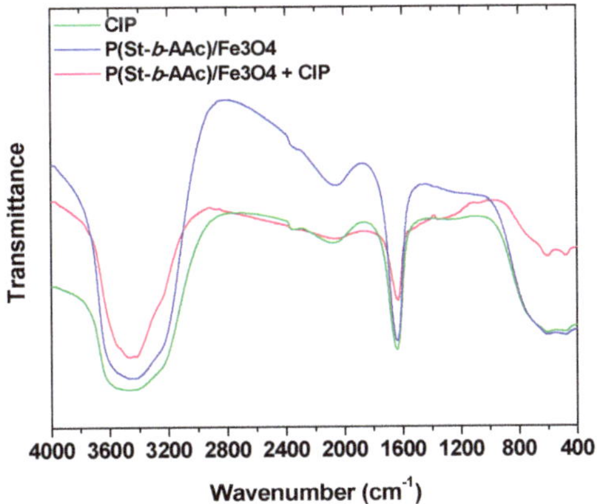

Figure 6. FTIR spectra of CIP, and P(St-*b*-AAc)/Fe$_3$O$_4$ (before and after adsorption).

3.2. Data Analysis

For the efficacy evaluation of antibiotic removal of ciprofloxacin using composite P(St-*b*-AAc), a composite design (one of the response surface methods) was used, and the effects of initial antibiotic concentration parameters, pH, adsorbance dose and reaction time were investigated. The response rate is presented in Table 2. The validity of the presented models was analyzed by ANOVA.

Table 2. Operational parameters and their efficacy in the removal of Ciprofloxacin antibiotic in different methods of performing runs.

Removal (%)	D (Time (min))	C (Nano(g))	B (pH)	A (Initial Concentration)	Run
73.11	1	1	1	−1	1
62.45	0	0	0	0	2
79.04	1	1	1	−1	3
99.5	0	0	0	−2	4
71.06	0	0	0	0	5
65.45	0	0	0	0	6
62	1	−1	0	1	7
89.73	0	2	1	0	8
60.91	1	1	1	1	9
57.03	−1	1	1	1	10
45.76	−1	−1	1	1	11
55.32	1	−1	1	1	12
66.7	−1	1	−1	1	13
66.6	0	0	0	0	14
72.8	1	−1	1	−1	15
53.68	−1	−1	−1	1	16
58.52	−1	−1	1	−1	17
67.5	−1	1	1	−1	18
57.74	−2	0	0	0	19
81.32	−1	1	−1	−1	20
56.45	0	0	−2	0	21
45.2	0	0	0	2	22
67.51	0	0	0	0	23
70.75	0	0	0	0	24
76.63	1	−1	−1	−1	25
51.12	0	−2	0	0	26
40.65	0	0	2	0	27
71.59	2	0	0	0	28
84.24	1	1	−1	−1	29
72.2	−1	−1	−1	−1	30

In Table 3, the parameters A, B, C and D are the main effect of independent variables, which are the initial concentration of ciprofloxacin, pH, adsorbent dose, and contact time, respectively. The variable AB represents the effect of the initial concentration of ciprofloxacin (factor A) and pH (factor B), and variable A^2 represents the square effect of factor A on the desired response.

Table 3. ANOVA for Response Surface Quadratic Model.

Source	Sum of Squares	p Value Prob > F	F Value	Mean Square	df	
Model	4492.14	0.0001	8.06	320.87	14	
A-Nano	353.88	0.0093	8.89	353.88	1	significant
B-pH	113.35	0.1121	2.85	113.35	1	
C-Concentration	822.02	0.0004	20.66	822.02	1	
D-Time	70.22	0.2039	1.76	70.22	1	
AB	1.32	0.8579	0.033	1.32	1	
AC	27.36	0.4199	0.69	27.36	1	
AD	12.83	0.5785	0.32	12.83	1	
BC	4.88	0.731	0.12	4.88	1	
BD	16.16	0.5335	0.41	16.16	1	
CD	1	0.8762	0.025	1	1	
A^2	159.99	0.0633	4.02	159.99	1	
B^2	318.55	0.0127	8.01	318.55	1	
C^2	150.96	0.0704	3.79	150.96	1	
D^2	2.42	0.8086	0.061	2.42	1	
Residual	596.83			39.79	15	
Lack of fit	541.99	0.0459	4.94	54.2	10	
Pure Error	54.83			10.97	5	significant
Cor. Total	5088.97				29	

The proposed model is presented as a modified model by removing non-significant variables via preserving the main effects of variables from the model for the antibiotic elimination efficacy in the following equation:

$$Y(\%) = 67.11 - 9.43X_1 - 4.7X_2 + 6.26X_3 + 3.71X_4 - 3.9X_2{}^2 \tag{3}$$

In this regard, X_1, X_2, X_3 and X_4 are coded values of the initial concentrations of antibiotics, pH, adsorbent dose and reaction time. The linear regression is another test that was used to validate the model [48]. In this test, the coefficient of determination ($R^2 = 0.8753$), the adjusted coefficient of determination ($R^2{}_{adj} = 0.8493$) and the prediction coefficient ($R^2{}_{pred} = 0.7955$) were calculated and reported. Also, in each model, there is very little difference between the values of R^2, $R^2{}_{adj}$ and $R^2{}_{pred}$ is observed.

The Effect of Variables on the Process

In order to study the effects of each variable and the interactions or duplicate effects of variables on the response generated by the model, the graphs were based on the polynomial model of the model, using the test design software. According to Equation (3), the initial concentration of antibiotics has the most significant effect on the removal process, with a coefficient equal to 9.94 and the reaction time smallest effect than other parameters with a coefficient of 3.71. The effect of independent variables on the efficacy of antibiotic removal is shown in Figures 7–9. Figure 7 shows the effect of the initial concentration of antibiotic and the pH of the solution. As shown from Figure 7, with the increase of antibiotic concentration, the removal efficiency decreases. In particular, with an increase of the antibiotic composition from 16.25 to 25 mg/L, the removal efficiency is reduced from 76.13 to 57.34%, respectively. The ideal efficiency was found to be at pH 6, while at pH > 6 and/or pH < 6, the efficiency is reduced.

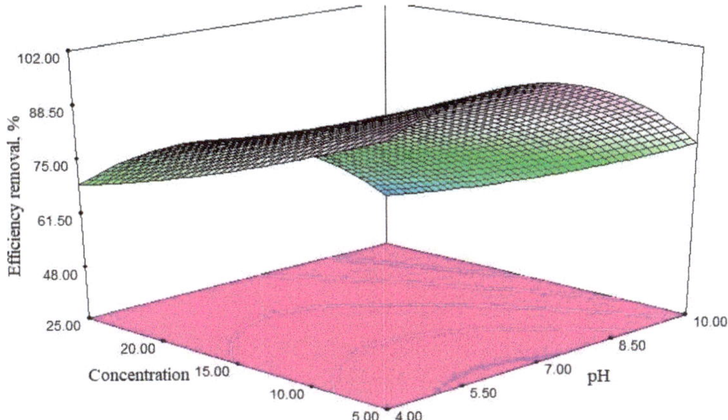

Figure 7. The simultaneous effect of two variables; initial concentration of antibiotic and pH of solution; adsorbent dose of 2 mg/L and reaction time of 37.5 min.

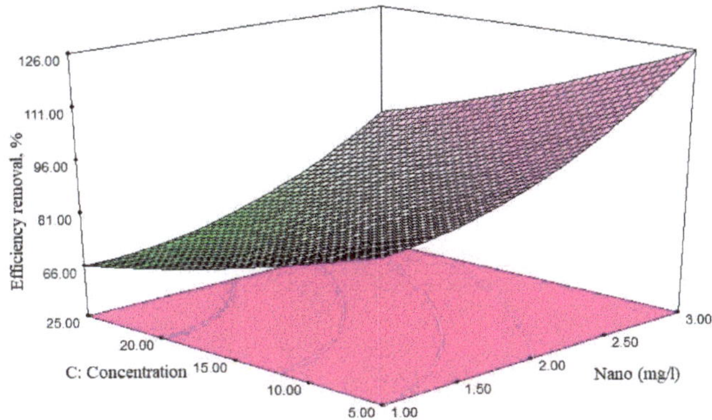

Figure 8. The simultaneous effect of two primary antibiotic and adsorbent dose variables: pH = 7 and reaction time of 37.5 min.

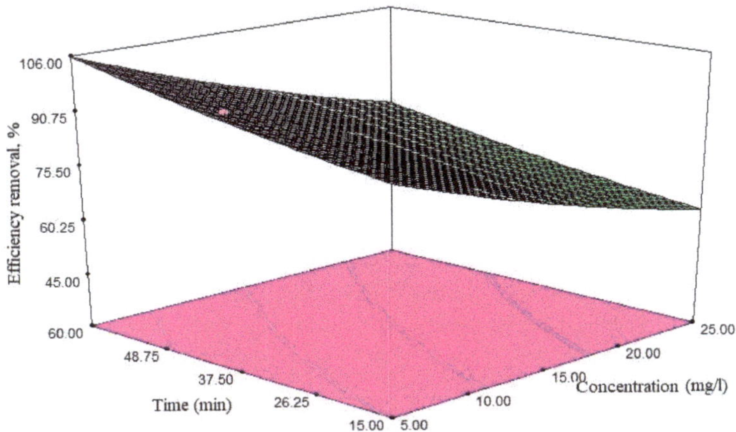

Figure 9. The simultaneous effect of the two initial variables of antibiotic concentration and reaction time: pH = 7 and the adsorbent dose is 2 mg/L.

The effect of the initial concentration of antibiotics, and the amount of adsorbent, are presented in Figure 8. By increasing the amount of adsorbent effluent, the efficiency increases; for example, when the antibiotic concentration is used at a minimum level and the adsorbent content is 5.5 mg/L, the removal efficiency is approximately 76%. If the concentration is constant and the adsorbent amount is equal to 2 mg/L, the removal efficiency is 95.9%.

The effect of initial CIP concentration and the reaction time are shown in Figure 9. According to the graph, when the antibiotic concentration reaches maximum, and in after 26.5 min, 70.11% of the antibiotic is removed. But when the response time reaches the 2+ level (equal to 48.75 min), removal efficiency is increased to 98.77%.

4. Discussion

The contact time is an important factor that directly influences the whole process. In the present work, for a concentration of 5 mg/L, the adsorption process reaches equilibrium at about 37 min, and then shows a relatively stable trend. The effect of the pollutant's initial concentration is affecting a lot the adsorption process. In this paper, the pollutant's initial concentration was studied, ranging from 5 to 50 mg/L. As shown in Figure 7, the initial CIP concentration had a negative effect on the elimination efficiency, and by increasing the ciprofloxacin concentration from 16.25 to 25 mg/L, the elimination efficiency decreased from 84 to 57%. The decrease in removal efficiency when increasing initial concentration can be explained by the fact that the active sites are constant with a constant amount of adsorbent dose, but as the concentration of the adsorbent increases, the pollutant molecules (in the medium—water) saturate the available adsorption sites, thereby, the removal efficiency is lowered [49]. Bajpai et al. observed that by increasing the initial concentration of ciprofloxacin from 10 to 20 mg/L, the adsorption capacity increased from 3.74 to 11.32 mg/g [50].

4.1. Effect of pH Solution

In the purification processes, including adsorption, pH plays an important role. The Solution's pH can affect the adsorbent's surface load, the degree of ionization of various pollutants, the separation of functional groups on active adsorbent sites, as well as the structure of the antibiotic molecule; in effect, the solution's pH affects the chemical environment of the aqueous and adsorption surface bonds. The pH changes were applied to the range of 4–10, and its effect on the removal efficiency was then analyzed. The removal process had the highest percentage at pH 6.2–7, while with the increase of pH, the removal efficiency decreased.

The effect of pH on the ciprofloxacin molecule has shown that in pH less than 6.2, the surface of the molecule appears cationic and positive due to the protonation of amino groups. At pH values higher than 8.6, the ciprofloxacin molecule is converted into anionic form, due to the loss of the proton from the carboxylic group in the antibiotic structure. In the range of 6.2 to 8.6, the deprotonation of carboxyl groups leads to negative carboxylate production. However, the amino group of proteins has a positive charge. In other words, it has a positive and a negative "head". The stabilization and behavior of ciprofloxacin molecule from 6.2 to 7.8 have also been investigated [51]. Since the pH value at pH_{pzc} at the isoelectric absorption point is 7.5, and is negatively charged at higher pH values, given that at pH values above 7.5, both the adsorbent and the antibiotic molecule are both negatively charged. At a pH of less than 6.2, the adsorbent and the antibiotic have positive charge, so in this range, the adsorption process occurs slower and reaches at minimum removal rate at pH = 6.2-6.8, because the unnamed bands reach the maximum electrostatic gravity.

4.2. Effect of Adsorbent's Dose

Based on the findings of this study, the adsorbent dose was the most important factor affecting the efficiency of ciprofloxacin elimination. The study of the effect of adsorbent mass on adsorption processes is one of the most important issues to be considered. Adsorption dose was applied to the range of 1 to 3 mg/L, and its effect on the effectiveness of ciprofloxacin antibiotic removal was measured.

Depending on the results obtained using constant concentrations of antibiotics, the increase in the dose of the adsorbent improves the removal efficiency. As shown in Figure 8, when the concentration of antibiotic is constant and equal to 16.25 mg/L, and the amount of adsorbent is 1.5 mg/L, the removal efficiency is 75.97%—and when the amount of adsorbent reaches 2 mg/L, the removal efficiency is improved, reaching 95.91%; at a constant concentration of antibiotic, by increasing the dose of adsorbent, the ratio of active sites on the adsorbent's surface is high relative to the adsorbing molecules (pollutants), resulting in increased elimination efficiency. On the contrary, in low adsorbent amounts, the ratio of active sites to the adsorbent molecules is lower, and the adsorption decreases.

On the other hand, with the increase of adsorbent above the optimal amount, the adsorption capacity decreased below the maximum level of 15.25 mg/g, which is also due to the fact that by increasing the adsorbent dose, the total capacity of the active sites present in the adsorbent level is completely covered. If not, its adsorption capacity is reduced. This can be the use of available surface in the form of unsaturated attributed adsorbent. The results show that the adsorption pattern in the non-saturable adsorbent form causes undesirable use of existing spaces; this issue is very important in the design of the process economics, particularly in scaling-up.

In this study, 5 mg/L of antibiotic and 2 mg/L of adsorbent were introduced as the optimum amount, at maximum efficiency, with application of 2 mg/L of adsorbent, despite the increase in adsorbent content, other increase in cleavage removal efficiency has not shown any increase. In other words, the removal rate remains constant. It can be concluded that this amount of adsorbent adsorbs all the antibiotics in the solution. Therefore, the antibiotic concentration in the solution is so low that it is no longer "able to be adsorbed" easily. A study by Peasant et al. also showed that with the increase in the adsorbent dose (chitosan/zeolite composite), the dye removal increases, due to the increasing number of adsorption sites, while the increase of adsorbent's dose reduces the adsorption capacity (from the maximum of 17.77 mg/g) [52].

4.3. Effect of Contact Time

An important issue when using the adsorption system is providing an effective contact time under specific conditions. In this paper, contact time was applied to the range of 15 to 60 min, and its effect on the ciprofloxacin antibiotic removal. Figure 7 shows that the adsorption process reaches equilibrium at different times. For a concentration of 5 mg/L, the adsorption process reaches equilibrium at about 37 min, and then shows a relatively stable trend. By increasing contact time, the probability of colliding with adsorbent molecules is also increased, and the efficiency of removal increased. Chang et al. (2012) obtained the equilibrium time for tetracycline removal by Monte Myrnolite for 8 h [53]. In another study by Liu et al. who removed tetracycline using zeolite= by increasing contact time, resulted in the removal efficiency also increased, and the time of equilibrium was 120 min [54].

4.4. Kinetics and Adsorption Isotherms

The adsorption kinetics depends on the adsorbent chemical and physical properties, which influence the adsorption mechanism. In this study, we have used different kinetic and isotherm adsorption models such as pseudo-first order, pseudo-second order, Langmuir, and Freundlich (Table 4).

The pseudo-first and pseudo-second order kinetic equations are shown in Figure 10. Adsorption kinetics were used to determine the control mechanism of adsorption processes. Thus, in this figure, the experimental points were not shown, and only theoretical ones are presented. Based on Table 5, the best fitting was achieved with pseudo-second order equation ($R^2 = 0.9984$).

Table 4. Equations used in this study.

Equation	Expression
Pseudo-first-order [55]	$$\log\left(q_e - q_t\right) = \log\left(q_e\right) - t\frac{k_1}{2.303}$$ • q_e is the amount of mass absorbed in equilibrium state (mg/g) • q_t is equal to (mg/g) the amount of mass absorbed at time t • k_1 is the equilibrium of the first-order kinetic velocity (min^{-1})
Pseudo-second-order [56]	$$\frac{t}{q_t} = \frac{1}{k_2 q_e^2} + t\frac{1}{q_e}$$ • k_2 is the constant of the equilibrium velocity of the quadratic kinetic equation (g mg^{-1} min^{-1})
Freundlich [57]	$$\ln\left(q_e\right) = \ln(K_F) + \frac{1}{n}\ln(C_e)$$ • C_e is the equilibrium concentration in the solution after adsorption (mg/L) • n and K_F are the Freundlich constants
Langmuir [58]	$$\frac{C_e}{q_e} = \frac{1}{q_m b} + \frac{C_e}{q_m}$$ • q_m represents absorption capacity (mg/g) • b is the Langmuir constant (L/mg)

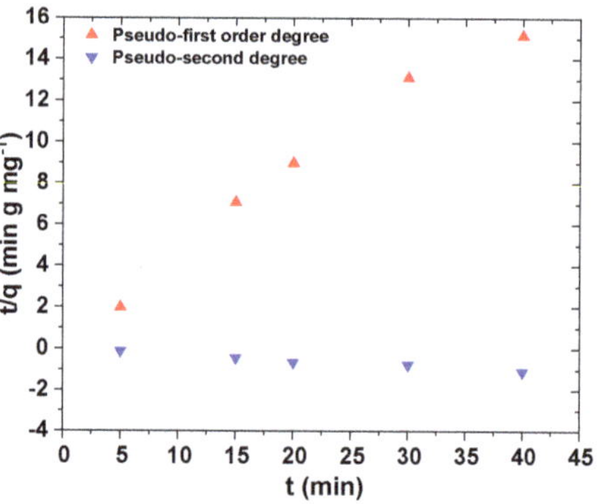

Figure 10. (Up triangle): pseudo-first order kinetic equation; (down triangle) pseudo-second order kinetic equation.

Table 5. Parameters and related kinetic coefficients.

Kinetic Constant Rate	R^2	Kinetic Model
0.0320 min^{-1}	0.7862	Pseudo-first order
1.91 g mg^{-1} min^{-1}	0.9984	Pseudo-second order

The isotherm of adsorption describes how the adsorbent and adsorbate interact. In this study, the experimental results were fitted to Freundlich and Langmuir isotherms. The Langmuir model is valid for single-layer adsorption on adsorbent surface, with limited and uniform adsorption locations, while the Freundlich isotherm is based on single-layer adsorption on heterogeneous adsorption sites with unequal and non-uniform energies. Figure 11 shows the relative isotherms.

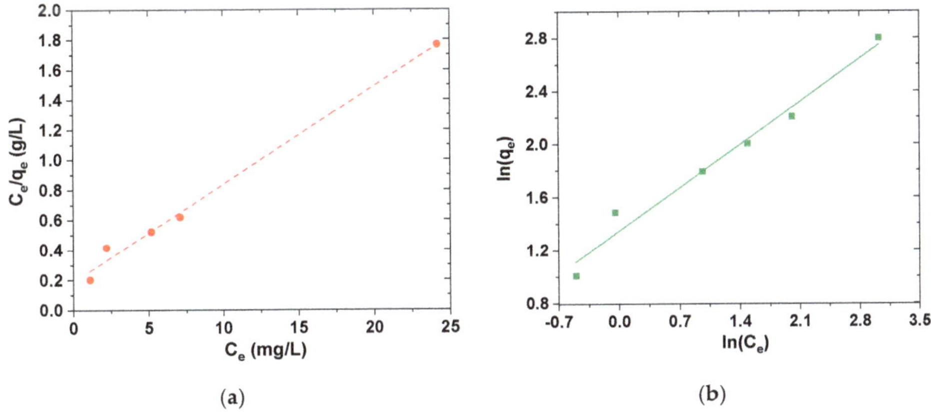

Figure 11. Equilibrium results fitted to (**a**) Langmuir isotherm and (**b**) Freundlich isotherm.

In Freundlich isotherm, when K_F increases, the adsorbent material adsorbed higher amounts of pollutant, and the value of n between 1 and 10 reflects the proper adsorption process. The parameters and coefficients are briefly summarized in Table 5. In this study, the calculated K_F value is 4.75, and the value of n is 2.79, which is within the specified range. Therefore, the adsorption of ciprofloxacin on the adsorbent is well fitted to Langmuir model (Table 6), but it is fact that the data may suggest the presence of non-specific or multi-type interactions between the adsorbate molecules and the adsorptive sites.

Table 6. Parameters and correlation coefficients of isotherm models.

q_m = 15.52 mg/g	b = 0.689 L/mg	R^2 = 0.9918	Langmuir isotherm
K_F = 4.75 (mg^{1-n} L^n/g)	n = 2.79	R^2 = 0.9845	Freundlich isotherm

A major concern regarding any synthesized adsorbent material is answering why this material was synthesized instead of another structure-type material? To respond, it is of fundamental importance to mention some facts. Nanoparticles have a unique combination of properties, such as small size, large surface area, catalytic potential, large number of active sites, high chemical reactivity; all of the above give nanoparticles high adsorption capacity [59]. Also, magnetic nanoadsorbents can be applied as cost-saving and effective materials to separate the materials (solid) from the liquid-phase (water) after the end of the adsorption process. Moreover, the relatively simple isolation of magnetic materials from the solution can aid to their regeneration and reuse [60]. Therefore, the magnetic nanoadsorbents can be good candidates for water/wastewater treatment. Based on the above, Poly(vinylimidazole-co-divinylbenzene) magnetic nanoparticles have been used for the adsorption of fluoroquinolones from aqueous environments [61]. Wang et al. also synthesized the easy to separate magnetic chalcogenide composite KMS-1/L-Cystein/Fe_3O_4 using L-cystein to connect KMS-1 and Fe_3O_4 nanoparticles for ciprofloxacin removal from aqueous solutions [62]. Table 7 shows a brief comparison of some other adsorbent materials tested for the removal of CIP. However, similar experimental conditions should be kept in order to compare two adsorbents (even for the treatment of the same pollutant). Parameters affecting adsorption are the contact time, the solutions' pH, the initial concentration of the pollutant, temperature, adsorbate volume, agitation speed, the solution's ionic strength, and adsorbent dosage. Any change to the abovementioned conditions will lead to different results, and the comparison can be made for adsorbent/adsorbate systems of the same study. Also, based on the interaction groups, a possible mechanism of adsorption is illustrated in Figure 12.

Table 7. CIP adsorption capacities comparison from aqueous solutions using various adsorbents.

Adsorbent	Q_m (mg/g)	Reference
Carbon nanotubes	135	[29]
Kaolinite	6.99	[29]
Bamboo-based carbon modified	153.17	[63]
Graphene oxide	379	[64]
Ca^{2+}-montmorillonite	330	[65]
Multi-walled nanotubes	194	[66]
Iron hydrous oxide	25.76	[67]
Aluminum hydrous oxide	14.72	[67]
Bentonite	147	[68]
Birnessite	80.96	[69]
Montmorillonite	137.7	[70]
Al-PILC	17.78	[71]
Polystyrene nanocomposites P(St-*b*-AAc)/Fe$_3$O$_4$	15.52	This study

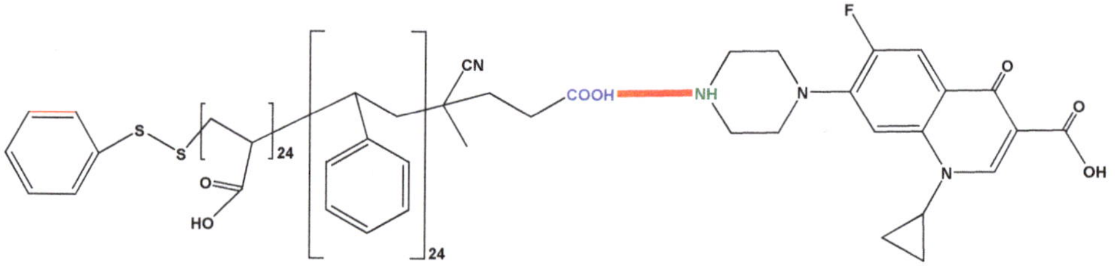

Figure 12. Possible adsorption interaction.

4.5. Aspects

It is known that activated carbon is a very popular adsorbent material, with the demand for virgin activated carbon expanding, since demand from water and wastewater treatment facilities has been steadily increasing. Together with the increase of wastewater treatment applications, the demand and production of activated carbon is also increasing. The largest quantities of activated carbon consumption are observed the U.S.A, Japan and then Europe [72]. Antibiotics are being detected in the aquatic environment. There are different ways for antibiotics to enter the aquatic environment with WWTP considered to be one of the main points of entrance. Even treated wastewater effluent can contain antibiotics, since WWTP cannot eliminate the presence of antibiotics. Compared to other tertiary treatments, adsorption can be a sustainable option for antibiotic removal from wastewaters. Activated carbon is used in the pharmaceutical for the removal of unwanted compounds [72]. Activated carbon possesses a plethora of disadvantages [73], such as high capital cost, ineffectiveness and non-selectivity against vat/disperse dyes. Furthermore, saturated carbon regeneration is expensive and leads to adsorbent loss. Depending on the demand, cost, and the nature of the pollutant to be adsorbed, the adsorbents are either disposed or regenerated for future use. Used adsorbents are considered hazardous waste, causing environmental and societal problems in various countries [74]. Heat accumulation and toxic adsorbates desorption could create hazardous conditions. In addition, odor can be caused by the dumping of adsorbents.

Since regeneration costs can be quite high, the reduction of consumption costs is the key to sustainable and industrial benefits. Substantial studies regarding the activated carbon-based adsorption of pollutants onto have been conducted, but research on regeneration methodologies remains limited [75]. Adsorbent regeneration capability cost analysis is necessary for the economic and environmental assessment of the adsorption process. For the spent adsorbent stabilizing or proper disposal seem to be difficult. The regeneration process of adsorbents from the points of view of sustainability and the environmental involves recovering valuable adsorbates, while reducing the need of virgin adsorbents, and this is extremely important. Studies on novel adsorbents, at full-scale adsorption systems, should be considered for potential industrial applications.

5. Conclusions

Antibiotics are still being detected in the effluents of WWTP, and adsorption seems to be a sustainable option for antibiotics removal from waters. Poly(St-*b*-AAc) diblock copolymers were prepared using the RAFT technique. This copolymer with acrylic acid group was adsorbed onto the surface of Fe_3O_4 nanoparticles, through the interaction with hydroxyl groups on the Fe_3O_4 nanoparticles' surface. A magnetic nanocomposite ranged in 30 nm was then prepared. The VSM analysis showed the saturation magnetization (26 emu/g for P(St-*b*-AAc)/Fe_3O_4). The removal process was performed using P(St-*b*-AAc)/Fe_3O_4 to remove ciprofloxacin antibiotic from synthetic sewage. The effects of parameters such as initial concentration of antibiotic, pH, soluble dose and reaction time were studied. The primary concentration of antibiotics with the highest negative effect and adsorbent dose showed the most positive effect in the removal process. The results also indicated that 97.5% of antibiotics were removed under optimal conditions, which include an initial antibiotic concentration of 5 mg/L, pH 7, and an adsorbent dose of 2 mg/L for 37.5 min. The adsorption of CIP was better fitted to Langmuir isotherm (R^2 = 0.9995), while the kinetics were better fitted to second-order kinetic equation (R^2 = 0.9973). Future work should include multi-component pharmaceutical adsorption with continuous adsorption of wastewaters, taking into account adsorbent regeneration.

Author Contributions: Methodology, L.M., A.G., G.F. and R.K.; A.R. and G.Z.K. writing—original draft preparation and supervision. All authors have read and agreed to the published version of the manuscript.

Funding: This research received no external funding.

Conflicts of Interest: The authors declare no conflict of interest.

References

1. Radjenovic, J.; Petrovic, M.; Barceló, J. Analysis of pharmaceuticals in wastewater and removal using a membrane bioreactor. *Anal. Bioanal. Chem.* **2006**, *387*, 1365–1377. [CrossRef] [PubMed]
2. González, O.; Bayarri, B.; Aceña, J.; Pérez, S.; Barceló, D. Treatment technologies for wastewater reuse: Fate of contaminants of emerging concern. *Handb. Environ. Chem.* **2015**, *20*, 95–100.
3. Patel, M.; Kumar, R.; Kishor, K.; Mlsna, T.; Pittman, C.U.; Mohan, D. Pharmaceuticals of emerging concern in aquatic systems: Chemistry, occurrence, effects, and removal methods. *Chem. Rev.* **2019**, *119*, 3510–3673. [CrossRef]
4. Chen, S.; Zhao, W. Adsorption of Pb2+ from aqueous solutions using novel functionalized corncobs via atom transfer radical polymerization. *Polymers* **2019**, *11*, 1715. [CrossRef]
5. Pham, T.D.; Vu, T.N.; Nguyen, H.L.; Le, P.H.P.; Hoang, T.S. Adsorptive removal of antibiotic ciprofloxacin from aqueous solution using Protein-Modified nanosilica. *Polymers* **2020**, *12*, 57. [CrossRef]
6. Ahmad, A.; Jamil, S.N.A.M.; Choong, T.S.Y.; Abdullah, A.H.; Mastuli, M.S.; Othman, N.; Jiman, N. Green flexible polyurethane foam as a potent support for Fe-Si adsorbent. *Polymers* **2019**, *11*, 2011. [CrossRef]
7. Maponya, T.; Ramohlola, K.E.; Kera, N.; Modibane, K.D.; Maity, A.; Katata-Seru, L.; Hato, M.J. Influence of magnetic nanoparticles on modified polypyrrole/M-Phenylediamine for adsorption of Cr(VI) from aqueous solution. *Polymers* **2020**, *12*, 679. [CrossRef]
8. Ren, L.; Yang, Z.; Huang, L.; He, Y.; Wang, H.; Zhang, L. Macroscopic poly schiff Base-Coated bacteria cellulose with high adsorption performance. *Polymers* **2020**, *12*, 714.
9. Shaipulizan, N.S.; Jamil, S.N.A.M.; Kamaruzaman, S.; Subri, N.N.S.; Adeyi, A.A.; Abdullah, A.H.; Abdullah, L. Abdullah preparation of ethylene glycol dimethacrylate (EGDMA)-Based terpolymer as potential sorbents for pharmaceuticals adsorption. *Polymers* **2020**, *12*, 423. [CrossRef]
10. Khan, M.A.; Siddiqui, M.R.; Otero, M.; Alshareef, S.A.; Rafatullah, M. Removal of rhodamine B from water using a solvent impregnated polymeric dowex 5wx8 resin: Statistical optimization and batch adsorption studies. *Polymers* **2020**, *12*, 500. [CrossRef] [PubMed]

11. Sudre, G.; Siband, E.; Gallas, B.; Cousin, F.; Hourdet, D.; Tran, Y. Responsive adsorption of N-Isopropylacrylamide based copolymers on polymer brushes. *Polymers* **2020**, *12*, 153. [CrossRef] [PubMed]

12. Zhang, W.; Yang, Z.-Y.; Cheng, X.-W.; Tang, R.-C.; Qiao, Y.-F. Adsorption, antibacterial and antioxidant properties of tannic acid on silk fiber. *Polymers* **2019**, *11*, 970. [CrossRef] [PubMed]

13. Guo, W.; Xia, T.; Pei, M.; Du, Y.; Wang, L. Bentonite modified by allylamine polymer for adsorption of amido black 10B. *Polymers* **2019**, *11*, 502. [CrossRef]

14. Kim, S.W.; Sohn, J.S.; Kim, H.K.; Ryu, Y.; Cha, S.W. Effects of gas adsorption on the mechanical properties of amorphous polymer. *Polymers* **2019**, *11*, 817. [CrossRef]

15. Wang, C.; Zhao, J.; Zhang, L.; Wang, C.; Wang, S. Efficient and selective adsorption of gold ions from wastewater with polyaniline modified by trimethyl phosphate: Adsorption mechanism and application. *Polymers* **2019**, *11*, 652. [CrossRef]

16. Huang, W.; Diao, K.; Tan, X.-C.; Lei, F.; Jiang, J.; Goodman, B.A.; Ma, Y.; Liu, S. Mechanisms of adsorption of heavy metal cations from waters by an amino bio-based resin derived from rosin. *Polymers* **2019**, *11*, 969. [CrossRef]

17. Kong, W.; Chang, M.; Zhang, C.; Liu, X.; He, B.; Ren, J. Preparation of Xylan-G-/P(AA-co-AM)/GO nanocomposite hydrogel and its adsorption for heavy metal ions. *Polymers* **2019**, *11*, 621. [CrossRef]

18. Sims, R.; Harmer-Bassell, S.; Gibson, C.T. The role of physisorption and chemisorption in the oscillatory adsorption of organosilanes on aluminium oxide. *Polymers* **2019**, *11*, 410. [CrossRef] [PubMed]

19. Othman, N.A.F.; Selambakkannu, S.; Abdullah, T.A.T.; Hoshina, H.; Sattayaporn, S.; Seko, N. Selectivity of copper by Amine-Based ion recognition polymer adsorbent with different aliphatic amines. *Polymers* **2019**, *11*, 1994. [CrossRef]

20. Dimitrakopoulou, D.; Rethemiotaki, I.; Frontistis, Z.; Xekoukoulotakis, N.; Venieri, D.; Mantzavinos, D. Degradation, mineralization and antibiotic inactivation of amoxicillin by UV-A/TiO2 photocatalysis. *J. Environ. Manag.* **2012**, *98*, 168–170. [CrossRef]

21. Hernando, M.D.; Mezcua, M.; Fernández-Alba, A.R.; Barcelo, D. Environmental risk assessment of pharmaceutical residues in wastewater effluents, surface waters and sediments. *Talanta* **2006**, *69*, 334–342. [CrossRef]

22. Ghamkhari, A.; Mohamadi, L.; Kazemzadeh, S.; Zafar, M.N.; Rahdar, A.; Khaksefidi, R. Synthesis and characterization of poly(Styrene-Block-Acrylic acid) diblock copolymer modified magnetite nanocomposite for efficient removal of penicillin G. *Compos. Part B Eng.* **2020**, *182*, 107643. [CrossRef]

23. Carabineiro, S.; Thavorn-Amornsri, T.; Pereira, M.; Figueiredo, J. Adsorption of ciprofloxacin on Surface-Modified carbonmaterials. *Water Res.* **2011**, *45*, 4583–4591. [CrossRef] [PubMed]

24. Avella, A.; Delgado, L.; Gorner, T.; Albasi, C.; Galmiche, M.; De Donato, P. Effect of cytostatic drug presence on extracellular polymeric substances formation in municipal wastewater treated by membrane bioreactor. *Bioresour. Technol.* **2010**, *101*, 518–526. [CrossRef] [PubMed]

25. De Witte, B.; Van Langenhove, H.; Demeestere, K.; Saerens, K.; De Wispelaere, P.; Dewulf, J. Ciprofloxacin ozonation in hospital wastewater treatment plant effluent: Effect of pH and H_2O_2. *Chemosphere* **2010**, *78*, 1142–1146. [CrossRef]

26. Sun, S.P.; Guo, H.Q.; Ke, Q.; Sun, J.H.; Shi, S.H.; Zhang, M.L. Degadation of antibiotic ciprofloxacin hydrochloride by photofenton oxidation process. *Sci. Eng.* **2009**, *26*, 753–759.

27. Liao, R.; Yu, Z.; Gao, N.; Peng, P. Oxidative transformation of ciprofloxacin in the presence of manganese oxide. *Eco Environ. Sci.* **2011**, *20*, 1143–1146.

28. Van Doorslaer, X.; Demeestere, K.; Heynderickx, P.M.; Van Langenhove, H.; Dewulf, J. UV-A and UV-C induced photolytic and photocatalytic degradation of aqueous ciprofloxacin and moxifloxacin: Reaction kinetics and role of adsorption. *Appl. Catal. Environ.* **2011**, *101*, 540–547. [CrossRef]

29. Carabineiro, S.A.C.; Thavorn-Amornsri, T.; Pereira, M.F.R.; Serp, P.; Figueiredo, J.L. Comparison between activated carbon, carbon xerogel and carbon nanotubes for the adsorption of the antibiotic ciprofloxacin. *Catal. Today* **2012**, *186*, 29–34. [CrossRef]

30. Mohammadi, L.; Bazrafshan, E.; Noroozifar, M.; Ansari-Moghaddam, A.; Barahuie, F.; Balarak, D. Adsorptive removal of benzene and toluene from aqueous environments by cupric oxide nanoparticles: Kinetics and isotherm studies. *J. Chem.* **2017**, *2017*, 1–10. [CrossRef]

31. Mohammadi, L.; Bazrafshan, E.; Noroozifar, M.; Ansari-Moghaddam, A.; Barahuie, F.; Balarak, D. Removing 2,4-Dichlorophenol from aqueous environments by heterogeneous catalytic ozonation using synthesized MgO nanoparticles. *Water Sci. Technol.* **2017**, *76*, 3054–3068. [CrossRef] [PubMed]

32. Do, Q.C.; Choi, S.; Kim, H.; Kang, S. Adsorption of lead and nickel on to expanded graphite decorated with manganese oxide nanoparticles. *Appl. Sci.* **2019**, *9*, 5375. [CrossRef]

33. Rojas, J.; Suarez, D.; Moreno, A.; Silva-Agredo, J.; Torres-Palma, R.A. Kinetics, isotherms and thermodynamic modeling of liquid phase adsorption of crystal violet dye onto Shrimp-Waste in its raw, pyrolyzed material and activated charcoals. *Appl. Sci.* **2019**, *9*, 5337. [CrossRef]

34. El-Azazy, M.; Dimassi, N.S.; El-Shafie, S.A.; Issa, A.A. Bio-Waste aloe vera leaves as an efficient adsorbent for titan yellow from wastewater: Structuring of a novel adsorbent using Plackett-Burman factorial design. *Appl. Sci.* **2019**, *9*, 4856. [CrossRef]

35. Tuomikoski, S.; Kupila, R.; Romar, H.; Bergna, D.; Kangas, T.; Runtti, H.; Lassi, U. Zinc adsorption by activated carbon prepared from lignocellulosic waste biomass. *Appl. Sci.* **2019**, *9*, 4583. [CrossRef]

36. Caban, R.T.; Vega-Olivencia, C.; Mina-Camilde, N. Adsorption of Ni2+ and Cd2+ from water by calcium alginate/spent coffee grounds composite beads. *Appl. Sci.* **2019**, *9*, 4531. [CrossRef]

37. Wang, S.; Wang, N.; Yao, K.; Fan, Y.; Li, W.; Han, W.; Yin, X.; Chen, D. Characterization and interpretation of Cd (II) adsorption by different modified rice straws under contrasting conditions. *Sci. Rep.* **2019**, *9*, 17868. [CrossRef]

38. Bui, T.X.; Choi, H. Adsorptive removal of selected pharmaceuticals by mesoporous silica SBA-15. *J. Hazard. Mater.* **2009**, *168*, 602–608. [CrossRef]

39. Lorphensri, O.; Intravijit, J.; Sabatini, D.; Kibbey, T.; Osathaphan, K.; Saiwan, C. Sorption of acetaminophen, 17α-Ethynyl estradiol, nalidixic acid, and norfloxacin to silica, alumina, and a hydrophobic medium. *Water Res.* **2006**, *40*, 1481–1491. [CrossRef]

40. Mestre, A.; Pires, J.; Nogueira, J.; Carvalho, A.P. Activated carbons for the adsorption of ibuprofen. *Carbon* **2007**, *45*, 1979–1988. [CrossRef]

41. Baccar, R.; Sarrà, M.; Bouzid, J.; Feki, M.; Blánquez, P. Removal of pharmaceutical compounds by activated carbon prepared from agricultural by-product. *Chem. Eng. J.* **2012**, *211*, 310–317. [CrossRef]

42. Fukahori, S.; Fujiwara, T.; Ito, R.; Funamizu, N. pH-Dependent adsorption of sulfa drugs on high silica zeolite: Modeling and kinetic study. *Desalination* **2011**, *275*, 237–242. [CrossRef]

43. Martucci, A.; Pasti, L.; Marchetti, N.; Cavazzini, A.; Dondi, F.; Alberti, A. Adsorption of pharmaceuticals from aqueous solutions on synthetic zeolites. *Microporous Mesoporous Mater.* **2012**, *148*, 174–183. [CrossRef]

44. Varanda, F.; De Melo, M.J.P.; Caço, A.I.; Dohrn, R.; Makrydaki, F.A.; Voutsas, E.; Tassios, D.; Marrucho, I.M. Solubility of antibiotics in different solvents. 1. hydrochloride forms of tetracycline, moxifloxacin, and ciprofloxacin. *Ind. Eng. Chem. Res.* **2006**, *45*, 6368–6374. [CrossRef]

45. Ghamkhari, A.; Massoumi, B.; Jaymand, M. Novel 'schizophrenic' diblock copolymer synthesized via RAFT polymerization: Poly (2-succinyloxyethyl methacrylate)-b-poly[(N-4-vinylbenzyl),N,N-diethylamine]. *Des. Monomers Polym.* **2016**, *20*, 190–200. [CrossRef]

46. Ghamkhari, A.; Ghorbani, M.; Aghbolaghi, S. A perfect Stimuli-Responsive magnetic nanocomposite for intracellular delivery of doxorubicin. *Artif. Cells Nanomed. Biotechnol.* **2018**, *46*, S911–S921. [CrossRef]

47. Li, H.; Zhang, D.; Han, X.; Xing, B. Adsorption of antibiotic ciprofloxacin on carbon nanotubes: pH dependence and thermodynamics. *Chemosphere* **2014**, *95*, 150–155. [CrossRef]

48. Yoosefian, M.; Ahmadzadeh, S.; Aghasi, M.; Dolatabadi, M. Optimization of electrocoagulation process for efficient removal of ciprofloxacin antibiotic using iron electrode; kinetic and isotherm studies of adsorption. *J. Mol. Liq.* **2017**, *225*, 544–553. [CrossRef]

49. Zhao, Y.; Tong, F.; Gu, X.; Gu, C.; Wang, X.; Zhang, Y. Insights into tetracycline adsorption onto goethite: Experiments and modeling. *Sci. Total Environ.* **2014**, *470*, 19–25. [CrossRef]

50. Bajpai, S.; Bajpai, M.; Rai, N. Sorptive removal of ciprofloxacin hydrochloride from simulated wastewater using sawdust: Kinetic study and effect of pH. *Water SA* **2012**, *38*, 673–682. [CrossRef]

51. El-Shafey, E.-S.I.; Al Lawati, H.A.J.; Al-Sumri, A.S. Ciprofloxacin adsorption from aqueous solution onto chemically prepared carbon from date palm leaflets. *J. Environ. Sci.* **2012**, *24*, 1579–1586. [CrossRef]

52. Dehghani, M.H.; Dehghan, A.; Alidadi, H.; Dolatabadi, M.; Mehrabpour, M.; Converti, A. Removal of methylene blue dye from aqueous solutions by a new chitosan/zeolite composite from shrimp waste: Kinetic and equilibrium study. *Korean J. Chem. Eng.* **2017**, *34*, 1699–1707. [CrossRef]

53. Chang, P.-H.; Li, Z.; Jean, J.-S.; Jiang, W.-T.; Wang, C.-J.; Lin, K.-H. Adsorption of tetracycline on 2:1 layered Non-Swelling clay mineral illite. *Appl. Clay Sci.* **2012**, *67*, 158–163. [CrossRef]

54. Liu, P.; Liu, W.-J.; Jiang, H.; Chen, J.-J.; Li, W.-W.; Yu, H.-Q. Modification of Bio-Char derived from fast pyrolysis of biomass and its application in removal of tetracycline from aqueous solution. *Bioresour. Technol.* **2012**, *121*, 235–240. [CrossRef]

55. Lagergren, S. About the theory of So-Called adsorption of soluble substances. *Handlingarl* **1898**, *24*, 1–39.

56. Ho, Y.; McKay, G. Pseudo-Second order model for sorption processes. *Process. Biochem.* **1999**, *34*, 451–465. [CrossRef]

57. Freundlich, H. Over the adsorption in solution. *Z. Phys. Chem.* **1906**, *57*, 385–470.

58. Langmuir, I. The adsorption of gases on plane surfaces of glass, mica and platinum. *J. Am. Chem. Soc.* **1918**, *40*, 1361–1403. [CrossRef]

59. Liu, S.; Ma, C.; Ma, M.-G.; Xu, F. Chapter 12—Magnetic nanocomposite adsorbents. In *Composite Nanoadsorbents*; Kyzas, G.Z., Mitropoulos, A.C., Eds.; Elsevier: Amsterdam, The Netherlands, 2019; pp. 295–316.

60. Malakootian, M.; Nasiri, A.; Mahdizadeh, H. Preparation of CoFe2O4/activated carbon@ chitosan as a new magnetic nanobiocomposite for adsorption of ciprofloxacin in aqueous solutions. *Water Sci. Technol.* **2018**, *78*, 2158–2170. [CrossRef]

61. Huang, X.; Wang, Y.; Liu, Y.; Yuan, D. Preparation of magnetic poly (Vinylimidazole-Codivinylbenzene) nanoparticles and their application in the trace analysis of fluoroquinolones in environmental water samples. *J. Sep. Sci.* **2013**, *36*, 3210–3219.

62. Wang, Y.X.; Gupta, K.; Li, J.R.; Yuan, B.; Yang, J.C.E.; Fu, M.L. Novel chalcogenide based magnetic adsorbent KMS-1/L-Cystein/Fe$_3$O$_4$ for the facile removal of ciprofloxacin from aqueous solution. *Coll. Surf. Physicochem. Eng. Asp.* **2018**, *538*, 378–386. [CrossRef]

63. Peng, X.; Hu, F.; Lam, F.L.-Y.; Wang, Y.; Liu, Z.; Dai, H. Adsorption behavior and mechanisms of ciprofloxacin from aqueous solution by ordered mesoporous carbon and bamboo-based carbon. *J. Colloid Interface Sci.* **2015**, *460*, 349–360. [CrossRef] [PubMed]

64. Chen, H.; Gao, B.; Li, H. Removal of sulfamethoxazole and ciprofloxacin from aqueous solutions by grapheme oxide. *J. Hazard. Mater.* **2015**, *282*, 201–207. [CrossRef]

65. Wang, C.-J.; Li, Z.; Jiang, W.-T.; Jean, J.-S.; Liu, C.-C. Cation exchange interaction between antibiotic ciprofloxacin and montmorillonite. *J. Hazard. Mater.* **2010**, *183*, 309–314. [CrossRef] [PubMed]

66. Yu, F.; Sun, S.; Han, S.; Zheng, J.; Ma, J. Adsorption removal of ciprofloxacin by Multi-Walled carbon nanotubes with different oxygen contents from aqueous solutions. *Chem. Eng. J.* **2016**, *285*, 588–595. [CrossRef]

67. Gu, C.; Karthikeyan, K.G. Sorption of the antimicrobial ciprofloxacin to aluminum and iron hydrous oxides. *Environ. Sci. Technol.* **2005**, *39*, 9166–9173. [CrossRef]

68. Genç, N.; Dogan, E.C.; Yurtsever, M. Bentonite for ciprofloxacin removal from aqueous solution. *Water Sci. Technol.* **2013**, *68*, 848–855. [CrossRef]

69. Jiang, W.-T.; Chang, P.-H.; Wang, Y.-S.; Tsai, Y.; Jean, J.-S.; Li, Z.; Krukowski, K. Removal of ciprofloxacin from water by birnessite. *J. Hazard. Mater.* **2013**, *250*, 362–369. [CrossRef]

70. Jalil, M.E.R.; Baschini, M.; Sapag, K. Influence of pH and antibiotic solubility on the removal of ciprofloxacin from aqueous media using montmorillonite. *Appl. Clay Sci.* **2015**, *114*, 69–76. [CrossRef]

71. Jalil, M.R.; Baschini, M.; Sapag, K. Removal of ciprofloxacin from aqueous solutions using pillared clays. *Materials* **2017**, *10*, 1345. [CrossRef]

72. Bansal, R.C.; Goyal, M. *Activated Carbon Adsorption*; Taylor & Francis: Boca Raton, FL, USA, 2005.

73. Konaganti, V.K.; Kota, R.; Patil, S.; Madras, G. Adsorption of anionic dyes on chitosan grafted poly (alkyl methacrylate)s. *Chem. Eng. J.* **2010**, *158*, 393–401. [CrossRef]

74. Miyake, Y. The soil purifying method that combine soil vapor extraction and activated carbon fiber. *J. Resour. Environ. (Sigen Kankyo Taisaku)* **1998**, *33*, 896–899.

75. Mohan, D.; Pittman, C.U., Jr. Arsenic removal from water/wastewater using Adsorbents—A critical review. *J. Hazard. Mater.* **2007**, *142*, 1–53. [CrossRef] [PubMed]

Article

Core−Shell Molecularly Imprinted Polymers on Magnetic Yeast for the Removal of Sulfamethoxazole from Water

Liang Qiu [1,2], **Guilaine Jaria** [2], **María Victoria Gil** [3], **Jundong Feng** [1], **Yaodong Dai** [1], **Valdemar I. Esteves** [2], **Marta Otero** [4,*] and **Vânia Calisto** [2]

[1]   Department of Materials Science and Technology, Nanjing University of Aeronautics & Astronautics, Nanjing 210016, China; Qiuliangyzl@126.com (L.Q.); jundongfeng@nuaa.edu.cn (J.F.); yd_dai@nuaa.edu.cn (Y.D.)

[2]   CESAM & Department of Chemistry, University of Aveiro, Campus Universitário de Santiago, 3810-193 Aveiro, Portugal; jaria.guilaine@ua.pt (G.J.); valdemar@ua.pt (V.I.E.); vania.calisto@ua.pt (V.C.)

[3]   Instituto de Ciencia y Tecnología del Carbono (INCAR-CSIC), C/ Francisco Pintado Fe 26, 33011 Oviedo, Spain; victoria.gil@incar.csic.es

[4]   CESAM & Department of Environment and Planning, Campus Universitário de Santiago, 3810-193 Aveiro, Portugal

*   Correspondence: marta.otero@ua.pt

Received: 28 May 2020; Accepted: 17 June 2020; Published: 20 June 2020

Abstract: In this work, magnetic yeast (MY) was produced through an in situ one-step method. Then, MY was used as the core and the antibiotic sulfamethoxazole (SMX) as the template to produce highly selective magnetic yeast-molecularly imprinted polymers (MY@MIPs). The physicochemical properties of MY@MIPs were assessed by Fourier-transform infrared spectroscopy (FT-IR), a vibrating sample magnetometer (VSM), X-ray diffraction (XRD), thermogravimetric analysis (TGA), specific surface area (S_{BET}) determination, and scanning electron microscopy (SEM). Batch adsorption experiments were carried out to compare MY@MIPs with MY and MY@NIPs (magnetic yeast-molecularly imprinted polymers without template), with MY@MIPs showing a better performance in the removal of SMX from water. Adsorption of SMX onto MY@MIPs was described by the pseudo-second-order kinetic model and the Langmuir isotherm, with maximum adsorption capacities of 77 and 24 mg g^{-1} from ultrapure and wastewater, respectively. Furthermore, MY@MIPs displayed a highly selective adsorption toward SMX in the presence of other pharmaceuticals, namely diclofenac (DCF) and carbamazepine (CBZ). Finally, regeneration experiments showed that SMX adsorption decreased 21 and 34% after the first and second regeneration cycles, respectively. This work demonstrates that MY@MIPs are promising sorbent materials for the selective removal of SMX from wastewater.

Keywords: antibiotics; emerging contaminants; pharmaceuticals; wastewater treatment; polymeric adsorbents; magnetization

---

1. Introduction

Antibiotics are intensively used as human and veterinary medicines for the treatment and prevention of infectious diseases [1]. Among them, sulfamethoxazole (SMX) is a sulfonamide bacteriostatic antibiotic that has been commonly used during the last 80 years to treat urinary tract infections due to its low cost and broad spectrum of activity to treat bacterial diseases [2,3]. However, the widespread and indiscriminate use of SMX, as of other antibiotics, constitutes a huge potential threat to human health and contaminates natural ecosystems by affecting aquatic and soil organisms [4,5].

Recently, SMX has been detected in effluents of sewage treatment plants (STP), and also in surface and groundwater [6,7]. Indeed, it is known that pharmaceuticals (including SMX) can reach the aquatic environment in their unchanged or transformed forms mainly through discharge of effluents from municipal STP [7]. According to the statistics, more than 20,000 tons of SMX enter the environment worldwide every year, resulting in concentrations that range from 0.001 to 5.0 $\mu g\ L^{-1}$ in untreated or treated wastewater [8–10]. Therefore, the problem of environmental contamination by SMX is of great concern as pathogen resistance is highly documented and has been induced even by low levels of antibiotics [11].

To solve the above-mentioned problems, substantial research efforts have been directed worldwide to develop sustainable treatments for the removal of antibiotics, including SMX, from contaminated waters, such as membrane separation, adsorption processes, photocatalysis, and chemical oxidation [12]. Among these treatments, adsorption-based processes have been highlighted to be efficient, easy to implement and, furthermore, avoid the generation of transformation products [13–15]. However, the application of these processes is quite challenging due to the characteristic features of contaminated wastewaters, namely, large discharge flux, complex composition, and very low antibiotic concentrations [16]. Increasing the adsorbent specificity has been proposed as a strategy to address these challenges and improve the efficiency of the adsorptive removal of antibiotics from such complex matrices [17].

Molecularly imprinted technology (MIT) involves the creation of tailor-made selective binding sites in a polymeric matrix with memory of the shape, size, and functional groups of the template. Thus, molecularly imprinted polymers (MIPs) have become increasingly attractive as adsorbent materials due to their capacity to selectively bind specific targets and to their promising characteristics, such as low cost, easy synthesis, high stability to harsh chemical and physical conditions, and excellent reusability [18,19]. In recent years, MIPs, whose application of the extraction and analysis of organic contaminants in environmental water samples is well-established [20], have been successfully used for the adsorptive removal of pharmaceuticals, including antibiotics, from contaminated water [21–24]. In the specific case of SMX adsorption by MIPs, few works have been published, with most of them aiming at the analytic quantification of this antibiotic. For example, Qin et al. [5] used Fe_3-O_4-chitosan MIPs for SMX selective extraction and determination in aqueous samples, with the produced materials having attained a maximum adsorption capacity of 4.32 mg g^{-1}. Zhao et al. [25] prepared core–shell MIPs on the surface of magnetic carbon nanotubes (MCNTs@MIP) for SMX, the resulting material having a maximum SMX adsorption capacity from aqueous solution of 864.9 $\mu g\ g^{-1}$. However, to the best of our knowledge, the removal of SMX from complex wastewaters using MIPs has just been assessed by Valtech et al. [19]. Among the materials produced by these authors [18], those having the largest maximum adsorption capacity (6.5×10^{-5} mol g^{-1} (16.5 mg g^{-1})) performed similarly to a commercial activated carbon in terms of removal, but presented higher selectivity toward SMX in the presence of other pharmaceuticals and better regeneration ability.

Despite the above-mentioned advantages and applications, the preparation of MIPs by conventional MIT has two main drawbacks: (1) The imprinted polymer matrices are thick and, thus, hold a small number of recognition sites per unit volume; and (2) the template molecules are deeply embedded in the matrix, so there is a diffusion barrier for them, the mass transfer rate is low, and binding to the recognition sites is somehow hampered [26]. Surface molecular imprinting has been proved to improve mass transfer, recognition, and binding ability relative to MIT [27]. Among solid-support substrates used for the surface molecular imprinting process, microbial nano-magnetic materials are alternative supporters that have many advantages compared to inorganic materials [28]: (1) They are easy to obtain and short generations can be artificially cultured [29]; (2) there are many surface chemical functional groups and so modification steps can be avoided, reducing secondary pollution; (3) cells can guide the regulation of the growth process of inorganic materials [30]; (4) microbial cells have a variety of structures and can provide a rich array of templates for nanomaterials by template-assisted synthesis; and (5) magnetic properties allow for a simple after-use separation of the materials.

Yeasts, which belong to the fungus kingdom, are relatively large eukaryotic and single-celled microorganisms (diameters typically measuring 2.0–4.0 μm). Their cell wall includes glucan, mannan, chitin protein, and a small amount of lipids, and it has many surface chemical groups such as carboxyl (–COOH), carbonyl (–C=O), amino (–NH$_2$), hydroxyl (–OH), and phosphoryl (–P=O) groups. Moreover, yeast is very cheap, easy to obtain, and environmentally friendly. These advantages make yeasts appropriate and widely used as supports for bio-nanocomposites [31].

In the above-described context, the objectives of this study were to: (1) Prepare a bio-nanocomposite of yeast-Fe$_3$O$_4$ (magnetic yeast, MY) using an in situ one-step preparation of nano-Fe$_3$O$_4$; (2) use MY as the core to synthesize magnetic yeast-molecularly imprinted polymers (MY@MIPs) by a surface-imprinted polymerization method with MIPs as the shell and SMX as the template molecule; (3) characterize the resulting materials by Fourier-transform infrared spectroscopy (FT-IR), a vibrating sample magnetometer (VSM), X-ray diffraction (XRD), thermogravimetric analysis (TGA), specific surface area (S_{BET}) determination, and scanning electron microscopy (SEM); (4) test the removal performance of MY@MIPs toward SMX and compare it with those of MY and MY@NIPs (magnetic molecularly imprinted polymers without template); and (5) explore the selective sorption capacity of MY@MIP in a real complex matrix (wastewater collected at a STP) and in the presence of other pharmaceuticals (diclofenac and carbamazepine).

2. Materials and Methods

2.1. Chemicals and Materials

Yeast cells (CICC 30225) were obtained from the China Center of Industrial Culture Collection (CICC). Iron salts used to produce MY were ferric chloride hexahydrate (FeCl$_3$·6H$_2$O) and ferrous chloride tetrahydrate (FeCl$_2$·4H$_2$O), purchased from Sigma-Aldrich (Stenheim, Germany). In addition, 2-vinyl pyridine (2-vpy), ethylene glycol dimethacrylate (EGDMA), acetonitrile (ACN), and azo-bis-isobutyronitrile (AIBN), which were also purchased from Sigma-Aldrich (Stenheim, Germany), were used for MIT. Other reagents used in this work included ammonium hydroxide, toluene (99.8%, Aldrich), ethanol (99.9%, Riedel-de Haën), methanol (99.99%, Fischer Chemical), and acetic acid (p.a., Merck). Ultrapure water was obtained from a Milli-Q water purification system (Millipore). SMX was purchased from TCI Europe (>98%); carbamazepine (CBZ; Sigma-Aldrich, 99%); diclofenac (DCF, TCI Europe, >98%). All solutions were stored at 4 °C immediately after preparation.

2.2. Materials Preparation

2.2.1. Preparation of Magnetic Yeast (MY)

Nano-Fe$_3$O$_4$ was loaded onto the yeast cell surface by a one-step method as described by Tian et al. [32]. Briefly, the yeast cells were cultured in ultrapure water with glucose. After reaching the exponential growth phase (6–10 h), the yeast cells were collected by centrifugation (4000 rpm). Then, collected cells (1.0 g) were suspended in 40 mL of 0.125 M FeCl$_3$ solution in a three-necked flask and stirred for 1 h at room temperature. After that, 0.6 g of FeCl$_2$·4H$_2$O was added under nitrogen atmosphere and stirred for another 1 h. The mixture was then heated in a water bath at 80 °C for 15 min, and the pH was adjusted to approximately 11 with 25% (*w/v*) ammonium hydroxide. Stirring was kept for 30 min and then stopped to age for 1 h. The resulting magnetic yeast (MY) was then washed, separated by applying a magnetic field, and then dried in an oven (35 °C, 4 h).

2.2.2. Preparation of Magnetic Yeast-Based Molecularly Imprinted Polymer (MY@MIPs)

MY was treated as the core and the MIPs as the shell. The process used for the production of MY@MIPs was as follows: 1 mg of SMX (template molecule) and 4 mmol of 2-vpy (monomer) were dissolved in 60 mL of ACN/toluene (3/1; *v/v*). This solution was then self-polymerized for 8 h at room temperature (25 °C). Subsequently, 100 mg of MY (polymer supporter), 0.36 mmol of AIBN (initiator), and EGDMA (crosslinker) were added into the polymerized solution (template:monomer:crosslinker, 1:4:20), which was ultrasonicated for 10 min. The mixture was heated and maintained at 60 °C for 24 h under stirring with nitrogen protection. At last, the MY@MIPs were washed with methanol/formic acid (9/1; *v/v*) for 12 h and purified for 24 h by a Soxhlet extraction method (the extraction solution was methanol). Meanwhile, the MY@NIPs were also produced by following the above-described procedure but in the absence of the template.

2.3. Characterization of MY, MY@MIPs, and MY@NIPs

Fourier-transform infrared spectra of the produced materials were obtained in a Shimadzu-IRaffinity-1 equipment, using an ATR module (FTIR-ATR), under a nitrogen purge. The measurements were recorded in the range 500–4000 cm^{-1}, 4.0 of resolution, 256 scans, and applying atmosphere and background correction.

A vibrating sample magnetometer (VSM EV9) with an oscillatory applied magnetic field (H) to a maximum of 22 kOe was used to determine the saturation magnetization (M_S). The M_S was calculated by plotting the magnetic moment versus the applied magnetic field, and it corresponded to the plateau value of the magnetic moment reached divided by the sample mass (10 mg). The sample was encapsulated in an acrylic cylindrical container (5.85 mm of diameter and 2.60 mm of height), which was coupled to the lineal motor of the VSM EV9 instrument, centered between the two polar heads of the electromagnet used to fluctuate the magnetic field. The instrument was calibrated with a disk of pure nickel (8 mm of diameter) using a procedure that establishes the determination of the magnetic field, applied at around 1 Oe, while the dispersion of the magnetic moment is inferior to 0.5%.

X-ray diffraction (XRD, 5–90°) was measured on a D8-Focus X-ray diffractometer (Bruker Optics) with a test rate of 10°·min^{-1}. The results were analyzed by Jade program (9.0) and Origin (9.0).

Thermogravimetric analysis (TGA) was performed in a thermogravimetric balance Setsys Evolution 1750, Setaram, TGA mode (S type sensor). The samples were heated at a heating rate of 10 °C min^{-1}, under nitrogen atmosphere, from room temperature to 105 °C and from 105 °C to 900 °C, maintaining constant temperature until total stabilization of the sample mass at the end of both stages (approximately 30 min).

The S_{BET} and micropore volume (W_0) were determined by nitrogen adsorption isotherms, acquired at 77 K using a Micromeritics Instrument, Gemini VII 2380, after outgassing the materials overnight at 120 °C. S_{BET} was calculated from the Brunauer–Emmett–Teller equation in the relative pressure range 0.01–0.1. Pore volume (V_p) was estimated from the amount of nitrogen adsorbed at a relative pressure of 0.99.

The surface morphology of the materials was analyzed by scanning electron microscopy (SEM) using a Hitachi S4100. The images were obtained at magnifications of 500, 3000, and 10,000×.

2.4. Adsorptive Removal of SMX by the Produced Materials

The produced materials (MY, MY@MIPs, MY@NIPs) were used as adsorbents for the removal of SMX under batch operation conditions. Summarizing, the materials were put in contact with a 5 mg L^{-1} SMX solution in polypropylene tubes, which were shaken in a head-over-head shaker (80 rpm) for a predetermined period of time at controlled temperature (32 °C). The corresponding adsorbent material was separated from the suspension liquid by an external magnetic field. At last, the concentration of SMX in the liquid phase was measured by micellar electrokinetic chromatography (MEKC), using a methodology adapted from Silva et al. (2019) [33]. The experiments were conducted in triplicate, and control experiments without adsorbent were run in parallel. The performance of the materials was evaluated by carrying out kinetic, equilibrium, pH, selectivity, and regeneration/reutilization studies, described in detail in the next subsections.

2.4.1. Kinetic Adsorption Studies in Ultrapure Water

In the kinetic study, tubes containing 250 mg of adsorbent material (MY, MY@NIPs, or MY@MIPs), together with 10 mL of a 5 mg L^{-1} SMX solution in ultrapure water, were incubated and shaken as described above. After shaking during defined periods of time (t, min), at intervals from 0 to 24 h, the materials were separated from the aqueous phase and the remaining SMX concentration in solution was measured by MEKC. At each time, the corresponding value of the adsorbed concentration (q_t, mg·g^{-1}) was determined as follows:

$$q_t = \frac{C_0 - C_t}{C_m} \tag{1}$$

where C_t (mg L^{-1}) is the residual SMX concentration at time t, C_0 is the initial SMX concentration (mg L^{-1}), and C_m is the adsorbent dosage (mg·L^{-1}).

When adsorption equilibrium was attained, the percentage of adsorption R (%) was determined as:

$$R\,(\%) = \frac{C_0 - C_e}{C_0} \times 100\% \tag{2}$$

where C_e (mg·L^{-1}) is the residual SMX concentration at equilibrium.

2.4.2. Equilibrium Adsorption Studies in Ultrapure Water

For the equilibrium studies, the corresponding adsorbent material (MY, MY@NIPs, or MY@MIPs), with doses ranging from 50 to 2000 mg L^{-1}, was added to 10 mL of a 5 mg L^{-1} solution of SMX in ultrapure water. Tubes with the mixtures were shaken for 16 h, which allowed equilibrium to be reached. The materials were recovered from the suspension by the application of a magnetic field and the residual concentration of SMX was determined by MEKC. Then, for the different doses of material, the adsorbed concentration at the equilibrium (q_e, mg·g^{-1}) was determined as follows:

$$q_e = \frac{C_0 - C_e}{C_m} \tag{3}$$

where C_e (mg L^{-1}) is the SMX concentration in the liquid phase at equilibrium.

2.5. Adsorptive Performance of MY@MIPs

From the results of the above-mentioned kinetic and equilibrium studies in ultra-pure water, the most efficient material for removal of SMX was MY@MIPs. Thus, in order to assess the practical application of this material, further studies were carried out on the adsorptive performance of MY@MIPs under different experimental conditions.

2.5.1. Kinetic and Equilibrium Adsorption Studies in STP Effluent

The kinetic and equilibrium procedures described in Section 2.4. were carried out using MY@MIPs for the adsorptive removal of SMX from a real matrix, namely the effluent from a STP. In this case, 5 mg L^{-1} solutions of SMX were prepared using a STP effluent instead of ultrapure water. The effluent was collected from an urban STP in Aveiro (Portugal) that is designed to serve 159,700 population equivalents. This STP consists of primary and biological treatment stages. For this work, water was collected at the outlet of the biological decanter, as this is the final treated effluent that is discharged from the STP into the aquatic environment. Immediately after collection, the effluent was filtered through 0.45 µm, 293 mm Supor® membrane disk filters (Gelman Sciences) and stored at 4 °C until use, which occurred within a maximum of 15 days. The collected effluent had a pH of 7.99, conductivity of 3.03 mS cm^{-1}, and total organic carbon content of 21.5 mg L^{-1}.

2.5.2. pH Study

Adsorption studies on the effect of pH were carried out at 32 °C with the initial conditions of $C_0 = 5$ mg L^{-1} in ultrapure water and $C_m = 300$ mg L^{-1}. Experiments were carried out at three different pHs, namely 4, 7, and 8 (pH was adjusted by adding HCl or NaOH, 1 M). After shaking during 16 h, MY@MIPs were separated from the liquid suspensions, the residual concentration of SMX was analyzed by MEKC, and the corresponding q_e (mg g^{-1}) at each pH was determined using Equation (3).

2.5.3. Selective Adsorption

To study the selective capacity of MY@MIPs toward SMX, diclofenac (DCF) and carbamazepine (CBZ) were used as competing species in the adsorption experiments. These pharmaceuticals were selected due to their high global frequency of occurrence in wastewater, surface water, and groundwater and their recalcitrant properties, with low removal rates after conventional STP treatments [34]. The concentration of DCF and CBZ in ultrapure water solution was the same as that of SMX (5 mg L^{-1}), the C_m was 300 mg L^{-1}, the incubation temperature was 32 °C, the pH was 4, and shaking was maintained during 16 h. Then, the residual concentration of SMX at equilibrium was analyzed and the corresponding q_e (mg g^{-1}) was determined with Equation (3).

2.5.4. Regeneration and Reutilization

In order to evaluate the adsorptive performance after regeneration, after SMX saturation in ultrapure water, MY@MIPs were regenerated and then tested for the adsorption of SMX in four subsequent cycles. For the regeneration, saturated MY@MIPs were washed by methanol/acetic acid (9/1, *v/v*) through Soxhlet extraction during 72 h. Then, the regenerated material was used in adsorption experiments as described in previous sections (shaking during 16 h at 32 °C with the initial conditions of $C_0 = 5$ mg L^{-1} in ultrapure water and $C_m = 300$ mg L^{-1}). The residual concentration of SMX at the equilibrium was analyzed and the corresponding R (%) was determined as for Equation (2).

3. Results

3.1. Preparation of MY

In this study, an in situ one-step method was carried out to load nano-Fe$_3$O$_4$ particles on the surface of yeast, which was used as a biological solid support. Under alkaline conditions, Fe^{2+} and Fe^{3+} co-precipitated on the surface of yeast and then Fe(OH)$_2$ or Fe(OH)$_3$ was converted to nano-Fe$_3$O$_4$ at 80 °C, according to the following chemical reactions:

$$Fe^{2+} + 2OH^- \rightarrow Fe(OH)_2\downarrow$$

$$Fe^{3+} + 3OH^- \rightarrow Fe(OH)_3\downarrow$$

$$Fe(OH)_2 + 2Fe(OH)_3 \xrightarrow{80\ ^\circ C} Fe(Fe_2O_4) \downarrow + 4H_2O$$

The observation of MY by a high-power optical microscope (Olympus CX22, Japan) clearly showed the loading of magnetic nanoparticles over yeast cells, as shown in Figure 1a. Meanwhile, Figure 1b represents the picture of MY at the actual size. Compared to other methods used to anchor Fe$_3$O$_4$ nanoparticles on the surface of yeast biomass, such as cross-linking or electrostatic-interaction-driven hypercoagulation, the one-step method applied here only took approximately 3.5 h, in opposition to the referred methods, which can take up to 13.5 and 6 h, respectively [35], without considering the time of washing and drying. Hence, the results suggest that the one-step method is an interesting synthesis option.

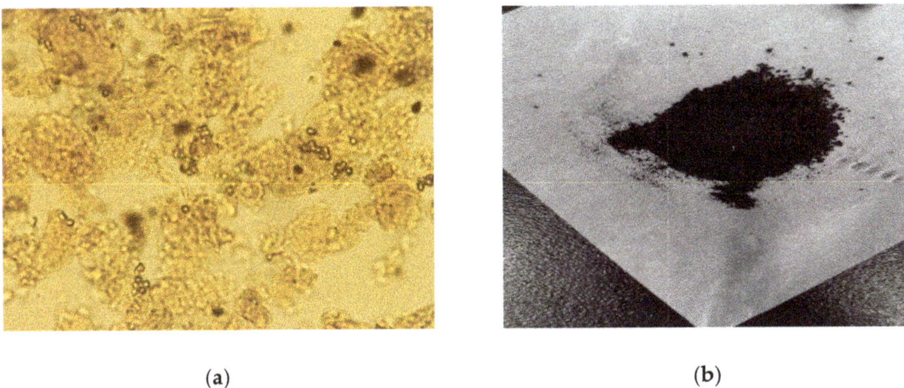

(a)   (b)

Figure 1. Optical microscopy photograph (500×) of magnetic yeast (MY) (**a**); actual size photograph of MY (**b**).

3.2. Preparation of MY@MIPs

MY was used as support of a MIP-based material for the selective adsorption of SMX. Compared to other support materials such as SiO_2, carbon nanotubes, or Fe_3O_4-SiO_2 used in the literature [36,37], MY particles can act as support without the need of an intermediate chemical modification step and can distinctly improve grafting efficiency.

The shell of MIPs was co-polymerized on the surface of MY. Hence, to synthetize MIPs with affinity, selectivity, and appreciable removal capacity toward the target compound, the monomer and crosslinker types and the ratio of the reagents should be taken into account. Normally, if the template molecule has an alkaline chemical group, the monomer should be methacrylate (MAA), but if it has an acidic group, the monomer should be vinyl pyrimidine (vpy) [38]. As SMX has an oxazole moiety that displays acidity, 2-vpy was chosen as it has both a hydrogen-bond acceptor (N atom of pyridine) and alkalinity [39]. In this work, the molar ratio of the mixture of template and monomer was 1:4, as the monomer and template were in dynamic equilibrium, and it is not useful to add the monomer indiscriminately. Indeed, an excessive monomer may increase the non-selective sites, resulting in a selectivity decrease. On the other hand, during the synthesis of MIPs, in order to immobilize the template into the polymer without changing the spatial configuration of pores in the polymer, this must have a high rigidity. Therefore, it was necessary to use a crosslinker for increasing rigidity, EDGMA being selected due to its appropriate cost and solubility. However, if the ratio of monomer to crosslinker is too high, it will make the extraction of the template difficult due to the excessive rigidity of the MIP. Considering the referred considerations and conclusions from other studies [40,41], the ratio of monomer and crosslinker was selected to be 1:5.

3.3. Characterization of MY, MY@MIPs, and MY@NIPs

FTIR spectra of the produced materials (MY, MY@MIPs, and MY@NIPs), which were obtained in order to shed some light about the chemical groups present on their surface, are depicted in Figure 2. At 548 cm^{-1}, a characteristic adsorption peak belonging to the Fe-O chemical bond was observed for all materials. Compared to MY, MY@MIPs and MY@NIPs had some new peaks. Among them were the absorption bands at 2363 and 2328 cm^{-1} (MY@MIPs) and at 2377 and 2337 cm^{-1} (MY@NIPs), which were attributed to the stretching vibrations of -CN or -NC, respectively. The peak at 1758 cm^{-1} (MY@MIPs) or at 1727 cm^{-1} (MY@NIPs) belongs to the stretching vibration of C=O in the EGDMA ester group and the carboxyl group, suggesting that EGDMA worked on the surface. Moreover, MY@MIPs had new peaks at 1118 and 955 cm^{-1}, which belonged to the symmetrical and asymmetric stretching vibration of C-O in EGDMA, respectively, and reflected that it had a cross-linking polymerization on the surface of MY@MIPs. In the spectra of MY@MIPs and MY@NIPs, adsorption peaks at 1350 or

1340 cm^{-1} were due to N–H bending vibrations, while this peak was very weak in MY, indicating the N–H bond of 2-vpy. Peaks at 1595 cm^{-1} (MY@MIPs) or 1572 cm^{-1} (MY@NIPs) were due to bending vibrations of N-H.

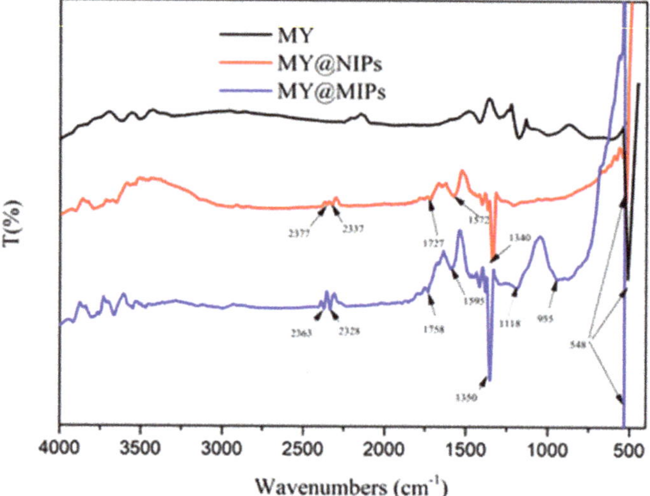

Figure 2. FTIR spectra of magnetic yeast (MY), magnetic yeast-molecularly imprinted polymers without template (MY@NIPs), and magnetic yeast-molecularly imprinted polymers (MY@MIPs).

The magnetic properties of the MY, MY@MIPs, and MY@NIPs were studied by VSM at room temperature, the M_S of each material being shown in Table 1. The M_S values were determined to be between 26 and 34 emu g^{-1}, which were compatible with good magnetization. Indeed, Figure S1, within Supplementary Information, shows that MY@MIPs can be easily separated by an external magnetic field, which is beneficial for the after-use separation of the saturated MY@MIPs from treated water, achieving one of the major goals of this study.

Table 1. Physical characterization of the produced materials.

Materials	S_{BET} (m^2 g^{-1})	V_p (cm^3 g^{-1})	D (nm)	M_S (emu g^{-1})
MY	38.8	0.11	5.71	26.1
MY@NIPs	39.2	0.11	5.41	24.2
MY@MIPs	43.2	0.11	5.06	34.1

N$_2$ adsorption at −196 °C; V_p = total pore volume; D = average pore diameter; M_S = saturation magnetization.

The XRD spectrum of MY in the 2θ range of 20 to 80° is shown in Figure 3, where the (220), (311), (400), (422), (511), and (440) planes of Fe$_3$O$_4$ may be observed at 2θ = 30.22°, 35.40°, 43.36°, 53.68°, 57.21°, and 62.43°. This pattern is consistent with the standard XRD data of Fe$_3$O$_4$ in the JCPDS-International Centre for Diffraction Data (JCPDS Card: PDF#75-0033). Therefore, XRD results evidenced that Fe$_3$O$_4$ was successfully loaded onto the yeast surface during the production of MY and that, subsequently, surface molecular imprinting did not change the crystalline structure of magnetic nanoparticles. Similar patterns confirming the effective loading of magnetite have been reported in the literature on magnetic MIPs (MMIPs), including MMIPs produced for melamine analysis in milk [42], PEGylated magnetic core–shell structure-molecularly imprinted polymers (PMMIPs) for the specific adsorption of bovine serum albumin (BSA) [43], or core–shell MMIPs for the selective adsorption of tetracycline [44].

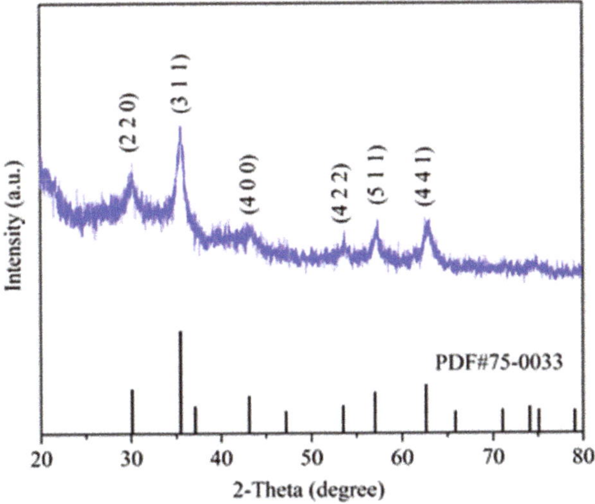

Figure 3. X-ray diffraction spectrum of MY.

The thermogravimetric (TG) and derivative thermogravimetric (DTG) curves of MY, MY@NIPs, and MY@MIPs are shown in Figure 4. All the materials evidenced three main weight loss peaks: The first at ~100 °C related to moisture; the second at ~300 °C related to the most thermolabile organic fraction; and the third centered at ~700 °C related to less thermolabile organic or inorganic fractions. For MY, a weight loss of approximately 62% was reached at 900 °C (Figure 4a); the second weight loss peak is particularly accentuated in this material, as it is the one with the highest amount of yeast per unit mass of material and, thus, yeast cells carbonized with increasing temperature. Meanwhile, MY@NIPs (Figure 4b) and MY@MIPs (Figure 4c) suffered, globally, a lower weight loss than MY, reaching 30% and 50% of weight loss, respectively, at 900 °C. This might be due to the introduction of less thermolabile structures in the composition of these materials (such as the magnetic nanoparticles and polymers).

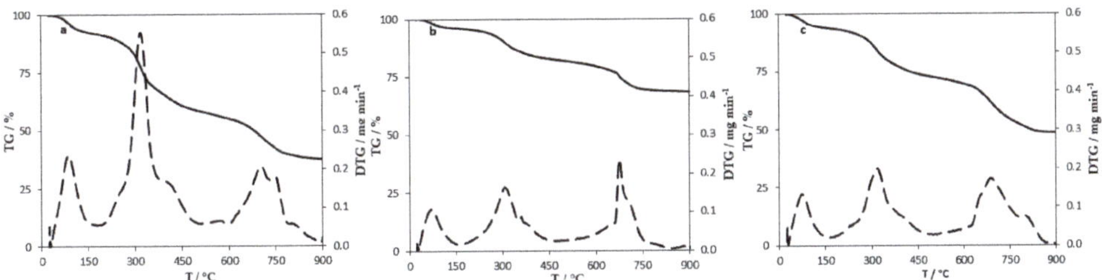

Figure 4. TG (full line) and DTG (dashed lines) curves of MY (**a**), MY@NIPs (**b**), and MY@MIPs (**c**).

The results of S_{BET} are shown in Table 1. The S_{BET} of each material was as follows: MY—38.8 m^2 g^{-1}, MY@MIPs—43.2 m^2 g^{-1}, and MY@NIPs—39.2 m^2 g^{-1}. Similar S_{BET} (47 m^2 g^{-1}) were determined for magnetic sorbents with a metal–organic framework core and MIP shell [45]. Meanwhile, lower S_{BET}, between 6 and 11 m^2 g^{-1}, have been measured for magnetic sorbents based on the iron oxide (Fe_3O_4) core and MIP shell [46,47]. Regarding the average pore diameter (*D*), it was 5.71, 5.41, and 5.06 nm respectively for MY, MY@MIPs, and MY@NIPs. Therefore, the three produced materials are mesoporous with no significant differences between them in terms of porosity.

The surface of MY, MY@MIPs, and MY@NIPs was examined by SEM (Figure 5). All the figures suggested that the particles (either MY, MY@NIPs, or MY@MIPs) were elliptical, which is due to the use of yeast as support, as it has been observed to have an ellipsoid shape with uniform size [48]. As it may be seen, MY@NIPs and MY@MIPs have a dispersed and comparatively smoother appearance than MY, which is rough-faced due to the magnetic nanoparticles coating the smooth-faced yeast [49]. Moreover, under 3000× magnification, results showed that MY@MIPs had a better dispersion compared to MY and MY@NIPs. Under 10,000×, MY@MIPs showed a bigger porosity than the other materials, which may benefit the adsorption of SMX and improve the mass transfer rate from the aqueous phase.

Globally, the characterization results demonstrated the successful loading of Fe_3O_4 on the yeast surface and the preparation of MIPs on the surface of MY.

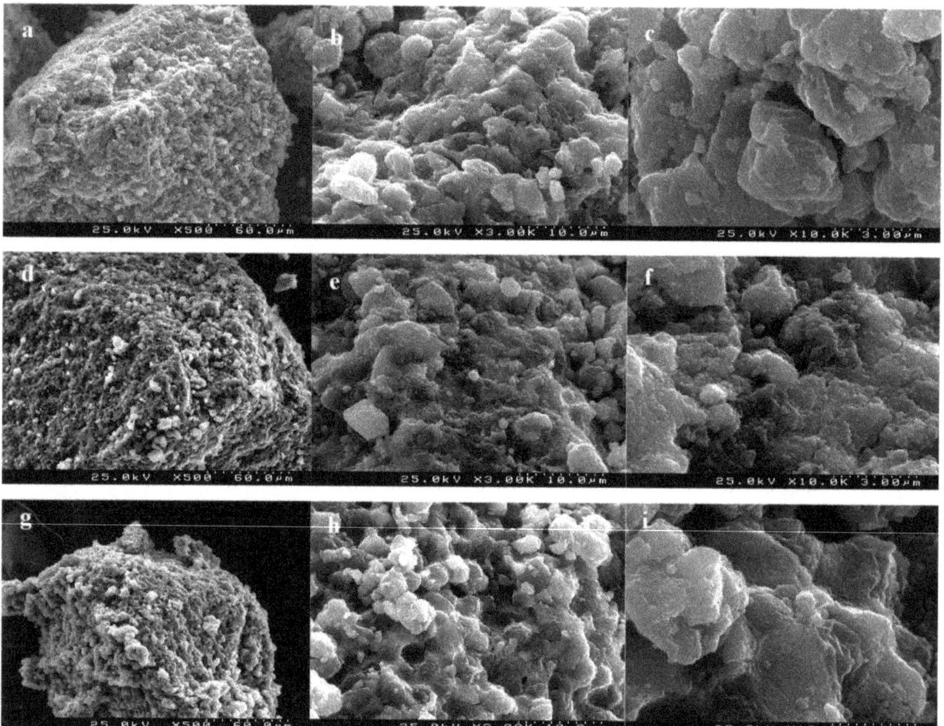

Figure 5. Scanning electron microscopy (SEM) images of MY (**a–c**), MY@NIPs (**d–f**), and MY@MIPs (**g–i**).

3.4. Adsorptive Removal of SMX by the Produced Materials

3.4.1. Adsorption Kinetics

The kinetic results on the adsorption of SMX onto the produced materials are shown in Figure 6, which evidences that, in all cases, the adsorbed concentration q_t (mg g^{-1}) rapidly increased until 360 min of contact and then slowly increased until becoming stable. Moreover, all the materials performed quite similarly from a kinetic point of view.

Comparing the results obtained here to those reported in the literature, it may be said that a shorter equilibrium time (around 20 min, at room temperature) was determined for the adsorption of SMX onto core–shell MIPs on the surface of magnetic carbon nanotubes (MCNTs@MIP) by Zhao et al. [25]. Meanwhile, using MIPs on the surface of yeast (yeast@MIPs), Wang et al. [16] found that (at 298 to 318 K) 200 min were necessary to attain equilibrium for the adsorption of ciprofloxacin (CIP), and Pan et al. [50] observed an equilibrium time around 375 min for the adsorption (at 303 K) of cephalexin. In any case, it has been noticed that surface-imprinting improves the binding kinetics as compared to traditionally imprinted materials, which take longer (usually around 12–24 h) to attain adsorption equilibrium [51].

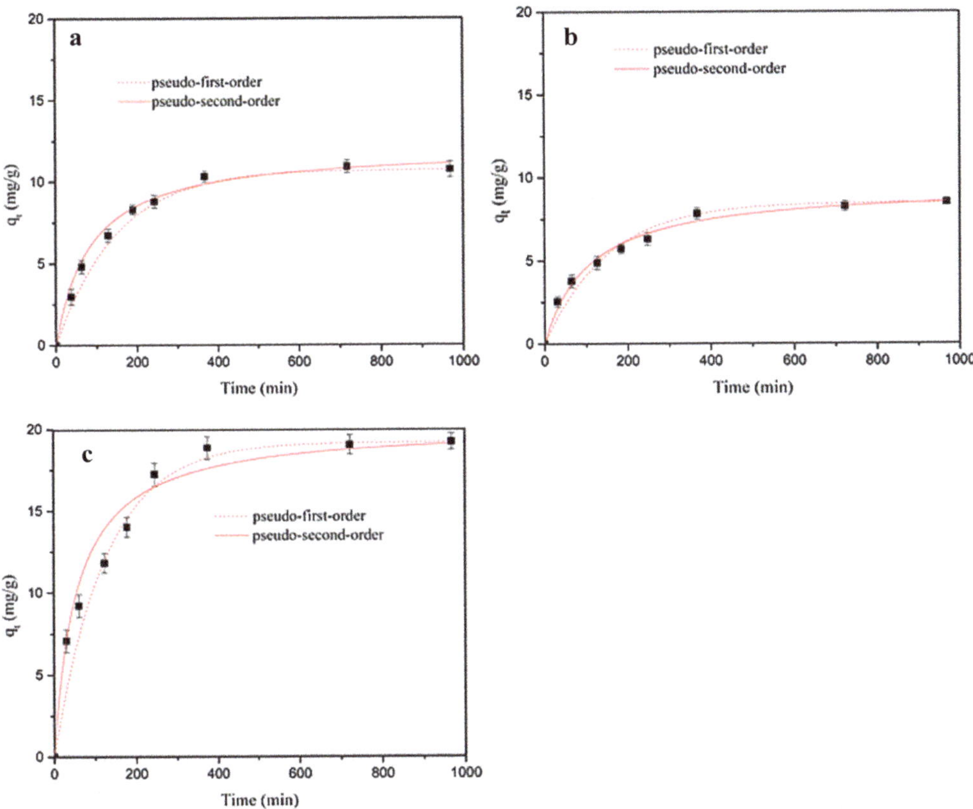

Figure 6. Experimental kinetic results together with pseudo-first- and pseudo-second-order model fittings for the adsorption of sulfamethoxazole (SMX) onto MY (**a**), MY@NIPs (**b**), and MY@MIPs (**c**) in ultrapure water.

Pseudo-first-order [52] and pseudo-second-order [53] kinetic models were applied to describe the adsorption kinetics of SMX onto the produced materials. The formulation of the models is as follows:

Pseudo-first-order

$$q_t = q_e \times \left(1 - e^{(-k_1 t)}\right) \tag{4}$$

Pseudo-second-order

$$q_t = \frac{k_2 \times q_e^2 \times t}{1 + k_2 \times q_e \times t} \tag{5}$$

where k_1 (min^{-1}) and k_2 (g mg^{-1} min^{-1}) are the pseudo-first-order and the pseudo-second-order rate constants.

The non-linear fitting kinetic parameters are summarized in Table 2. According to the correlation coefficient (R^2) and concordance between experimental and fitted q_e values, both models described the SMX adsorption onto the produced materials, with the pseudo-second-order model describing slightly better the results onto MY and MY@NIPs and the pseudo-first-order model onto MY@MIPs.

Table 2. Kinetic parameters corresponding to the adsorption of SMX onto MY, MY@NIPs, and MY@MIPs in ultrapure water.

Materials	Experimental	Pseudo-First Order Model			Pseudo-Second Order Model		
	q_e (mg g^{-1})	q_e (mg g^{-1})	k_1 (min^{-1})	R^2	q_e (mg g^{-1})	k_2 (g mg^{-1} min^{-1})	R^2
MY	10.6 ± 0.7	11.1 ± 0.9	0.993 ± 0.003	0.931	12.5 ± 0.8	0.009 ± 0.005	0.994
MY@NIPs	8.2 ± 0.2	8.1 ± 0.1	0.993 ± 0.002	0.978	9.5 ± 0.2	0.009 ± 0.005	0.988
MY@MIPs	19 ± 1	19 ± 1	0.992 ± 0.001	0.966	21 ± 1	0.001 ± 0.004	0.952

3.4.2. Adsorption Isotherm

Equilibrium results on the adsorption of SMX onto the produced materials are shown in Figure 7. With the aim of describing these results, four isotherm models were used: Langmuir [54] and Freundlich [55] isotherm models for the adsorption of SMX onto MY@MIPs; BET isotherm [56] for the adsorption onto MY@NIPs; and Zhu–Gu isotherm [57] for the adsorption onto MY. The equations of these models are as follows:

Langmuir isotherm

$$q_e = \frac{q_m \times b \times C_e}{1 + b \times C_e} \tag{6}$$

Freundlich isotherm

$$q_e = k_f \times C_e^{\frac{1}{n}} \tag{7}$$

BET isotherm

$$q_e = \frac{q_m \times c \times C_e}{(1 - c \times C_e) \times (1 - c \times C_e + c \times C_e)} \tag{8}$$

Zhu–Gu isotherm

$$q_e = \frac{q_m \times \left(g \times C_e \times \left(\frac{1}{r} + e \times C_e^{r-1}\right)\right)}{(1 + g \times C_e) \times \left(1 + e \times C_e^{r-1}\right)} \tag{9}$$

where q_m is the maximum adsorption capacity (mg g^{-1}); b (L mg^{-1}) is the Langmuir equilibrium constant; k_f (mg g^{-1} (mg L^{-1})$^{-1/n}$) is the Freundlich constant; n is the degree of non-linearity in the Freundlich isotherm; c is the BET constant, related to the energy of adsorption in the first adsorbed layer; g is the Zhu–Gu constant related to the first adsorption step (the first layer of molecules on the materials); e is the Zhu–Gu constant related to the subsequent layers adsorbed; and r is the aggregation number in the Zhu–Gu isotherm.

Experimental results on the equilibrium of SMX adsorption onto the produced materials are depicted in Figure 7 together with fittings to the above-mentioned isotherm models. From Figure 7, it is evident that, contrarily to the adsorption onto MY@MIPs, in the case of MY and MY@NIPs, the q_e did not tend to stabilization. Furthermore, in the C_e range between 0 and 3 mg L^{-1}, a lower q_e occurred for MY@NIPs than for MY. This may be related to the presence of chemical groups on the surface of MY, which were able to bind SMX groups, but became inaccessible in MY@NIPs due to molecular imprinting. In addition, a first stage with stabilization of q_e at C_e around 3 mg L^{-1} may be observed in the MY isotherm, which could be associated with the saturation of the chemical adsorption sites. In the case of MY@MIPs, the isotherm showed an increase in q_e with C_e with a stabilization trend from $C_e \sim 3$ mg L^{-1}. Furthermore, it should be noted that, at relatively low C_e, the q_e values determined for MY@MIPs are higher than those for MY and MY@NIPs, which points to their larger affinity for SMX.

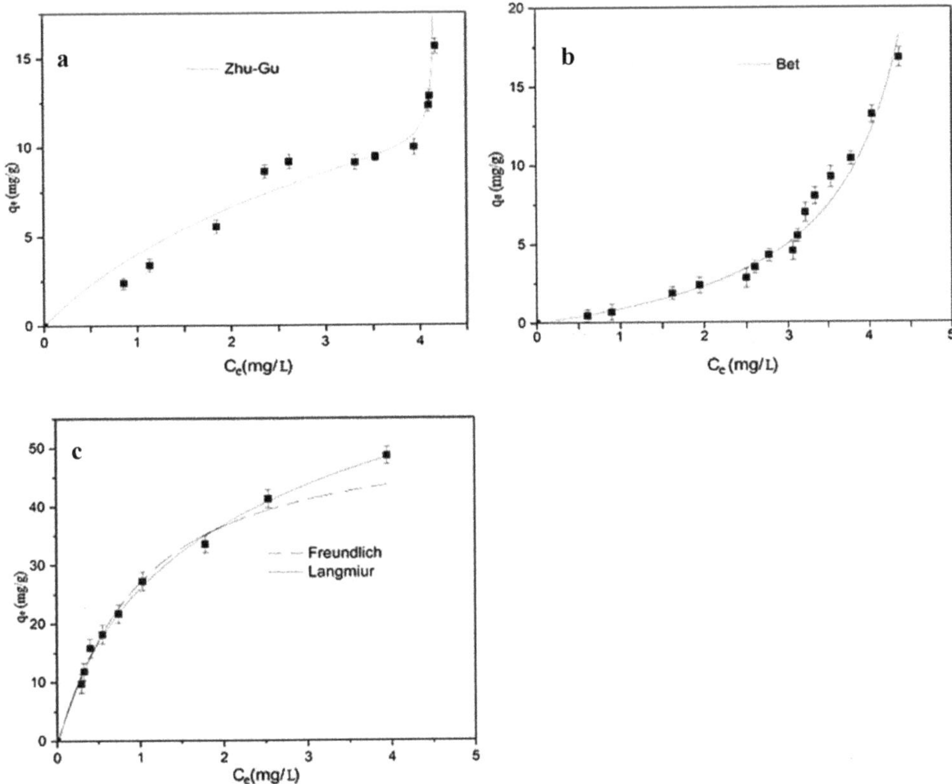

Figure 7. Experimental equilibrium results together with fittings to the considered models for the adsorption isotherm of SMX onto MY (**a**), MY@NIPs (**b**), and MY@MIPs (**c**) in ultrapure water.

The fitted equilibrium parameters are shown in Table 3 together with the correlation coefficients of the fittings (R^2). In the case of SMX adsorption onto MY@MIPs, the Langmuir isotherm provided the best fitting of equilibrium results with a higher R^2 than the Freundlich model. From the results of non-linear fittings for the different isotherm models, the values of q_m for MY, MY@NIPs, and MY@MIPs were, respectively, 23 ± 1, 3.8 ± 0.3, and 77 ± 3 mg g^{-1}. These values indicate that molecularly imprinted polymers with the template resulted in a substantial increase in the monolayer adsorption capacity, SMX adsorbing onto the surface of MY@MIPs in a homogeneous distribution by occupying specific sites. Similarly, equilibrium results on the adsorption of SMX onto the MCNTs@MIP produced by Zhao et al. [25] also fitted the Langmuir isotherm, but with a considerably lower q_m (0.87 mg g^{-1}). Indeed, compared to other materials used for the adsorption of SMX (Table 4), MY@MIPs are competitive in terms of SMX adsorption capacity.

Table 3. Equilibrium parameters corresponding to the adsorption of SMX onto MY, MY@NIPs, and MY@MIPs from ultrapure water.

Materials	Isotherm Model	Parameters	Fitted Values
MY	Zhu–Gu	q_m (mg g^{-1})	23 ± 1
		r	0.834 ± 0.005
		e	0.775 ± 0.005
		R^2	0.957
MY@NIPs	BET	q_m (mg g^{-1})	3.8 ± 0.3
		c	0.205 ± 0.008
		R^2	0.979
MY@MIPs	Freundlich	k_f (mg g^{-1} (mg L^{-1})$^{-1/n}$)	26 ± 1
		$1/n$	0.575 ± 0.003
		R^2	0.965
	Langmuir	q_m (mg g^{-1})	77 ± 3
		b (L mg^{-1})	0.498 ± 0.003
		R^2	0.998

Table 4. Maximum Langmuir adsorption capacities (q_m, mg g^{-1}) of different MIPs used for the adsorption of SMX.

Adsorbent (mg)	q_m (mg g^{-1})	Experimental Conditions	References
MIPs (100)	16.5	pH = 3; Time = 15 min; room temperature; C_{SMX} = 7500 µmol/L; V = 15 mL	[19]
Fe$_3$O$_4$-chitosan MIPs (10)	4.32	pH = 4; Time = 30 min; room temperature; C_{SMX} = 200 µg/mL; V = 10 mL	[5]
Magnetic carbon MIPs (15)	0.87	pH = 4; Time = 60 min; room temperature; C_{SMX} = 8 µg/mL; V = 20 mL	[25]
Monolithic MIPs (200)	0.02	pH = 3; Time = 30 min; room temperature; C_{SMX} = 4 µmol/L; V = 10 mL	[58]
MY@MIPs (250)	77	pH = 4; Time = 360 min; room temperature; C_{SMX} = 5 mg/L; V = 10 mL	This study

C_{SMX} = Initial concentration of SMX.

3.4.3. Kinetic and Equilibrium Adsorption Studies from STP Effluent

In order to assess the practical applicability of MY@MIPs, kinetic and equilibrium experiments were carried out in a real matrix, namely the effluent from a STP. The obtained results together with fittings to the considered kinetic and equilibrium models are in shown in Figure 8, and the fitted parameters are depicted in Table 5. As it may be seen, the pseudo-second-order and the Langmuir isotherm models were those that best described the kinetic and equilibrium experimental results, respectively. On the other hand, it is evident in Figure 8 that, under identical experimental conditions, the adsorption velocity was slower and the q_e values were lower for the STP effluent than they were for ultrapure water. This was confirmed by the parameters in Table 5, especially by the comparatively lower q_m (24 ± 2 mg g^{-1}) than in ultrapure water, which might be related to interferences due to the complex composition of the STP effluent.

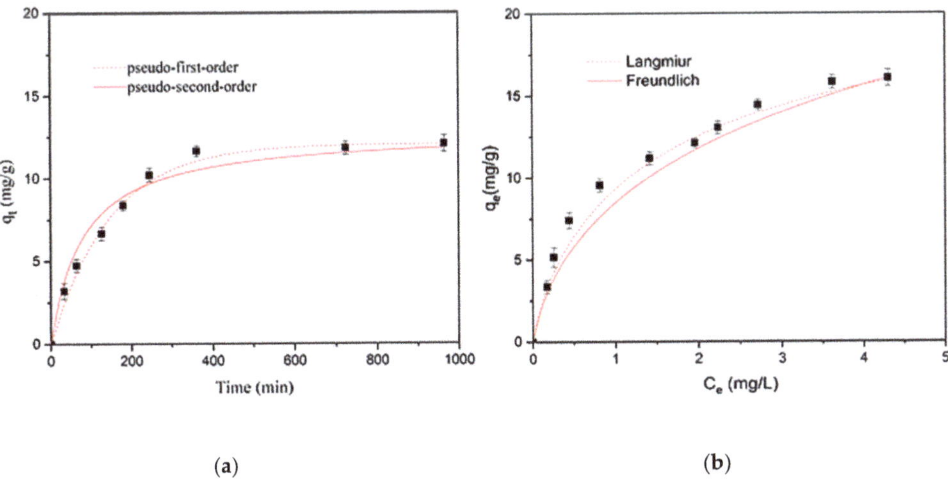

(a) (b)

Figure 8. Experimental results together with fittings to the considered models for the adsorption kinetics (**a**) and adsorption equilibrium isotherm (**b**) of SMX onto MY@MIPs in sewage treatment plant (STP) effluent.

Table 5. Kinetic and equilibrium parameters corresponding to the adsorption of SMX onto MY@MIPs from the STP effluent.

Adsorption	Models	Parameters	Fitted Values	Experimental
Kinetics	Pseudo-first order	q_e (mg g^{-1}) k_1 (min^{-1}) R^2	11.9 ± 0.3 0.471 ± 0.004 0.978	12.3 ± 0.2
	Pseudo-second order	q_e (mg g^{-1}) k_2 (g mg^{-1} min^{-1}) R^2	13.8 ± 0.7 0.009 ± 0.005 0.979	12.3 ± 0.2
Equilibrium	Freundlich	k_f (mg g^{-1} (mg L^{-1})$^{-1/n}$) $1/n$ R^2	8.5 ± 0.4 0.387 ± 0.008 0.979	
	Langmuir	q_m (mg g^{-1}) b (L mg^{-1}) R^2	24 ± 2 0.609 ± 0.004 0.989	

3.5. pH Study

Results from the study of pH effects on the adsorption of SMX onto MY@MIPs are shown in Figure 9. Under identical experimental conditions, except for the pH, decreasing q_e values were obtained at pH 4 > pH 7 > pH 9, thus indicating that SMX adsorption onto MY@MIPs was favored under acidic conditions. This may be related to the pH influence on the status of not only the adsorbate (by protonation/deprotonation) but also the adsorbent (by surface charge). For SMX, the pK_a values are 1.97 and 6.16 (Table S1, as Supplementary Information), which means that SMX is mostly positively charged (protonated NH$_2$ groups, NH$_3^+$ groups) at pH < 1.97 but predominantly negatively charged (deprotonated NH groups, N−) at pH > 6.16. Therefore, at the experimental pH 4, adsorption of SMX in the non-ionic form was favored, while at pH 7 and 9, SMX was mostly present in the anionic form, which partially hindered its adsorption. Indeed, the decrease in SMX adsorption from wastewater (with pH > 7) has already been related to electrostatic repulsion between the negatively charged SMX and the negatively charged surface of the waste-based adsorbents [59]. Moreover, the monomer used in the synthesis of MIPs was 2-vpa and the pK_a of pyridine was 5.21, which is, therefore, negatively charged when pH > 5.21. Thus, at the experimental pH 7, electrostatic repulsion forces between SMX and MY@MIPs cannot be disregarded, these increasing at pH 9.

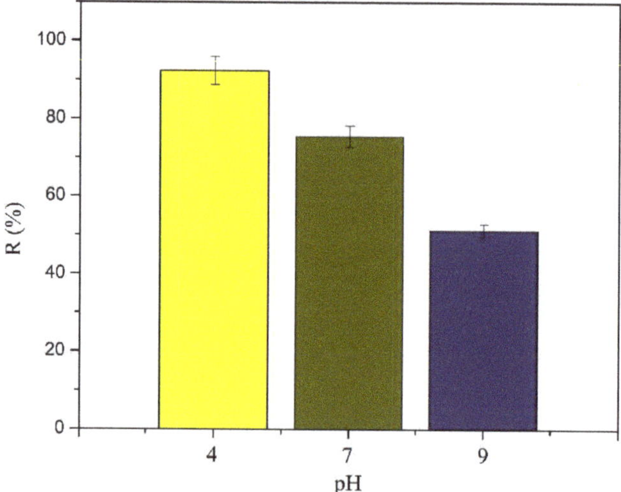

Figure 9. Effect of pH on the percentage of SMX adsorption (R (%)) onto MY@MIPs.

3.6. Selective Adsorption

In order to find out the selectivity of MY@MIPs toward SMX, its adsorption was compared to those of DCF and CBZ from their single solution and then from their ternary solution. The values of percentage of adsorption (R (%)) for the single adsorption of each pharmaceutical are shown in Figure 10a and for adsorption from their ternary solution in Figure 10b.

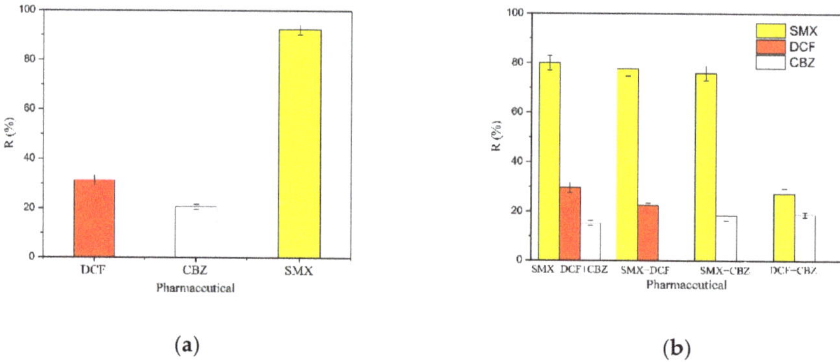

(a) (b)

Figure 10. Percentage of adsorption (R (%)) of SMX, diclofenac (DCF), and carbamazepine (CBZ) onto MY@MIPs from single solution (**a**) and ternary solution (**b**).

The results in Figure 10a evidence that MY@MIPs have a larger R (%) for SMX than for DCF or CBZ. Furthermore, under identical experimental conditions but from the ternary solution of the considered pharmaceuticals (Figure 10b), selective adsorption of SMX onto MY@MIPs occurred. Indeed, the R (%) determined for SMX from the ternary solution was just slightly lower than from its single solution, which points to the selectivity of MY@MIPs. Moreover, the results reflected that the adsorption of MY@MIPs was SMX > DCF > CBZ.

In this work, 2-vpy was the monomer and it was combined through $-NH_2$. Moreover, considering the structure and properties of SMX, DCF, and CBZ, which are depicted in Table S1, they all have $-NH_2$ and/or $-NH$ groups. However, SMX has two amino groups: $-NH$ and $-NH_2$, which is probably the main reason for its selective adsorption onto MY@MIPs under the presence of DCF and CBZ. For DCF and CBZ, the pKa was 4 and 15.96 (Table S1), respectively. Meanwhile, the pK_a value of 2-vpy (monomer) is 5.21, which may explain why MY@MIPs had a better removal ability for DCF than for CBZ. Selectivity toward SMX was also verified by Zhao et al. [25], who prepared MCNTs@MIP by using SMX as the template molecule and copolymerization of vinyl end groups on the surface of MCNTs [25]. These authors demonstrated the selective adsorption of SMX under the presence of other sulfonamides (SAs), namely sulfamethazine (SMZ), sulfamerazine (SMR), sulfadimethoxine (SDM), and sulfameter (SME). Still, the adsorbed concentration of SMX from the quinary solution was lower than from its single solution, which was ascribed to the close structure of the other SAs, which, therefore, could competitively occupy the imprinted sites.

3.7. Regeneration and Reutilization

Saturation of the produced MY@MIPs with SMX was carried out as described in Section 2.4.1. At that moment, R (%) calculated by Equation (2) was 92 ± 4%. Then, saturated MY@MIPs was regenerated as indicated in Section 2.5.4. and reused for the adsorption of SMX until saturation. A total of four regeneration/reutilization cycles were performed and the R (%) calculated for each of them are shown in Table 6.

Table 6. SMX adsorption onto MY@MIPs in subsequent cycles after regeneration.

Cycles	R (%)
SMX saturation	92 ± 4
1	73 ± 3
2	61 ± 2
3	58 ± 1
4	55 ± 2

As it may be seen, after cycles 1 and 2, the R (%) values decreased to 73 ± 3 and 61 ± 2%, respectively. Such decreases (21 and 34%, respectively, in cycle 1 and 2) indicate that the regeneration procedure affected the adsorption sites on the surface of MY@MIPs, which lost efficiency in the removal of SMX. After cycle 2, just a slight decrease in R (%) occurred, its value being similar in cycles 3 and 4 (58 ± 1 and 55 ± 2%, respectively). Therefore, deterioration of MY@MIPs adsorptive properties was not progressive with successive regenerations but occurred initially, with the performance remaining stable after cycle 2. MIPs sorbents are known to be easily regenerated by washing with organic solvents, with mixtures of methanol and acetic acid having been successfully employed to remove adsorbed pharmaceuticals [60]. Using the same regeneration agent as in this work, namely methanol/acetic acid (9/1, *v/v*), Dai et al. [61] regenerated MIPs synthesized for the adsorption of diclofenac and carried out thirty cycles with ≥95% recovery. Likewise, Duan et al. [62] also used this mixture for the regeneration of a multitemplate MIP, which was used in twenty regeneration/reutilization cycles, giving ≥95% removal of ibuprofen, naproxen, ketoprofen, diclofenac, and clofibric acid. Wang et al. [48], who used MIPs on the surface of yeast (yeast@MIPs), desorbed ciprofloxacin using the same mixture with losses of only about 8.5% of initial capacity after five cycles. Therefore, the relatively larger deterioration of the adsorptive performance observed in the present work may be related to the fact that magnetic yeast was used here as MIPs support. Thus, further work is to be carried out on the regeneration of the produced MY@MIPs, to maintain a high R (%) upon cyclic operation.

4. Conclusions

This work developed an efficient strategy to prepare yeast-Fe_3O_4 (magnetic yeast, MY) and then used molecularly imprinted technology (MIT) to modify MY. The characterization of the produced magnetic yeast-molecularly imprinted polymers (MY@MIPs) showed that elliptical and monosized imprinted polymeric nanospheres with a surface area of about 43.2 m^2 g^{-1} were successfully produced. Sulfamethoxazole (SMX) adsorption studies using MY@MIPs indicated that the equilibrium was attained in 360 min either in ultrapure water or in a sewage treatment plant (STP) effluent. The Langmuir isotherm model provided the best fitting of equilibrium results and pointed to the monolayer and favorable adsorption of SMX onto MY@MIPs. In addition, the fitted parameters of the Langmuir isotherm model indicated that the maximum SMX adsorption capacity of MY@MIPs was 77 and 24 mg g^{-1} in ultrapure water and STP effluent, respectively. The pH study pointed out that hydrogen binding was underneath the SMX adsorption onto MY@MIPs. Moreover, MY@MIPs showed successful selective adsorption of SMX from ternary solution under competition by other pharmaceuticals, namely diclofenac (DCF) and carbamazepine (CBZ). Finally, regeneration implied a reduction in SMX removal by MY@MIPs in the first two cycles, then tending to stabilization. Overall, it may be concluded that the MIPs-coated magnetic yeast designed here could be an alternative adsorbent for the selective removal of SMX from complex matrices such as wastewaters.

Supplementary Materials: The following are available online at http://www.mdpi.com/2073-4360/12/6/1385/s1, Table S1: Physico-chemical properties of the pharmaceuticals used in this study (Source: Drugbank), Figure S1: MY@MIPs in the presence (**a**) and absence (**b**) of an external magnetic field.

Author Contributions: Conceptualization, J.F., Y.D., V.I.E., M.O. and V.C.; methodology, L.Q., J.F., Y.D., V.I.E., M.O. and V.C.; materials characterization, L.Q. and M.V.G.; experimental work, L.Q. and G.J.; data analysis, Q.L., V.I.E., M.O. and V.C.; writing, L.Q., M.O. and V.C.; supervision, J.F., Y.D. and V.I.E. All authors have read and agreed to the published version of the manuscript.

Funding: This work is a contribution to the research project WasteMAC (POCI-01-0145-FEDER-028598) funded by FCT – Fundação para a Ciência e a Tecnologia, I.P., through national funds, and the co-funding by the FEDER, within the PT2020 Partnership Agreement and Compete 2020. Thanks are due to FCT/ Ministério da Ciência, Tecnologia e Ensino Superior (MCTES), for the financial support to CESAM (UIDP/50017/2020+UIDB/50017/2020), through national funds. The work was also sponsored by the Natural Science Foundation of Jiangsu Province (SBK201404182); the environmental protection scientific research subject in Jiangsu province (Grant NO.2016003); and the A Project Fund by the Priority Academic Program Development of Jiangsu Higher Education Institutions (PAPD). Marta Otero and Vânia Calisto are thankful to FCT for the Investigator Program (IF/00314/2015) and for the Scientific Employment Stimulus (CEECIND/00007/2017), respectively. Guilaine Jaria thanks for her FCT PhD grant (SFRH/BD/138388/2018) supported by the National Funds and FSE, POCH (Programa Operacional Capital Humano), and the European Union. María V. Gil acknowledges support from a Ramón y Cajal grant (RYC-2017-21937) of the Spanish Government, co-financed by the European Social Fund (ESF).

Conflicts of Interest: The authors declare no conflict of interest. Furthermore, funders had no role in the design of the study; in the collection, analyses, or interpretation of data; in the writing of the manuscript, or in the decision to publish the results.

References

1. Guo, J.; Sinclair, C.J.; Selby, K.; Boxall, A.B.A. Toxicological and ecotoxicological risk-based prioritization of pharmaceuticals in the natural environment. *Environ. Toxicol. Chem.* **2016**, *35*, 1550–1559. [CrossRef] [PubMed]

2. Andreozzi, R.; Caprio, V.; Ciniglia, C.; De Champdoré, M.; Giudice, R.L.; Marotta, R.; Zuccato, E. Antibiotics in the Environment: Occurrence in Italian STPs, Fate, and Preliminary Assessment on Algal Toxicity of Amoxicillin. *Environ. Sci. Technol.* **2004**, *38*, 6832–6838. [CrossRef] [PubMed]

3. Ashfaq, M.; Khan, K.N.; Rehman, M.S.U.; Mustafa, G.; Nazar, M.F.; Sun, Q.; Iqbal, J.; Mulla, S.I.; Yu, C.-P. Ecological risk assessment of pharmaceuticals in the receiving environment of pharmaceutical wastewater in Pakistan. *Ecotoxicol. Environ. Saf.* **2017**, *136*, 31–39. [CrossRef] [PubMed]

4. Hsu, J.-T.; Chen, C.-Y.; Young, C.-W.; Chao, W.-L.; Li, M.-H.; Liu, Y.-H.; Lin, C.-M.; Ying, C. Prevalence of sulfonamide-resistant bacteria, resistance genes and integron-associated horizontal gene transfer in natural water bodies and soils adjacent to a swine feedlot in northern Taiwan. *J. Hazard. Mater.* **2014**, *277*, 34–43. [CrossRef]

5. Qin, S.; Su, L.; Wang, P.; Gao, Y. Rapid and selective extraction of multiple sulfonamides from aqueous samples based on Fe3O4–chitosan molecularly imprinted polymers. *Anal. Methods* **2015**, *7*, 8704–8713. [CrossRef]

6. Thiebault, T. Sulfamethoxazole/Trimethoprim ratio as a new marker in raw wastewaters: A critical review. *Sci. Total Environ.* **2020**, *715*, 136916. [CrossRef] [PubMed]

7. Yang, Y.; Ok, Y.S.; Kim, K.-H.; Kwon, E.E.; Tsang, Y.F. Occurrences and removal of pharmaceuticals and personal care products (PPCPs) in drinking water and water/sewage treatment plants: A review. *Sci. Total Environ.* **2017**, *596*, 303–320. [CrossRef] [PubMed]

8. Lu, Z.; Na, G.; Gao, H.; Wang, L.; Bao, C.; Yao, Z. Fate of sulfonamide resistance genes in estuary environment and effect of anthropogenic activities. *Sci. Total Environ.* **2015**, *527*, 429–438. [CrossRef] [PubMed]

9. Na, G.; Lu, Z.; Gao, H.; Zhang, L.; Li, Q.; Li, R.; Yang, F.; Huo, C.; Yao, Z. The effect of environmental factors and migration dynamics on the prevalence of antibiotic-resistant Escherichia coli in estuary environments. *Sci. Rep.* **2018**, *8*, 1663. [CrossRef]

10. Andreozzi, R. Carbamazepine in water: Persistence in the environment, ozonation treatment and preliminary assessment on algal toxicity. *Water Res.* **2002**, *36*, 2869–2877. [CrossRef]

11. Dmitrienko, S.G.; Kochuk, E.V.; Apyari, V.V.; Tolmacheva, V.V.; Zolotov, Y.A. Recent advances in sample preparation techniques and methods of sulfonamides detection—A review. *Anal. Chim. Acta* **2014**, *850*, 6–25. [CrossRef] [PubMed]

12. Jia, A.; Hu, J.; Wu, X.; Peng, H.; Wu, S.; Dong, Z. Occurrence and source apportionment of sulfonamides and their metabolites in Liaodong Bay and the adjacent Liao River basin, North China. *Environ. Toxicol. Chem.* **2011**, *30*, 1252–1260. [CrossRef] [PubMed]

13. Ramos, A.M.; Otero, M.; Rodrigues, A.E. Recovery of Vitamin B12 and cephalosporin-C from aqueous solutions by adsorption on non-ionic polymeric adsorbents. *Sep. Purif. Technol.* **2004**, *38*, 85–98. [CrossRef]

14. Jaria, G.; Lourenço, M.A.; Silva, C.P.; Ferreira, P.; Otero, M.; Calisto, V.; Esteves, V.I. Effect of the surface functionalization of a waste-derived activated carbon on pharmaceuticals' adsorption from water. *J. Mol. Liq.* **2020**, *299*, 112098. [CrossRef]

15. Silva, C.P.; Jaria, G.; Otero, M.; Esteves, V.I.; Calisto, V. Waste-based alternative adsorbents for the remediation of pharmaceutical contaminated waters: Has a step forward already been taken? *Bioresour. Technol.* **2018**, *250*, 888–901. [CrossRef]

16. Wang, C.; Zhang, S.; Guo, F.; Ge, Y.; Wang, Y.; Li, H.; Hu, J.; Liu, H. Local Environment Structure in Positively Charged Porous Ionic Polymers for Ultrafast Removal of Sulfonamide Antibiotics. *Ind. Eng. Chem. Res.* **2019**, *58*, 16629–16635. [CrossRef]

17. Lu, Y.C.; Mao, J.H.; Zhang, W.; Wang, C.; Cao, M.; Wang, X.D.; Wang, K.Y.; Xiong, X. A novel strategy for selective removal and rapid collection of triclosan from aquatic environment using magnetic molecularly imprinted nano–polymers. *Chemosphere* **2020**, *238*, 124640. [CrossRef] [PubMed]

18. Chen, L.; Xu, S.; Li, J. Recent advances in molecular imprinting technology: Current status, challenges and highlighted applications. *Chem. Soc. Rev.* **2011**, *40*, 2922–2942. [CrossRef]

19. Valtchev, M.; Palm, B.S.; Schiller, M.; Steinfeld, U. Development of sulfamethoxazole-imprinted polymers for the selective extraction from waters. *J. Hazard. Mater.* **2009**, *170*, 722–728. [CrossRef]

20. Speltini, A.; Maraschi, F.; Govoni, R.; Milanese, C.; Profumo, A.; Malavasi, L.; Sturini, M. Facile and fast preparation of low-cost silica-supported graphitic carbon nitride for solid-phase extraction of fluoroquinolone drugs from environmental waters. *J. Chromatogr. A* **2017**, *1489*, 9–17. [CrossRef]

21. Bajpai, S.K.; Jhariya, S. Selective Removal of Amikacin From Simulated Polluted Water Using Molecularly Imprinting Polymer (MIP). *J. Macromol. Sci. Part A* **2015**, *52*, 901–911. [CrossRef]

22. Hou, L.; Han, X.; Wang, N. High performance of molecularly imprinted polymer for the selective adsorption erythromycin in water. *Colloid Polym. Sci.* **2020**, in press. [CrossRef]

23. Ou, H.; Chen, Q.; Pan, J.; Zhang, Y.; Huang, Y.; Qi, X. Selective removal of erythromycin by magnetic imprinted polymers synthesized from chitosan-stabilized Pickering emulsion. *J. Hazard. Mater.* **2015**, *289*, 28–37. [CrossRef] [PubMed]

24. Zhang, X.; Gao, X.; Huo, P.; Zhu, Z. Selective adsorption of micro ciprofloxacin by molecularly imprinted functionalized polymers appended onto ZnS. *Environ. Technol.* **2012**, *33*, 2019–2025. [CrossRef] [PubMed]

25. Zhao, Y.; Bi, C.; Hea, X.; Chen, L.; Zhangad, Y. Preparation of molecularly imprinted polymers based on magnetic carbon nanotubes for determination of sulfamethoxazole in food samples. *RSC Adv.* **2015**, *5*, 70309–70318. [CrossRef]

26. Bao-Jiao, G.; Wang, J.; An, F.; Liu, Q. Molecular imprinted material prepared by novel surface imprinting technique for selective adsorption of pirimicarb. *Polymer* **2008**, *49*, 1230–1238. [CrossRef]

27. Guan, W.-S.; Lei, J.-R.; Wang, X.; Zhou, Y.; Lu, C.-C.; Sun, S.-F. Selective recognition of beta-cypermethrin by molecularly imprinted polymers based on magnetite yeast composites. *J. Appl. Polym. Sci.* **2013**, *129*, 1952–1958. [CrossRef]

28. Liu, B.; Chen, W.; Peng, X.; Cao, Q.; Wang, Q.; Wang, N.; Meng, X.; Yu, G. Biosorption of lead from aqueous solutions by ion-imprinted tetraethylenepentamine modified chitosan beads. *Int. J. Boil. Macromol.* **2016**, *86*, 562–569. [CrossRef]

29. Qiu, L.; Feng, J.; Dai, Y.; Chang, S. Biosorption of strontium ions from simulated high-level liquid waste by living Saccharomyces cerevisiae. *Environ. Sci. Pollut. Res.* **2018**, *25*, 17194–17206. [CrossRef]

30. Gai, Q.-Q.; Qu, F.; Liu, Z.-J.; Dai, R.; Zhangad, Y. Superparamagnetic lysozyme surface-imprinted polymer prepared by atom transfer radical polymerization and its application for protein separation. *J. Chromatogr. A* **2010**, *1217*, 5035–5042. [CrossRef]

31. Qiu, L.; Feng, J.; Dai, Y.; Chang, S. Mechanisms of strontium's adsorption by Saccharomyces cerevisiae: Contribution of surface and intracellular uptakes. *Chemosphere* **2019**, *215*, 15–24. [CrossRef] [PubMed]

32. Tian, Y.; Ji, C.; Zhao, M.; Xu, M.; Zhang, Y.; Wang, R. Preparation and characterization of baker's yeast modified by nano-Fe3O4: Application of biosorption of methyl violet in aqueous solution. *Chem. Eng. J.* **2010**, *165*, 474–481. [CrossRef]

33. Silva, C.P.; Jaria, G.; Otero, M.; Esteves, V.I.; Calisto, V. Adsorption of pharmaceuticals from biologically treated municipal wastewater using paper mill sludge-based activated carbon. *Environ. Sci. Pollut. Res.* **2019**, *26*, 13173–13184. [CrossRef]

34. Mestre, A.; Carvalho, A.P. Photocatalytic Degradation of Pharmaceuticals Carbamazepine, Diclofenac, and Sulfamethoxazole by Semiconductor and Carbon Materials: A Review. *Molecules* **2019**, *24*, 3702. [CrossRef] [PubMed]

35. Qiu, Y.; Guo, H.; Guo, C.; Zheng, J.; Yue, T.; Yuan, Y. One-step preparation of nano-Fe3O4 modified inactivated yeast for the adsorption of patulin. *Food Control.* **2018**, *86*, 310–318. [CrossRef]

36. Inanan, T.; Tüzmen, N.; Akgöl, S.; Denizli, A. Selective cholesterol adsorption by molecular imprinted polymeric nanospheres and application to GIMS. *Int. J. Boil. Macromol.* **2016**, *92*, 451–460. [CrossRef]

37. Clausen, D.N.; Pires, I.M.R.; Tarley, C.R.T. Improved selective cholesterol adsorption by molecularly imprinted poly(methacrylic acid)/silica (PMAA–SiO2) hybrid material synthesized with different molar ratios. *Mater. Sci. Eng. C* **2014**, *44*, 99–108. [CrossRef]

38. Qin, S.; Su, L.; Wang, P.; Deng, S. Mixed templates molecularly imprinted solid-phase extraction for the detection of sulfonamides in fish farming water. *J. Appl. Polym. Sci.* **2014**, *132*, 41491. [CrossRef]

39. Qin, S.; Deng, S.; Su, L.; Wang, P. Simultaneous determination of five sulfonamides in wastewater using group-selective molecularly imprinted solid-phase extraction coupled with HPLC-DAD. *Anal. Methods* **2012**, *4*, 4278. [CrossRef]

40. Zheng, N.; Li, Y.; Wen, M.-J. Sulfamethoxazole-imprinted polymer for selective determination of sulfamethoxazole in tablets. *J. Chromatogr. A* **2004**, *1033*, 179–182. [CrossRef]

41. Xu, W.; Wang, Y.; Huang, W.; Yu, L.; Yang, Y.; Liu, H.; Yang, W. Computer-aided design and synthesis of CdTe@SiO2core-shell molecularly imprinted polymers as a fluorescent sensor for the selective determination of sulfamethoxazole in milk and lake water. *J. Sep. Sci.* **2017**, *40*, 1091–1098. [CrossRef] [PubMed]

42. Qi, Y.; Li, G.; Wei, C.; Zhao, L.; Gong, B. Preparation of Magnetic Molecularly Imprinted Polymer for Melamine and its application in milk sample analysis by HPLC. *J. Biomed. Sci.* **2016**, *5*, 1–10. [CrossRef]

43. Li, X.; Liu, H.; Deng, Z.; Chen, W.; Li, T.; Zhang, Y.; Zhang, Z.; He, Y.; Tan, Z.; Zhong, S. PEGylated Thermo-Sensitive Bionic Magnetic Core-Shell Structure Molecularly Imprinted Polymers Based on Halloysite Nanotubes for Specific Adsorption and Separation of Bovine Serum Albumin. *Polymers* **2020**, *12*, 536. [CrossRef] [PubMed]

44. Dai, J.; Zhou, Z.; Zhao, C.; Wei, X.; Dai, X.; Gao, L.; Cao, Z.; Yan, Y. Versatile Method to Obtain Homogeneous Imprinted Polymer Thin Film at Surface of Superparamagnetic Nanoparticles for Tetracycline Binding. *Ind. Eng. Chem. Res.* **2014**, *53*, 7157–7166. [CrossRef]

45. Parvinizadeh, F.; Daneshfar, A. Fabrication of a magnetic metal–organic framework molecularly imprinted polymer for extraction of anti-malaria agent hydroxychloroquine. *New J. Chem.* **2019**, *43*, 8508–8516. [CrossRef]

46. Rezaei, M.; Rajabi, H.R.; Rafiee, Z. Selective and rapid extraction of piroxicam from water and plasma samples using magnetic imprinted polymeric nanosorbent: Synthesis, characterization and application. *Colloids Surf. A Physicochem. Eng. Asp.* **2020**, *586*, 124253. [CrossRef]

47. Wang, X.; Pei, Y.; Hou, Y.; Pei, Z. Fabrication of Core-Shell Magnetic Molecularly Imprinted Nanospheres towards Hypericin via Click Polymerization. *Polymers* **2019**, *11*, 313. [CrossRef]

48. Wang, J.; Daib, J.; Meng, M.; Song, Z.; Pan, J.; Yan, Y.; Li, C. Surface molecularly imprinted polymers based on yeast prepared by atom transfer radical emulsion polymerization for selective recognition of ciprofloxacin from aqueous medium. *J. Appl. Polym. Sci.* **2013**, *131*, 40310. [CrossRef]

49. Li, X.; Pan, J.; Dai, J.; Dai, X.; Xu, L.; Wei, X.; Hang, H.; Li, C.; Liu, Y. Surface molecular imprinting onto magnetic yeast composites via atom transfer radical polymerization for selective recognition of cefalexin. *Chem. Eng. J.* **2012**, *198*, 503–511. [CrossRef]

50. Pan, J.; Hang, H.; Li, X.; Zhu, W.; Meng, M.; Dai, X.; Daib, J.; Yan, Y. Fabrication and evaluation of temperature responsive molecularly imprinted sorbents based on surface of yeast via surface-initiated AGET ATRP. *Appl. Surf. Sci.* **2013**, *287*, 211–217. [CrossRef]

51. Li, L.; He, X.; Chen, L.; Zhangad, Y. Preparation of Core-shell Magnetic Molecularly Imprinted Polymer Nanoparticles for Recognition of Bovine Hemoglobin. *Chem. Asian J.* **2009**, *4*, 286–293. [CrossRef] [PubMed]

52. Lagergren, S. Zur theorie der sogenannten adsorption gelöster stoffe. *Water Res.* **1996**, *30*, 1143–1148.

53. Ho, Y.; McKay, G. Pseudo-second order model for sorption processes. *Process. Biochem.* **1999**, *34*, 451–465. [CrossRef]

54. Langmuir, I. The Adsorption of Gases on Plane Surfaces of Glass, Mica and Platinum. *J. Am. Chem. Soc.* **1918**, *40*, 1361–1403. [CrossRef]

55. Freundlich, H. *Über die Adsorption in Lösungen. Habilitationsschrift durch welche. zu haltenden Probevorlesung "Kapillarchemie und Physiologie" einladet Dr. Herbert Freundlich*; W. Engelmann: Leipzig, Germany, 1906.

56. Brunauer, S.; Emmett, P.H.; Teller, E. Adsorption of Gases in Multimolecular Layers. *J. Am. Chem. Soc.* **1938**, *60*, 309–319. [CrossRef]

57. Gu, T.; Zhu, B.-Y. The S-type isotherm equation for adsorption of nonionic surfactants at the silica gel—water interface. *Colloids Surf.* **1990**, *44*, 81–87. [CrossRef]

58. Liu, X.; Ouyang, C.; Zhao, R.; Shangguan, D.; Chen, Y.; Liu, G. Monolithic molecularly imprinted polymer for sulfamethoxazole and molecular recognition properties in aqueous mobile phase. *Anal. Chim. Acta* **2006**, *571*, 235–241. [CrossRef]

59. Jaria, G.; Calisto, V.; Silva, C.P.; Gil, M.V.; Otero, M.; Esteves, V.I. Fixed-bed performance of a waste-derived granular activated carbon for the removal of micropollutants from municipal wastewater. *Sci. Total Environ.* **2019**, *683*, 699–708. [CrossRef]

60. Madikizela, L.; Tavengwa, N.; Pakade, V. Molecularly Imprinted Polymers for Pharmaceutical Compounds: Synthetic Procedures and Analytical Applications. In *Recent Research in Polymerization*; IntechOpen: London, UK, 2018; pp. 47–67.

61. Dai, C.; Zhou, X.; Zhang, Y.; Liu, S.-G.; Zhang, J. Synthesis by precipitation polymerization of molecularly imprinted polymer for the selective extraction of diclofenac from water samples. *J. Hazard. Mater.* **2011**, *198*, 175–181. [CrossRef]

62. Duan, Y.-P.; Dai, C.; Zhang, Y.; Chen, L.-. Selective trace enrichment of acidic pharmaceuticals in real water and sediment samples based on solid-phase extraction using multi-templates molecularly imprinted polymers. *Anal. Chim. Acta* **2013**, *758*, 93–100. [CrossRef]

Article

Activated Carbon Microsphere from Sodium Lignosulfonate for Cr(VI) Adsorption Evaluation in Wastewater Treatment

Keyan Yang [1], Jingchen Xing [1], Pingping Xu [1], Jianmin Chang [1,*], Qingfa Zhang [2] and Khan Muhammad Usman [3]

[1]   College of Material Science and Technology, Beijing Forestry University, 35 Qinghua East Road, Haidian District, Beijing 100083, China; ykysdut@163.com (K.Y.); jingchenx612@163.com (J.X.); m18813102213@163.com (P.X.)

[2]   School of Agricultural and Food Engineering, Shandong University of Technology, 266 Xincun West Road, Zibo 255000, China; zhangqingfacll@126.com

[3]   Department of Biological Systems Engineering, Washington State University, Richland, WA 99354, USA; muhammadusman.khan@wsu.edu

*   Correspondence: cjianmin@bjfu.edu.cn; Tel.: +86-010-6233-7733

Received: 19 December 2019; Accepted: 15 January 2020; Published: 19 January 2020

Abstract: In this study, activated carbon microsphere (SLACM) was prepared from powdered sodium lignosulfonate (SL) and polystyrene by the Mannich reaction and $ZnCl_2$ activation, which can be used to remove Cr(VI) from the aqueous solution without adding any binder. The SLACM was characterized and the batch experiments were conducted under different initial pH values, initial concentrations, contact time durations and temperatures to investigate the adsorption performance of Cr(VI) onto SLACM. The results indicated that the SLACM surface area and average pore size were 769.37 m^2/g and 2.46 nm (the mesoporous material), respectively. It was found that the reduced initial pH value, the increased temperature and initial Cr(VI) concentration were beneficial to Cr(VI) adsorption. The maximum adsorption capacity of Cr(VI) on SLACM was 227.7 mg/g at an initial pH value of 2 and the temperature of 40 °C. The adsorption of SLACM for Cr(VI) mainly occurred during the initial stages of the adsorption process. The adsorption kinetic and isotherm experimental data were thoroughly described by Elovich and Langmuir models, respectively. SL could be considered as a potential raw material for the production of activated carbon, which had a considerable potential for the Cr(VI) removal from wastewater.

Keywords: activated carbon microsphere; sodium lignosulfonate; Cr(VI); adsorption

1. Introduction

Water pollution has become a global issue because of its increasing impact on human and animal health. The presence of heavy metals in industrial water is not only causing severe damage to human and animal health, but it is also damaging aquatic life [1–3]. The effluents from a variety of industrial processes such as metallurgy, petroleum refining, batteries, and electroplating are responsible for introducing these heavy metals into aquatic environments [4,5]. These heavy metals include chromium (Cr), cadmium (Cd), lead (Pb), mercury (Hg), and other heavy metal ions and compounds [6–8]. These heavy metals are insistent and non-degradable in nature, but they are soluble in the aquatic environment; therefore, they can be easily absorbed into living cells [9]. At the same time, contamination of water by heavy metals is also considered one of the main reasons for the non-availability of potable water in developing and underdeveloped countries [10]. Among all of the heavy metals, Cr is one of the main 16 toxic metals considered detrimental for human health [11]. Cr(III) is a human micronutrient, while Cr(VI) is extremely toxic and is a strong oxidizing agent [12]. The presence of Cr(VI) in water can cause severe diseases, such as kidney circulation, dermatitis and lung cancer [13,14]. Therefore, removal of Cr(VI) from aquatic environments is indispensable for public health, as well as for the protection of the environment and aquatic life. Moreover, strict environmental mechanisms and the enactment of legal standards is necessary to avoid excessive discharge of Cr(VI) into potable water sources [15]. The Environmental Protection Agency (EPA) has recommended that the maximum concentration of Cr(VI) in drinking water not exceed more than 0.05 mg/L [12].

Many techniques, such as adsorption, membrane filtration, electrodialysis ion exchange, reduction, reverse osmosis and biological removal have been developed to remove Cr(VI) from aquatic environments [13,16,17]. Among them, adsorption is one of the most effective methods and is suitable for use in developing countries by using adsorbents [18]. Adsorption is a green and low-cost wastewater treatment technique, especially useful for heavy metal ions and hydrophilic compounds such as chromium, lead, ammonium ions and antibiotics [10,19–21]. The adsorbent generally has an abundant pore structure, while simultaneously having chemical functional groups on the surface, and it can be easily modified through increased surface charge [18]. In terms of removing chromium ions from wastewater, the adsorbent used by the adsorption method has a lower cost than the ionic exchange resins used by membrane filtration and the membranes used by electrodialysis ion exchange; furthermore, adsorption is easier to perform than reduction and reverse osmosis, and it has a higher processing efficiency and more extensive application fields than biological removal. In general, compared to other methods, activated carbon absorption has been proved to be a preferable technique for the removal of Cr(VI) from wastewater due to its higher efficiency, lower operating cost, higher adsorption capacity, easier operation, and non-hazardous technique. Thus, the activated carbon adsorption process has been developed and applied extensively [21,22].

Currently, the raw material to produce activated carbon comes from conventional fossil fuels such as coal and petroleum. With the development of renewable energy sources, the exploration of novel and effective adsorbents based on renewable sources has also received more attention than ever before due to the growing concerns about environment and increasing cost of fossil fuels [23]. Keeping in mind the environmental and industrialization concerns, it is necessary to develop activated carbons from low-cost and abundant precursors. During the last few years, many researchers have used adsorbents made of different types of renewable biomasses such as tobacco stems [24], longan seed [4], rice straw [25], juniperus procera sawdust, avocado kernel seeds and papaya peels [26], activated carbon derived from leucaena leucocephala [27], and sterculia guttata shell [28], and the results of these studies indicates that these adsorbents are helpful in achieving sufficient removal of Cr(VI) from wastewater. However, the cost of these adsorbents was found to be higher, due to which these adsorbents are not being used at commercial scale.

Therefore, a novel, cost-effective and efficient activated carbon is highly needed for proper and cost-effective treatment of wastewater at larger scales. The SL is an inevitable by-product of the paper and pulp industry. Currently, most of the SL generated by the paper and pulp industry is burnt or dumped into open lands and drains, which is not only causing an increase environmental pollution, but also wastes a valuable resource. It is necessary to achieve high-value utilization of SL, rather than wasting it. In addition to high carbon content, SL contains oxygenic functional groups such as carboxyl and phenol hydroxyl [29]. Moreover, some researchers have also found that SL and its derivatives have remarkable electrochemical performance and the potential to adsorb certain heavy metals, including toluene, and some other pollutants [30,31]. However, the problem associated with SL as an adsorbent is its high solubility in water, due to which it is difficult to remove powder SL from aqueous solution after the adsorption process, which can cause secondary pollution. That is a common reason for limiting powder SL application as adsorbing material. To solve the separation problem of powder SL after adsorption from solution, magnetic lignosulfonate was fabricated based on magnetic separation technique [32]. However, beyond that, changing the macroscopic character of SL from powder to granular to prepare granular activated carbon is a valuable research direction for solving the separation problem, as well as for improving the adsorption properties of SL. The activated carbon prepared by conventional methods from SL is also powdered. However, the powdered activated carbon is very difficult to apply in wastewater treatment, because it is dispersed in the water and very hard to separate after adsorption. If the powdered activated carbon is to be applied commercially, then it must be prepared into granular activated carbon by means of a binder. However, the application of binder is costly and environmentally unfriendly. Therefore, a new preparation method for activated carbon microspheres from SL needs to be developed. Because the SL has good reaction activity, it could be used to react with basis materials such as polystyrene to generate large particles of adsorption material with easy recovery and separation, in order to realize high-value industrial utilization of SL as an adsorption material. The SL also has lower cost than many other fossil fuels as a kind of renewable biomass resource with vast reserves. Therefore, the SL that is considered to be waste in the paper and pulp industry and is generally dumped into the open environment, causing an increase in environmental pollution, can be used as a renewable precursor to substitute traditional fossil fuels for the production of activated carbon, which can then be used as a low-cost adsorbent for the removal of heavy metals from polluted aquatic environments.

Keeping in mind the acute need for low-cost and efficient renewable adsorbents in the wastewater treatment sector and the promising behavior of adsorbents prepared from biomass, in this study the SLACM was developed, aiming to remove Cr(VI) from wastewater. The preparation of SLACM was done in two steps, i.e., the preparation of activated carbon precursor from powdered SL and polystyrene by Mannich reaction without the addition of any binder, and the $ZnCl_2$ activation of the precursor. The effects of the adsorption process on the adsorption capacity of SLACM for Cr(VI) were investigated according to batch experiments carried out under different initial pH, initial concentration, contact time, and temperature. The prepared SLACM was characterized with respect to its pore volume, surface area, and sorption efficiency, and the transformation of structure and properties before and after Cr(VI) adsorption was compared. On the basis of the ability to remove Cr(VI) in the solution experiments, adsorption kinetics, and isotherm fitting, the potential of using SLACM for the purification of wastewater was evaluated.

2. Materials and Methods

2.1. Materials

The SL was purchased from Shanghai Macklin Biochemical Co., Ltd., Shanghai, China. The SL was composed of 41.63% carbon, 28.32% oxygen, and 24.39% sodium, as well as small amounts of silicon. The chloromethylated polystyrene was used as the base material in the preparation. chloromethylstyrene-divinylbenzene-styrene copolymer (CMPS) was purchased from Tianjin Xingnan macromolecule technology Co., Ltd., Tianjin, China. 1,3-diaminopropane ($C_3H_{10}N_2$, MW 74.13) was purchased from Shanghai Macklin Biochemical Co., Ltd. Tetrahydrofuran (C_4H_8O) was purchased from Beijing Chemical Co., Ltd. Beijing, China. Formaldehyde (HCHO) was purchased from Xilong Scientific Co., Ltd., Guangdong, China. $ZnCl_2$ was purchased from Shanghai Macklin Biochemical Co., Ltd., Shanghai, China. Methyl orange was purchased from shanghai D&B biological science and technology Co., Ltd., Shanghai, China. Deionized water was used for all the experiments. All chemicals and materials were of analytical grade and were used without further purification.

2.2. Preparation of SLACM

The preparation of SLACM was done in two main steps, i.e., preparation of activated carbon precursor from powdered SL and polystyrene by Mannich reaction and the $ZnCl_2$ activation of the precursor. The preparation process diagram of SLACM was shown in Figure 1. The substrate material CMPS need to be amination pretreated before reaction. The CMPS was first swollen 2 h with tetrahydrofuran (3 mL/g) and then amination treated for 12 h in the 50 °C with 1,3-diaminopropane (5 mL/g). After that, the SL and amino CMPS ware generate large particles of adsorbent resin microsphere (ARM) by Mannich reaction in solution catalyzed by formaldehyde at a mass ratio of 1:1. The Mannich reaction of SL and amino CMPS continuously proceed for 12 h in the 90 °C water bath. The ARM samples are shown in Figure 2a. The ARM was oxidized for 30 min at 180 °C before the impregnation process to obtain pre-oxidized adsorbent resin microsphere (PARM) as an activated carbon precursor. A suitable amount of PARM was added to the activator aqueous solution (50 wt.%), and the solution was stirred continuously at room temperature to prepare a uniform solution. $ZnCl_2$ was used as an activating agent during the impregnation process with an impregnation ratio of 1:1 of activated carbon precursors to activator respectively. After 12 h of the impregnation process, the samples were dried at 105 °C for 6 h in an oven. The impregnated activated carbon precursors were put into a porcelain boat in a horizontal tube furnace and heated with nitrogen (99.99%) protected. During the activation process, the temperature was initially increased from room temperature to the activation temperature of 600 °C by using a heating rate of 10 °C/min, and then the temperature was kept constant at 600 °C for 2 h. After this, the samples were cooled to room temperature under nitrogen atmosphere. Finally, the samples were washed 2–3 times with rare hydrochloric acid (0.5 mol/L) and deionized water in sequence until the pH of filtrate became neutral and then dried at 105 °C for 12 h to obtain dry SLACM. The SLACM samples are shown in Figure 2b.

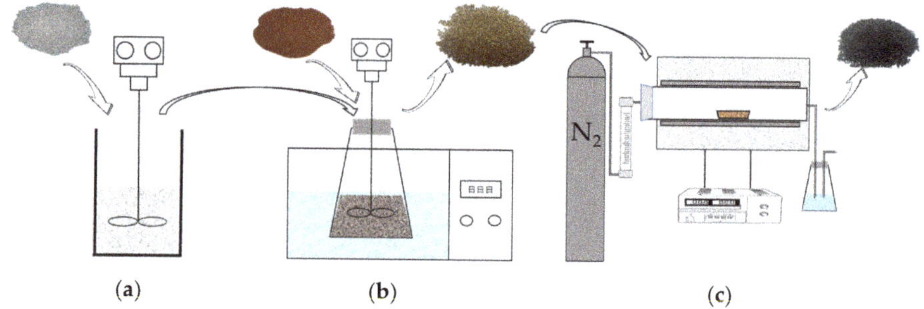

(a)   (b)   (c)

Figure 1. The preparation process diagram of SLACM: (**a**) Amination of CMPS; (**b**) Mannich reaction of SL and ACMPS; (**c**) Activation processes.

(a)   (b)

Figure 2. (**a**) The ARM sample picture; (**b**) The SLACM sample picture.

2.3. Characterizations of SLACM

The surface morphology of SLACM samples was observed by scanning electron microscopy (SEM, Hitachi S-4800, Hitachi, Japan) at an accelerating voltage 5.00 kV and the element chemistry configuration was analyzed by SEM with energy dispersive spectrometry (EDS) system.

The microstructures of the SLACM samples was characterized by transmission electron microscope (TEM, FEI TF30, Hillsboro, OR, USA). A physisorption analyzer (Quantachrome autosorb-IQ, Boynton Beach, FL, USA) was used to determine the specific surface area, pore volume and other pore characteristics of SLACM samples. The SLACM samples pulverized to carbon powder were treated by vacuum degassing 3 h at 200 °C at first, and then N_2 adsorption and desorption isotherms were measured at 77 K. The specific surface area and pore volume of SLACM samples were analyzed by the BET method according to the N_2 adsorption and desorption isotherms. The BET equation was applied to N_2 adsorption data from $0.06 < P/P_0 < 0.25$ to determine surface area and total pore volume was measured at $P/P_0 = 0.985$. The surface functional groups of SLACM were characterized before and after adsorption by using a Fourier transform infrared spectrometer (FT-IR, Thermo Scientific Nicolet iS10, San Jose, CA, USA) in the range from 4000 to 400 cm^{-1} at 4 cm^{-1} resolutions with 64 scans. The crystalline structure of SLACM specimen was investigated by X-ray diffraction (XRD, Bruker D8 ADVANCE diffractometer) with Cu Kα radiation ($\lambda = 0.15417$ nm) in the range of $2\theta = 10$–90 at a step rate of 0.02. The mass change of activated carbon precursors was measured by the thermogravimetric analysis (TG) (STA449C, Selb, Germany) under an N_2 atmosphere from 30 to 800 °C at the rate of 10 °C/min.

2.4. Adsorption Experiment

All batch experiments were carried out in 250 mL Erlenmeyer flasks with 100 mg SLACM and 50 mL Cr(VI) solution in order to study the adsorption capacity of adsorbent on Cr(VI). Several experimental parameters which may affect the absorption efficiency of SLACM were studied, including initial pH, contact time, temperature and initial concentration of Cr(VI). When the SLACM and Cr(VI) solution was taken in the flasks, HCl solution (0.1 mol/L) and NaOH solution (0.1 mol/L) was used to adjusted the initial pH of the solution to setting value. The Erlenmeyer flasks were then placed on a 150-rpm shaking table to determine the adsorption capacity of SLACM according to setting different contact time and temperature. After the designated time of adsorption, the mixture was separated using a 0.45 μm filter. The filtrate absorbance was measured by UV-Vis spectrophotometer, and Cr(VI) concentration was calculated according to the diphenyl carbohydrazide spectrophotometric method (Chinese National Standards GB/T 7467-87). All the adsorption experiments were repeated 3 times and the average values were used as final results.

The adsorption capacity q_t and q_e of Cr(VI) onto SLACM were calculated as follows:

$$q_t = \frac{(C_i - C_t)V}{M},\tag{1}$$

$$q_e = \frac{(C_i - C_e)V}{M},\tag{2}$$

where q_t (mg/g) is the adsorption capacity at time t, q_e (mg/g) is the adsorption capacity at equilibrium, C_i (mg/L) is the initial concentration of Cr(VI), C_t (mg/L) is the concentration of Cr(VI) at time t, C_e (mg/L) is the concentration of Cr(VI) at equilibrium, V (L) is the volume of suspension of Cr(VI), M (g) is the dry weight of the SLACM.

2.4.1. Effects of Initial pH on Adsorption

The effects of initial pH on the Cr(VI) adsorption were studied by adding SLACM to flasks with 150 mg/L Cr(VI) solution under the similar experimental conditions. The pH value of the solution was then adjusted to 1.0, 2.0, 3.0, 4.0, 5.0, 6.0, 7.0, 8.0 and 9.0 by using HCl solution (0.1 mol/L) and NaOH solution (0.1 mol/L). All the samples were shaken in a shaking table for 12 h (30 °C, 150 rpm).

2.4.2. Adsorption Kinetics

The adsorption kinetics of Cr(VI) on the SLACM was studied by using a contact times ranging from 2 min to 24 h with initial Cr(VI) concentration of 150 mg/L and pH value 2. The agitation speed and the contact temperature of the shaking table were kept constant at 150 rpm and 30 °C, respectively.

2.4.3. Adsorption Isotherm

Adsorption isotherms are usually used to describe the distribution of metal ions between the liquid phase and the solid phase [33]. Adsorption isotherms were performed for 20, 30 and 40 °C by using initial Cr(VI) concentrations, pH, contact time, and a shaking speed of 100 to 600 mg/L, 2, 12 hand 150 rpm, respectively.

3. Results and Discussion

3.1. Characterization

The pore structure of activated carbon is an important factor influencing adsorption capacity. The specific surface area and pore volume of SLACM sample was evaluated by the BET method according to Nitrogen adsorption isotherms measured at 77 K. The porous structure parameters of SLACM are shown in Table 1. It is clear from the table that the BET surface area of SLACM is 769.37 m^2/g and the total pore volume is 0.47 cm^3/g. The chromium ions could be trapped in the pores of SLACM due to the radiuses of chromium ions were less than the pore size of the adsorbent [33]. The well-developed BET surface area and the total pore volume mean that the absorption performance of SLACM is good and it can be used as a substitute to conventional adsorbents. The average pore size of SLACM was found to be 2.46 nm, and thus it belongs to the mesoporous materials.

Table 1. Porous structure parameters of SLACM.

Sample	S_{BET} (m^2/g)	S_{mic} (m^2/g)	V_{tot} (cm^3/g)	V_{mic} (cm^3/g)	D_p (nm)
SLACM	769.37	639.28	0.47	0.26	2.46

The surface microstructure and fracture micro appearance of SLACM was characterized using SEM and TEM, the surface microstructure of SLACM, fracture micro appearance of crushed SLACM before and after adsorption, the TEM images of SLACM before and after adsorption are presented in Figure 3. As shown in Figure 3a, the SLACM was a regular sphere and the diameter of most particles was around 0.5–1.5 mm. Due to its larger size, the activated carbon can be separated out more easily from solution after adsorbing than the powdery SL. Thus, the recycling properties as well as convenience of use of the activated carbon were improved significantly. There were a lot of different size clearances and cracks present on the SLACM surface, as shown in Figure 3b, and these were the main channels through which the adsorbate could enter inside [10]. The well-developed pore structure should be responsible for the adsorption of Cr(VI) onto SLACM. It was also found by the fracture micro appearance of inner side of SLACM that the abundant pores and cavity structure confirmed the high specific surface areas of adsorbent, this could also be supported by the porous structure parameters of SLACM. The TEM image confirms that the SLACM before adsorption is hollow, and this observation is in line with SEM fracture surface image. As can be seen from Figure 3d,f, the pores of SLACM were blocked, and obvious attachment appeared on the sample surface after Cr(VI) adsorption. The EDS spectrum of the SLACM before and after Cr(VI) adsorption and EDS elemental mapping patterns of C, N, O and Cr of SLACM after Cr(VI) adsorption were shown in Figure 4. It was observed that carbon, nitrogen and oxygen were the major elements in SLACM and that carbon, nitrogen, oxygen and chromium were the major elements on SLACM after Cr(VI) adsorption. The elements of nitrogen and oxygen have a greater promoting effect in the adsorption which could enhance the adsorption efficiency of SLACM for Cr(VI) [34]. The SLACM after adsorption showed distinct peaks of chromium present on the surface and the EDS mapping reveals that the chromium was evenly distributed over SLACM surface. Therefore, EDS analysis confirmed the adsorption of Cr(VI) onto SLACM.

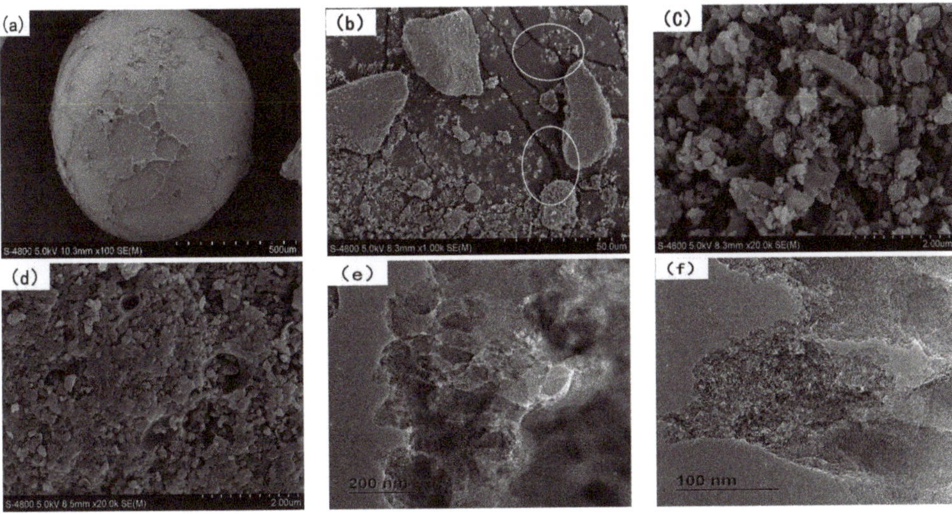

Figure 3. (**a**,**b**) SEM images of SLACM surface before Cr(VI) adsorption of 100 and 1 k; (**c**) SEM image of SLACM fracture surface before Cr(VI) adsorption of 20 k; (**d**) SEM image of SLACM fracture surface after Cr(VI) adsorption of 20 k; (**e**,**f**) TEM images of SLACM before and after Cr(VI) adsorption.

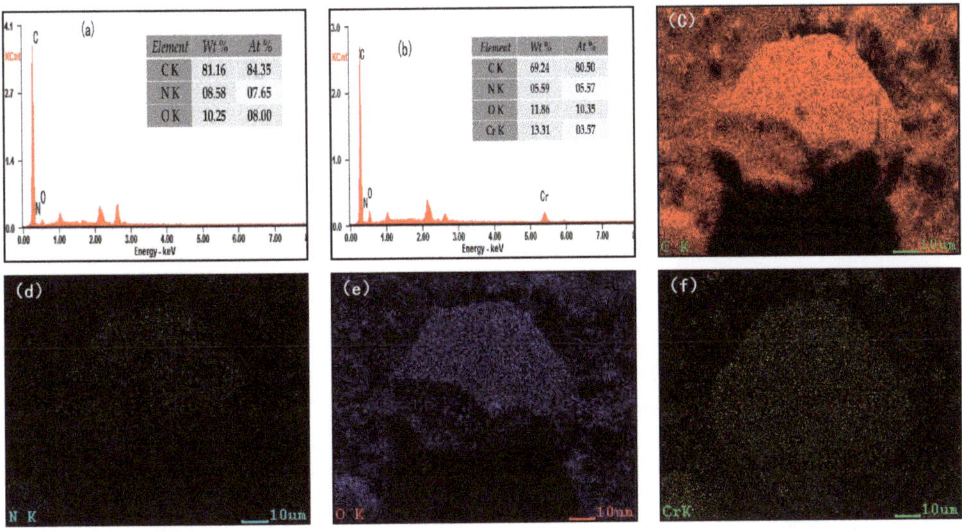

Figure 4. (**a**) EDS spectrum of SLACM before Cr(VI) adsorption; (**b**) EDS spectrum of SLACM after Cr(VI) adsorption; (**c**–**f**) EDS elemental mapping patterns of C, N, O and Cr after Cr(VI) adsorption.

The thermal behavior of the CMPS, ARM and PARM was analyzed under nitrogen atmosphere. The TG and derivative thermogravimetric (DTG) curves of SL, CMPS, ARM and PARM are illustrated in Figure 5. The SL has a high residual mass rate after heating and a low peak value of DTG than others. A high residual mass rate is positive for preparing of SLACM. Compared with CMPS, the maximum weight loss temperature (T_{max}) of AR was reduced more significantly. This is due to the destruction of part of long chain structure or aromatic substitution during the graft polymerization of SL and CMPS, the protection of lignin was break down [35]. For these reasons, the residual mass rate of ARM is much less than that of SL and CMPS. The residual mass rate of PARM has a larger increase compared to ARM, probably because of partial structural carbonization or pyrolysis in ARM during the oxidation process [36]. Keeping in mind the larger increase in the residual mass rate of PARM, it can be concluded that a higher yield SLACM can be prepared by using PARM as raw material compare to ARM. It is also clear from Figure 5 that the residual mass rate of all the samples did not decrease when the temperature was higher than 600 °C, which suggests that the activation temperature should not be increased to higher than 600 °C during the preparation of SLACM. Thus, the activation temperature for SLACM was found to be lower than activated carbon production from other studies in the literature [33,37–39], which can save a large amount of energy during production process.

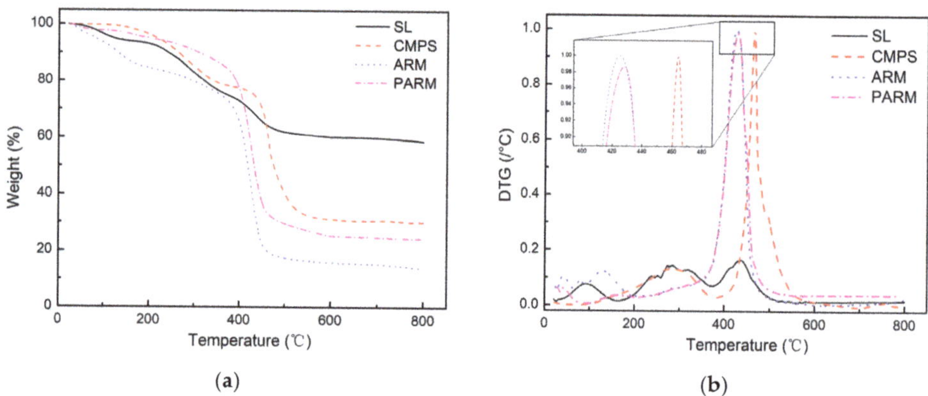

Figure 5. (**a**) The TG curves of SL, CMPS, ARM and PARM; (**b**) The DTG curves of SL, CMPS, ARM and PARM.

The structural variation of the SLACM samples and other materials were analyzed by FT-IR and the recorded spectra are shown in Figure 6a. The wide peak at 3423 cm^{-1} corresponding to stretching of aliphatic and phenolic –OH groups was observed in all of the samples, especially SL. The peaks at 1593 and 1444 cm^{-1} were typical of aromatic C–C stretching and C–H bending in SL. The peak intensity of SL at 3423 and 1593 cm^{-1} was much stronger than ARM, which meant aromatic phenolic -OH groups in SL were combined with aminated CMPS after Mannich reaction. After the pre-oxidation, new peaks appeared in PARM. The peak around 2929 cm^{-1} was attributed to the stretching vibration of saturated hydrocarbon $–CH_3$, $–CH_2–$ and the methyl or methylene in aliphatic groups. The peak at 1660 cm^{-1} was attributed to the stretching vibration of amide –CO–N [31]. This may be because the –NH oxidizes in AR and reaction produced -CO-N groups. The peak at 1103 cm^{-1} was attributed to the antisymmetric stretching vibration of S–O in SO_3^{3-} groups, and the SO_3^{3-} group was from SL [32]. This indicates that SL was successfully grafted onto aminated CMPS. After activation, the peak amount of SLACM was significantly less than PARM and the peak intensity was obviously weaker than PARM. This was because some chemical bonds of PARM were broken down at higher temperature, such as alken-CH_3, aliphatic ether $–C–O–$, and the benzene ring replacing op (CH), and forming graphitized carbon materials [31]. Some small weak peaks appeared at 3423 and 1593 cm^{-1} on the SLACM curve. These results show that SLACM contains a variety of functional groups including phenols, alcohols, alkenes, amines, etc. The main functional groups are in good agreement with biomass-activated carbon from others [31,40]. The oxygen-containing functional groups of SLACM determine the surface acidity-basicity and the adsorption performances. After adsorption of Cr(VI), these peaks of SLACM-Cr were not obvious, indicating an interaction between the Cr (VI) and SLACM.

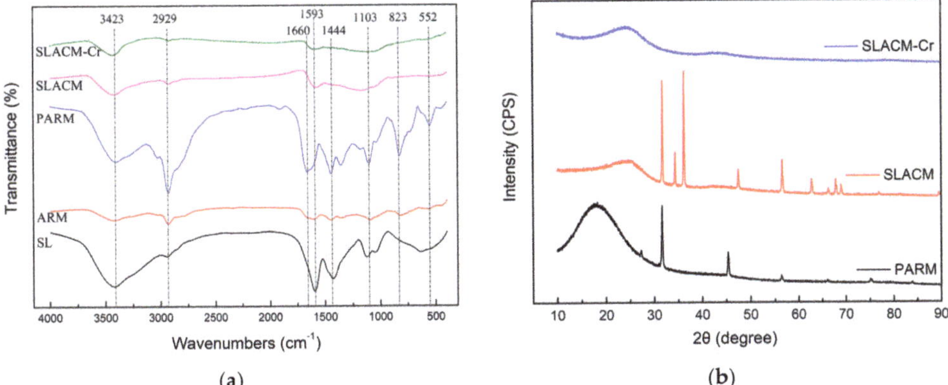

Figure 6. (**a**) The FT-IR spectra of samples; (**b**) The XRD curves of samples.

The XRD spectra representing the crystalline structures of PARM, SLACM, SLACM-Cr (after adsorption of Cr(VI)) are shown in Figure 6b. The results show that PARM has an amorphous wide peak at 18.28°. The wide peak of the SLACM decreased and moved to the right compared to PARM. This illustrates that the distance of crystal face of SLACM decreased and the graphitization degree increased after activation at high temperature [33,41]. The diffraction peaks of SLACM occurred at 31.72°, 34.35°, 36.2°, 56.54°. This is because the crystal structure of samples changed at higher temperature, main performance for small particle graphite structure and undefined structure. The diffraction peaks of SLACM-Cr disappeared after the Cr(VI) adsorption. This was due to fact that the Cr(VI) was adsorbed successfully by SLACM as a physical or chemical form and the adsorbate Cr(VI) damaged the crystal structures.

3.2. Effect of Initial pH on Adsorption

The initial pH of the solution is an important parameter that influences the adsorption capacity of the adsorbent for metal ions [42]. The effect of initial pH on the adsorption capacity of SLACM for Cr(VI) was tested, and the results are shown in Figure 7a. It is clear from the results that adsorption capacity of SLACM for Cr(VI) is strongly dependent on the initial pH of the solution. Moreover, the adsorption capacity of SLACM continuously decreases as the pH increases from 2.0 to 9.0. This result is comparable with other consequences reported in the literature [43]. Cr(VI) exist in the form of oxyanions in solution, and the existence form of Cr(VI) ion depends mainly on the solution pH value. The predominant Cr(VI) existence forms are dichromate ($Cr_2O_7^{2-}$) and hydrogen chromate ($HcrO^{4-}$) when the pH of the aqueous solutions was in range of 2.0–6.0. When the pH exceeded 7.0, chromate (CrO_4^{2-}) was the main Cr(VI) species [33,37]. The initial pH of the solution simultaneously affects the adsorbent surface charge and degree of ionization [44]. At lower pH, the functional groups on the surface of the SLACM were protonated by plentiful hydrogen ions, and there was a strong electrostatic interaction between the negatively charged Cr(VI) and the positively charged SLACM. The strong electrostatic interaction was very favorable for adsorbing dichromate and hydrogen chromate. Cr(VI) could be reduced to Cr(III) with the SLACM surface charge provided by the oxygen-containing functional groups, such as -C-O- and -OH. The Cr(VI) was adsorbed by complexation or reduced may be described by the following reactions (Equations (3)–(6)) [13,33]. The adsorbed Cr(VI) was reduced to Cr(III) by the electron-donor groups of the porous carbon from corn straw was also proved by Ma et al. [33]. Therefore, the highest adsorption capacity of SLACM for Cr(VI) was the 75.25 mg/g when the initial pH of solution was 2.0. As the initial pH of the solution increased, the protonation degree of the SLACM surface functional groups decreased and gradually converted to negatively charged. There exists competitive adsorption between OH- and chromate, which interferes with the binding site on the adsorbents; the OH- content in solution increased, and the competitive adsorption was found to be stronger at higher pH values. Electrostatic repulsion and competitive adsorption caused a decrease in the adsorption capacity of SLACM for Cr(VI); therefore, the adsorption capacity of SLACM for Cr(VI) dropped to 19.42 mg/g when the pH was raised to 9.0. Moreover, Cr(VI) and the –COOH group of SLACM may have an ion exchange reaction (Equation (7)). These results indicate that the Cr(VI) removal behavior was affected by the initial pH of solution, and pH 2.0 was applied to the following adsorption experiments to obtain the optimal adsorption performance in this study.

$$RO^- + Cr^{6+} \rightarrow RO^- \dots Cr^{6+} \tag{3}$$

$$3ROH + Cr_2O_7^{2-} + 4H^+ \rightarrow 3RO + HCrO_4^- + Cr^{3+} + 3H_2O \tag{4}$$

$$3ROH + HCrO_4^- + 4H^+ \rightarrow 3RO + Cr^{3+} + 4H_2O \tag{5}$$

$$3C-O- +3HCrO_4^- + 5H^+ \rightarrow 3C-O + 3R \dots Cr^{3+} + 4H_2O \tag{6}$$

$$3R-COOH + Cr^{6+} \rightarrow R-COO^- \dots Cr^{6+} + H^+ \tag{7}$$

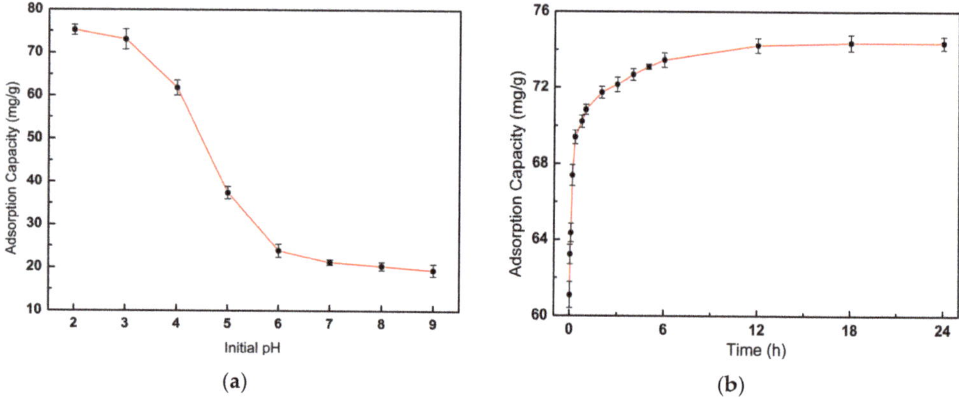

Figure 7. (**a**) Effect of initial pH on adsorption capacity of SLACM for Cr(VI); (**b**) Effect of time on adsorption capacity of SLACM for Cr(VI).

3.3. Adsorption Kinetics

To evaluate the kinetic behavior of adsorption process of SLACM for Cr(VI), the adsorption time of SLACM in solution was set as 2 min to 24 h. The adsorption velocity of SLACM for Cr(VI) was different at different time points during the adsorption process, but eventually it all balanced out. The adsorption kinetics were determined in order to study the relationship between the adsorption capacity and adsorption time during the adsorption process, and were significant in explaining the adsorption process in view of the order of the rate constants. The relationship of adsorption capacity of SLACM for Cr(VI) and adsorption time are shown in Figure 7b. As can be seen, the adsorption capacity of SLACM for Cr(VI) increased rapidly in the first hour and was 70.84 mg/g after 1 h. The increase in adsorption capacity slowed down between 1 and 6 h, and then the adsorption capacity gradually reached equilibrium. The equilibrium adsorption capacity was 74.38 mg/g. The pseudo first-order (PFO), pseudo second-order (PSO) and Elovich models were applied to the experimental data to evaluate the adsorption kinetics, in order to understand the controlling mechanisms of the Cr(VI) sorption on SLACM [7,27]. These models are given as the following equations, respectively.

$$\text{PFO}: q_t = q_e\left(1 - e^{-k_1 t}\right), \tag{8}$$

$$\text{PSO}: q_t = \frac{q_e^2 k_2 t}{1 + q_e k_2 t}, \tag{9}$$

$$\text{Elovich}: q_t = \frac{1}{\beta}\ln(1 + \alpha\beta t),\tag{10}$$

where q_t (mg/g) is the adsorption capacity at t time, q_e (mg/g) is the adsorption capacity at t time at equilibrium, t (h) is adsorption capacity time, k_1 (min^{-1}) is the PFO equilibrium rate constant, k_2 (g/mg·min) is the PSO equilibrium rate constant, α and β are Elovich constants. The adsorption kinetic constants and dynamic fitting parameters were summarized in Table 2. The fitted results in Table 2 show that the Elovich model is in better agreement with the experimental results when explaining adsorption rate than the PFO and PSO models, indicating the better suitability of Elovich model for describing the adsorption kinetics of Cr(VI) onto SLACM. Therefore, the adsorption type of Cr(VI) onto SLACM was determined to be predominantly chemisorption, which was consistent with heavy metal ion adsorption results using different solid adsorbents, as described in other literature studies [7]. The adsorption of SLACM includes mass transfer of Cr(VI) to the external and internal surface of adsorbent, particle diffusion, active site adsorption of adsorbent and other multi-stages adsorption process [40]. The adsorption process can be divided into several stages. The Cr(VI) in solution firstly diffused to the SLACM surface then it entered inside the adsorbent through pores. The adsorbate was adsorbed by pore structures and the active sites on external and internal surface of SLACM during these processes. The available active sites on the surface was sufficient and the pore structure was not completely filled during the initial stages of adsorption process, the mass transfer resistance of Cr(VI) was small. This was due to the presence of a greater driving force provided by the higher initial concentration of Cr(VI), which increased the probability of collisions between Cr (VI) and the active sites. Therefore, the adsorption process has a high adsorption rate at the beginning. The active sites on the surface were inhibited and the pore structure was filled gradually through the process of adsorption. Therefore, the interactions between Cr(VI) and active sites decreased and the mass transfer resistance of adsorbate increased. As shown in the curve in Figure 7b, the adsorption rate decreased and the adsorption capacity of SLACM for Cr(VI) gradually reached to the equilibrium state.

Table 2. Adsorption kinetics model parameters of SLACM for Cr(VI).

Model	Parameters	Value
PFO	q_e	71.1
	K_L	106.75
	R^2	38.83
PSO	q_e	72.03
	K_2	3.59
	R^2	73.12
Elovich	α	5.81
	β	0.54
	R^2	95.63

3.4. Adsorption Isotherm

SLACM was brought into contact with different concentrations of the solution (50–600 mg/L) in order to evaluate the effect of initial concentration on the adsorption capacity of SLACM for Cr(VI). Adsorption isotherms were used to evaluate the adsorption type and study the adsorption behavior under different equilibrium condition. The Langmuir model and Freundlich model were used to fit equilibrium experimental data of Cr(VI) adsorption on SLACM at temperatures of 20, 30 and 40 °C [37]. The isotherm models can be explained as follows.

This is an example of an equation:

$$\text{Langmuir}: q_e = \frac{K_L C_e q_m}{1 + K_L C_e},\tag{11}$$

$$R_L = \frac{1}{1 + K_L C_0},\tag{12}$$

$$\text{Freundlich}: q_e = K_F C_e^{\frac{1}{n}},\tag{13}$$

where q_e (mg/g) is the maximum adsorption capacity, C_e (mg/L) is the equilibrium Cr(VI) concentration, q_m (mg/g) is monolayer adsorption capacity, K_L (L/mg) is the Langmuir affinity constant, K_F (L/mg) is the Freundlich constant related to the adsorption capacity of the adsorbent and n is an empirical constant related to the adsorption intensity which varies with the heterogeneity of material.

The experimental data and non-linear fitting curves are shown in Figure 8. The fitting results and correlation coefficients are shown in Table 3. The Langmuir isotherm model is generally applicable to monolayer adsorption onto a homogeneous surface with a finite number of identical and equivalent sites. Meanwhile, it was supposed that the interaction forces between different adsorbates, between adsorbed molecules and adsorbent surface-active sites does not exist. The Freundlich isotherm model is generally applied to multilayer adsorption with interaction forces between different adsorbates, between adsorbed molecules and adsorbent surface-active sites [7]. As can be seen from the results in Table 3, Langmuir adsorption model has a higher correlation coefficient, and it can be better fitted to the adsorption type of SLACM for Cr(VI) than Freundlich model. Therefore, the isotherm model fitting results indicate that the adsorption process of SLACM for Cr(VI) was monolayer adsorption. The separation factor (R_L) is an important parameter for Langmuir isotherm model, which can be used to evaluate whether the adsorption process is favorable ($0 < R_L < 1$), linear ($R_L = 1$), or unfavorable ($R_L > 1$) [37]. In this test Cr(VI) concentration range (50–600 mg/L), the R_L values decreased from 0.1 to 0.0092 at 20 °C, from 0.0833 to 0.0075 at 30 °C, and from 0.0588 to 0.0052 at 40 °C. All of the R_L values were greater than 0 but less than 1, and which were decreases with higher Cr(VI) initial concentration and higher adsorption temperature. The results indicate that the adsorption process of Cr(VI) onto SLACM was favorable in the concentration range studied, the higher Cr(VI) initial concentration and higher temperature was beneficial for adsorption. The equilibrium data fitted well with the Langmuir isotherm models, and the maximum adsorption capacity of Cr(VI) on SLACM was around 227.7 mg/g initial pH 2, 40 °C. The Cr(VI) adsorption capacities of the other adsorbents from biomass waste are listed in Table 4. within comparison with others, it can be seen that the SLACM had excellent adsorption capacity for the Cr(VI) removal from aqueous solution, and thus it could be a promising adsorbent.

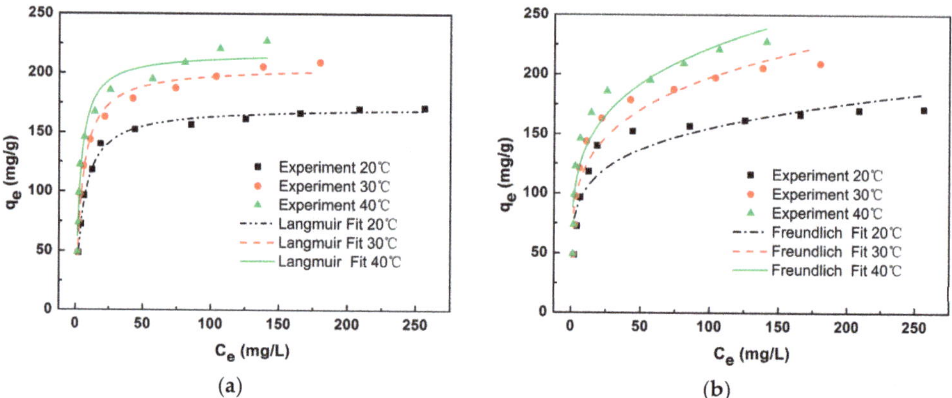

Figure 8. (**a**) Adsorption isotherm of Cr(VI) onto SLACM and Langmuir isotherm models fitting curves; (**b**) Adsorption isotherm of Cr(VI) onto SLACM and Freundlich isotherm models fitting curves.

Table 3. Adsorption isotherm parameters and correlation coefficients of SLACM for Cr(VI).

Temp °C	Langmuir			Freundlich		
	q_m mg/g	K_L L/mg	R^2	K_F L/mg	$\frac{1}{n}$ L/mg	R^2
20	172.41	0.18	99.22	67.80	0.18	84.45
30	206.13	0.22	98.59	75.09	0.21	89.59
40	218.19	0.32	96.27	85.05	0.21	89.03

Table 4. The Cr(VI) adsorption capacities of the other adsorbents from biomass waste. Data from [33,34,39,44,45].

Adsorbent Precursor	Activating Agend	q_{max} (mg/g)	pH	Temp °C	Reference
Mixed waste tea	/	94.34	2.0	30	[44]
Coffee ground	/	87.72	2.0	30	[44]
Algal bloom	CO_2	96	1.0	20	[39]
Glucose	KOH	332.53	2.0	25	[45]
Corn straw	KOH	175.44	3.0	25	[33]
Glucose monohydrate	H_2O	117.2	2.0	30	[34]
SL	$ZnCl_2$	227.7	2.0	40	This work

4. Conclusions

Activated carbon microsphere was prepared from powdered SL without adding any kind of binder. The preparation process consisted of activated carbon precursor preparation by Mannich reaction and $ZnCl_2$ activation. The results of the study indicated that the SLACM had excellent adsorption capacity for the removal of Cr(VI) from aqueous solution. It was found that the SL had a good Mannich reaction with aminated CMPS in creating ARM, and the ARM with good thermal stability was beneficial to the SLACM preparation. SLACM comprises tiny spherical particles with a well-developed porous structure. As a mesoporous material, the surface area and average pore size of spherical SLACM with a porous structure are 769.37 m^2/g and 2.46 nm, respectively. The adsorption capacity of SLACM for Cr(VI) is greatly dependent on solution pH, and the acidity of the solution is beneficial to Cr(VI) adsorption. The adsorption kinetic results showed that the adsorption of SLACM for Cr(VI) mainly occurred during the initial stages of the adsorption process; the Elovich model was in better agreement with the experimental data to describe adsorption process than the PFO and PSO models. In the range of test temperatures, the adsorption capacity increased with increasing temperature. The equilibrium data fitted well with the Langmuir isotherm models, and the maximum adsorption capacity of Cr(VI) on SLACM was around 227.7 mg/g at an initial pH value of 2 and a temperature of 40 °C. Therefore, the SL could be considered to be a potential raw material for the production of activated carbon in removing the Cr(VI) from wastewater. The SLACM prepared in this work provides the possibility of wastewater treatment in the future.

Author Contributions: Conceptualization, K.Y. and J.C.; methodology, K.Y. and J.X.; formal analysis, K.Y. and P.X.; investigation, K.Y. and J.X.; resources, J.C.; writing—original draft preparation, K.Y. and J.X.; writing—review and editing, K.Y., Q.Z. and K.M.U.; project administration, J.C.; funding acquisition, J.C. All authors have read and agreed to the published version of the manuscript.

Funding: This research was funded by National Key R&D Program of China, grant number 2017YFD0601004.

References

1. Sessarego, S.; Rodrigues, S.C.G.; Xiao, Y.; Lu, Q.; Hill, J.M. Phosphonium-enhanced chitosan for Cr(VI) adsorption in wastewater treatment. *Carbohydr. Polym.* **2019**, *211*, 249–256. [CrossRef]

2.   Vilela, P.B.; Dalalibera, A.; Duminelli, E.C.; Becegato, V.A.; Paulino, A.T. Adsorption and removal of chromium (VI) contained in aqueous solutions using a chitosan-based hydrogel. *Environ. Sci. Pollut. Res.* **2018**, *26*, 28481–28489. [CrossRef] [PubMed]

3.   Hussein, H.; Ibrahim, S.F.; Kandeel, K.; Moawad, H. Biosorption of heavy metals from waste water using *Pseudomonas* sp. *Electron. J. Biotechnol.* **2004**, *7*, 38–46. [CrossRef]

4.   Yang, J.; Yu, M.; Chen, W. Adsorption of hexavalent chromium from aqueous solution by activated carbon prepared from longan seed: Kinetics, equilibrium and thermodynamics. *J. Ind. Eng. Chem.* **2015**, *21*, 414–422. [CrossRef]

5.   Fazlzadeh, M.; Rahmani, K.; Zarei, A.; Abdoallahzadeh, H.; Nasiri, F.; Khosravi, R. A novel green synthesis of zero valent iron nanoparticles (NZVI) using three plant extracts and their efficient application for removal of Cr(VI) from aqueous solutions. *Adv. Powder Technol.* **2017**, *28*, 122–130. [CrossRef]

6.   Miretzky, P.; Cirelli, A.F. Cr(VI) and Cr(III) removal from aqueous solution by raw and modified lignocellulosic materials: A review. *J. Hazard. Mater.* **2010**, *180*, 1–19. [CrossRef]

7.   Norouzi, S.; Heidari, M.; Alipour, V.; Rahmanian, O.; Fazlzadeh, M.; Mohammadi-moghadam, F.; Nourmoradi, H.; Goudarzi, B.; Dindarloo, K. Preparation, characterization and Cr(VI) adsorption evaluation of NaOH-activated carbon produced from Date Press Cake; an agro-industrial waste. *Bioresour. Technol.* **2018**, *258*, 48–56. [CrossRef]

8.   Tan, X.; Liu, S.; Liu, Y.; Gu, Y.; Zeng, G.; Hu, X.; Wang, X.; Liu, S.; Jiang, L. Biochar as potential sustainable precursors for activated carbon production: Multiple applications in environmental protection and energy storage. *Bioresour. Technol.* **2017**, *227*, 359–372. [CrossRef]

9.   Tripathi, A.; Rawat Ranjan, M. Heavy Metal Removal from Wastewater Using Low Cost Adsorbents. *J. Bioremediat. Biodegrad.* **2015**, *6*, 1–5. [CrossRef]

10. Ilangovan, M.; Guna, V.; Olivera, S.; Ravi, A.; Muralidhara, H.B.; Santosh, M.S.; Reddy, N. Highly porous carbon from a natural cellulose fiber as high efficiency sorbent for lead in waste water. *Bioresour. Technol.* **2017**, *245*, 296–299. [CrossRef]

11. Qiang, W.; Jiaoyan, S.; Mengzhu, S. Characteristic of adsorption, desorption and oxidation of Cr (III) on birnessite. *Energy Procedia* **2011**, *5*, 1104–1108. [CrossRef]

12. Parlayici, Ş.; Pehlivan, E. Comparative study of Cr(VI) removal by bio-waste adsorbents: Equilibrium, kinetics, and thermodynamic. *J. Anal. Sci. Technol.* **2019**, *10*, 15. [CrossRef]

13. Zhang, J.; Chen, S.; Zhang, H.; Wang, X. Removal behaviors and mechanisms of hexavalent chromium from aqueous solution by cephalosporin residue and derived chars. *Bioresour. Technol.* **2017**, *238*, 484–491. [CrossRef] [PubMed]

14. Zhang, L.; Mu, C.; Zhong, L.; Xue, J.; Zhou, Y.; Han, X. Recycling of Cr (VI) from weak alkaline aqueous media using a chitosan/triethanolamine/Cu (II) composite adsorbent. *Carbohydr. Polym.* **2019**, *205*, 151–158. [CrossRef]

15. Garg, V.K.; Gupta, R.; Kumar, R.; Gupta, R.K. Adsorption of chromium from aqueous solution on treated sawdust. *Bioresour. Technol.* **2004**, *92*, 79–81. [CrossRef]

16. Aghababaei, A.; Ncibi, M.C.; Sillanpää, M. Optimized removal of oxytetracycline and cadmium from contaminated waters using chemically-activated and pyrolyzed biochars from forest and wood-processing residues. *Bioresour. Technol.* **2017**, *239*, 28–36. [CrossRef]

17. Hernandez-Ramirez, O.; Holmes, S.M. Novel and modified materials for wastewater treatment applications. *J. Mater. Chem.* **2008**, *18*, 2751. [CrossRef]

18. Pham, T.D.; Tran, T.T.; Le, V.A.; Pham, T.T.; Dao, T.H.; Le, T.S. Adsorption characteristics of molecular oxytetracycline onto alumina particles: The role of surface modification with an anionic surfactant. *J. Mol. Liq.* **2019**, *287*, 110900. [CrossRef]

19. Pham, T.; Bui, T.; Nguyen, V.; Bui, T.; Tran, T.; Phan, Q.; Pham, T.; Hoang, T. Adsorption of Polyelectrolyte onto Nanosilica Synthesized from Rice Husk: Characteristics, Mechanisms, and Application for Antibiotic Removal. *Polymers* **2018**, *10*, 220. [CrossRef]

20. Pham, T.D.; Do, T.T.; Ha, V.L.; Doan, T.H.Y.; Nguyen, T.A.H.; Mai, T.D.; Kobayashi, M.; Adachi, Y. Adsorptive removal of ammonium ion from aqueous solution using surfactant-modified alumina. *Environ. Chem.* **2017**, *14*, 327. [CrossRef]

21. Qi, W.; Zhao, Y.; Zheng, X.; Ji, M.; Zhang, Z. Adsorption behavior and mechanism of Cr(VI) using Sakura waste from aqueous solution. *Appl. Surf. Sci.* **2016**, *360*, 470–476. [CrossRef]

22. Kazakis, N.; Kantiranis, N.; Kalaitzidou, K.; Kaprara, E.; Mitrakas, M.; Frei, R.; Vargemezis, G.; Vogiatzis, D.; Zouboulis, A.; Filippidis, A. Environmentally available hexavalent chromium in soils and sediments impacted by dispersed fly ash in Sarigkiol basin (Northern Greece). *Environ. Pollut.* **2018**, *235*, 632–641. [CrossRef] [PubMed]

23. Rangabhashiyam, S.; Balasubramanian, P. The potential of lignocellulosic biomass precursors for biochar production: Performance, mechanism and wastewater application—A review. *Ind. Crop. Prod.* **2019**, *128*, 405–423.

24. Li, W.; Zhang, L.; Peng, J.; Li, N.; Zhu, X. Preparation of high surface area activated carbons from tobacco stems with K2CO3 activation using microwave radiation. *Ind. Crop. Prod.* **2008**, *27*, 341–347. [CrossRef]

25. Gao, H.; Liu, Y.; Zeng, G.; Xu, W.; Li, T.; Xia, W. Characterization of Cr(VI) removal from aqueous solutions by a surplus agricultural waste—Rice straw. *J. Hazard. Mater.* **2008**, *150*, 446–452. [CrossRef]

26. Mekonnen, E.; Yitbarek, M.; Soreta, T.R. Kinetic and thermodynamic studies of the adsorption of Cr(VI) onto some selected local adsorbents. *South Afr. J. Chem.* **2015**, *68*, 45–52. [CrossRef]

27. Malwade, K.; Lataye, D.; Mhaisalkar, V.; Kurwadkar, S.; Ramirez, D. Adsorption of hexavalent chromium onto activated carbon derived from Leucaena leucocephala waste sawdust: Kinetics, equilibrium and thermodynamics. *Int. J. Environ. Sci. Technol.* **2016**, *13*, 2107–2116. [CrossRef]

28. Rangabhashiyam, S.; Selvaraju, N. Adsorptive remediation of hexavalent chromium from synthetic wastewater by a natural and ZnCl$_2$ activated Sterculia guttata shell. *J. Mol. Liq.* **2015**, *207*, 39–49. [CrossRef]

29. Doherty, W.O.S.; Mousavioun, P.; Fellows, C.M. Value-adding to cellulosic ethanol: Lignin polymers. *Ind. Crop. Prod.* **2011**, *33*, 259–276. [CrossRef]

30. Xi, Y.; Yang, D.; Qiu, X.; Wang, H.; Huang, J.; Li, Q. Renewable lignin-based carbon with a remarkable electrochemical performance from potassium compound activation. *Ind. Crop. Prod.* **2018**, *124*, 747–754. [CrossRef]

31. Zhang, N.; Shen, Y. One-step pyrolysis of lignin and polyvinyl chloride for synthesis of porous carbon and its application for toluene sorption. *Bioresour. Technol.* **2019**, *284*, 325–332. [CrossRef]

32. Geng, J.; Gu, F.; Chang, J. Fabrication of magnetic lignosulfonate using ultrasonic-assisted in situ synthesis for efficient removal of Cr(VI) and Rhodamine B from wastewater. *J. Hazard. Mater.* **2019**, *375*, 174–181. [CrossRef]

33. Ma, H.; Yang, J.; Gao, X.; Liu, Z.; Liu, X.; Xu, Z. Removal of chromium (VI) from water by porous carbon derived from corn straw: Influencing factors, regeneration and mechanism. *J. Hazard. Mater.* **2019**, *369*, 550–560. [CrossRef]

34. Kumar, S.; Sivaranjanee, R.; Saravanan, A. Carbon sphere: Synthesis, characterization and elimination of toxic Cr(VI) ions from aquatic system. *J. Ind. Eng. Chem.* **2018**, *60*, 307–320.

35. Fang, W.; Weisheng, N.; Andong, Z.; Weiming, Y. Enhanced anaerobic digestion of corn stover by thermo-chemical pretreatment. *Int. J. Agric. Biol. Eng.* **2015**, *8*, 84–90.

36. Wang, F.; Zhang, D.; Wu, H.; Yi, W.; Fu, P.; Li, Y.; Li, Z. Enhancing biogas production of corn stover by fast pyrolysis pretreatment. *Bioresour. Technol.* **2016**, *218*, 731–736. [CrossRef]

37. Xu, Z.; Yuan, Z.; Zhang, D.; Huang, Y.; Chen, W.; Sun, Z.; Zhou, Y. Cr(VI) removal with rapid and superior performance utilizing cost-efficient waste-polyester-textile-based mesoporous carbon: Behavior and mechanism. *J. Mol. Liq.* **2019**, *278*, 496–504. [CrossRef]

38. Martins, A.C.; Pezoti, O.; Cazetta, A.L.; Bedin, K.C.; Yamazaki, D.A.S.; Bandoch, G.F.G.; Asefa, T.; Visentainer, J.V.; Almeida, V.C. Removal of tetracycline by NaOH-activated carbon produced from macadamia nut shells: Kinetic and equilibrium studies. *Chem. Eng. J.* **2015**, *260*, 291–299. [CrossRef]

39. Cui, Y.; Masud, A.; Aich, N.; Atkinson, J.D. Phenol and Cr(VI) removal using materials derived from harmful algal bloom biomass: Characterization and performance assessment for a biosorbent, a porous carbon, and Fe/C composites. *J. Hazard. Mater.* **2019**, *368*, 477–486. [CrossRef]

40. Pezoti, O.; Cazetta, A.L.; Bedin, K.C.; Souza, L.S.; Martins, A.C.; Silva, T.L.; Santos Júnior, O.O.; Visentainer, J.V.; Almeida, V.C. NaOH-activated carbon of high surface area produced from guava seeds as a high-efficiency adsorbent for amoxicillin removal: Kinetic, isotherm and thermodynamic studies. *Chem. Eng. J.* **2016**, *288*, 778–788. [CrossRef]

41. Mondal, S.; Derebe, A.T.; Wang, K. Surface functionalized carbon microspheres for the recovery of copper ion from refinery wastewater. *Korean J. Chem. Eng.* **2018**, *35*, 147–152. [CrossRef]

42. Jeon, C. Removal of Cr(VI) from aqueous solution using amine-impregnated crab shells in the batch process. *J. Ind. Eng. Chem.* **2019**, *77*, 111–117. [CrossRef]
43. Zhang, X.; Lv, L.; Qin, Y.; Xu, M.; Jia, X.; Chen, Z. Removal of aqueous Cr(VI) by a magnetic biochar derived from Melia azedarach wood. *Bioresour. Technol.* **2018**, *256*, 1–10. [CrossRef]
44. Cherdchoo, W.; Nithettham, S.; Charoenpanich, J. Removal of Cr(VI) from synthetic wastewater by adsorption onto coffee ground and mixed waste tea. *Chemosphere* **2019**, *221*, 758–767. [CrossRef]
45. Liang, H.; Song, B.; Peng, P.; Jiao, G.; Yan, X.; She, D. Preparation of three-dimensional honeycomb carbon materials and their adsorption of Cr(VI). *Chem. Eng. J.* **2019**, *367*, 9–16. [CrossRef]

MDPI

St. Alban-Anlage 66

4052 Basel

Switzerland

Tel. +41 61 683 77 34

Fax +41 61 302 89 18

www.mdpi.com

Polymers Editorial Office

E-mail: polymers@mdpi.com

www.mdpi.com/journal/polymers